ANALOG AND DIGITAL
COMMUNICATION
SYSTEMS

ANALOG AND DIGITAL COMMUNICATION SYSTEMS

MARTIN S. RODEN

Professor of Electrical Engineering
California State University, Los Angeles

PRENTICE-HALL, INC., *Englewood Cliffs, N.J. 07632*

Library of Congress Cataloging in Publication Data

RODEN, MARTIN S
   Analog and digital communication systems.

   Includes bibliographies and index.
   1. Telecommunication. 2. Telecommunication
systems. I. Title.
TK5101.R58   621.38'0413   78-16428
ISBN 0-13-032722-0

Editorial/production supervision and interior
design by Donald S. Rosanelli
Cover design by Katrine Stevens
Manufacturing buyer: Cathie Lenard

10 9 8 7 6 5 4 3 2 1

Printed in the United States of America

PRENTICE-HALL INTERNATIONAL, INC., *London*
PRENTICE-HALL OF AUSTRALIA PTY. LIMITED, *Sydney*
PRENTICE-HALL OF CANADA, LTD., *Toronto*
PRENTICE-HALL OF INDIA PRIVATE LIMITED, *New Delhi*
PRENTICE-HALL OF JAPAN, INC., *Tokyo*
PRENTICE-HALL OF SOUTHEAST ASIA PTE. LTD., *Singapore*
WHITEHALL BOOKS LIMITED, *Wellington, New Zealand*

To ROSE and LEONARD RODEN

CONTENTS

7   DIGITAL COMMUNICATION   *272*

8   INFORMATION THEORY AND CODING   *300*

APPENDIXES

PREFACE

Analog and Digital Communication Systems evolved from an earlier text by the same author, *Introduction to Communication Theory* (Pergamon Press, 1972). The book is intended as an introductory text for the study of analog and/or digital communication systems, with or without noise.

The text stresses a mathematical-systems approach to all phases of the subject matter. The mathematics used throughout is as elementary as possible but is carefully chosen so as not to contradict any more sophisticated approach which may eventually be required. An attempt is made to apply intuitive techniques prior to grinding through the mathematics. The style is informal, and the text has been thoroughly tested in the classroom with excellent success.

Chapter 1 lays the mathematical groundwork of signal analysis including discrete and continuous signals. Chapter 2 applies this to the study of linear systems with emphasis upon analog and digital filters. Chapter 3 introduces probability and random analysis and applies these to the analysis of narrowband noise. If it is desired to omit noise in a first approach to the subject, this chapter may be skipped. The factor of noise has been introduced as the last section of each remaining chapter. The student may therefore conveniently eliminate noise analysis without adversely affecting continuity. Chapter

4 is a thorough treatment of AM, including applications to broadcast radio and television. A final section studies noise in AM systems. Chapter 5 parallels Chapter 4 but for FM instead of AM. A section is included on broadcast FM and FM stereo. Chapter 6 explores analog forms of pulse modulation and introduces the concepts of time-division multiplexing and cross talk. Chapter 7 treats digital communication with emphasis upon PCM and delta modulation. A final section deals with noise and errors in digital communication. Chapter 8 introduces key factors from information and coding theory. Appendices cover a summary of symbols used in the text, a brief table of Fourier Transforms, a table of error functions, and answers to selected problems.

Problems and references are presented at the end of each chapter. Illustrative examples are given within each chapter.

It gives me a great deal of pleasure to acknowledge the assistance and support of the following people without whom this text would not have been possible:

To the many classes of students who were responsive during lectures and helped indicate the clearest approach to each topic.

To my colleagues at Bell Telephone Labs, Hughes Aircraft (Fullerton), and California State University, Los Angeles.

To Professor Roger Brandt, who, although he enjoys teaching basic communication, continually scheduled me to teach it since he knew I was writing this text.

To Dean Leslie Cromwell, who forgave me if my writing sometimes took precedence over my duties as Associate Dean.

To all who adopted the earlier text and responded to my request for an evaluation

To Mr. William Gehr of the Trident Shop without whom this text would have been delayed several years.

To Mr. Dennis J. E. Ross for guidance and for assistance with the diagrams.

Finally, to Prof. A. Papoulis, who played a key role during the formative years of my education.

INTRODUCTION
TO THE
STUDENT

The field of communications, as is true of many engineering areas, experiences a revolution several times each decade. Some important milestones of the past include

The television revolution: After less than four decades, television has become a way of life. The technology for wall-size or wrist-size TV is just around the corner.

The space revolution: This has been the catalyst for many innovations in long-distance communication.

The digital revolution: The tremendous emphasis upon digital electronics and processing has resulted in a rapid change in direction from analog to digital communication.

The microprocessor revolution: This is changing the shape of everything related to computing and control, including many phases of communication.

The recreation (consumer) revolution: Consumers first discovered calculators and digital watches. They are now expanding into sophisticated TV recording systems, CB radio, and video games. The potential impact upon the future of communication is tremendous.

There is every reason to expect this revolution rate to continue or, indeed, accelerate. As this text is being prepared, the first large-scale commercial application of two-way cable television is being implemented. Subscribers can request programming, or newspaper facsimile, or respond to surveys.

As resources become more scarce and gasoline more expensive, people will try to perform more and more daily functions without physically moving around. They will order groceries, hold business-related conferences, send documents from one terminal to another, and satisfy recreational needs without leaving the home.

The problems of urban centers are being attacked through communications. Indeed, adequate communication can make cities unnecessary. The potential impact of these developments is mind boggling, and the possibilities involving communications are most exciting.

All of the above should excite the reader to consider the field of communications as a career. But even if you decide not to approach the field any more deeply than this text, you will be a far more aware person having experienced even the small amount contained herein. If nothing more, the questions arising in modern communications as it relates to everything from the space program to home entertainment will take on a new meaning for you. The devices around you will no longer be mysterious. You will learn what magic force lights the "stereo" indicator on your receiver, the basics of how a video Ping-Pong game works, and the mechanism by which your voice can travel by telephone to any part of the globe in a moment's time.

Enjoy the book. Enjoy the subject. And, please, after studying this text, communicate any comments that you may have to me at California State University, Los Angeles, California 90032.

Los Angeles, California   MARTIN S. RODEN

SIGNAL ANALYSIS

1

By this stage of your education, you have probably heard of Fourier Series expansions. You have probably also learned that the earlier part of each course of study tends to be the least interesting. Certain groundwork must be laid prior to getting into the interesting applications.

We could start by presenting the rules for finding Fourier Series expansions and from there go directly into the business of signal analysis. We choose instead to take a more general approach. Since the principles introduced in this approach are not crucial to the understanding of *elementary* communication systems, their presentation at this time must be somehow justified.

The purpose of this section is to put the study of signals into proper perspective in the much broader area of applied mathematics. Signal analysis, and indeed most of communication theory, is a mathematical science. Probability theory and transform analysis techniques (a term which will have to be defined later) form the backbone of all communication theory. Both of these disciplines fall clearly within the realm of mathematics.

It is quite possible to study communication systems without relating the results to more general mathematical concepts, but this seems to be quite narrow-minded and tends to downgrade engineers (a phenomenon to which we need not contribute). More important than that, the narrow-minded approach reduces a person's ability to extend or modify existing results in order to apply them to new problems.

We shall therefore begin by examining some of the more common properties of orthogonal vector spaces. We do this with the intention of generalizing the results to apply to orthogonal function spaces. The generation of the Fourier Series representation of a function (our immediate goal) will prove to be a trivial application of the principles of orthogonal function spaces.

As an added bonus, the techniques and principles we develop will prove crucial to later studies in more advanced communication theory. For example, orthogonality is basic to detection problems (e.g., radar).

If words such as "space" and "orthogonal" make you uncomfortable, skip to Section 1.3 and start with the definition of a Fourier Series. This will cost nothing as far as basic communication theory is concerned.

1.1 ORTHOGONAL VECTOR SPACES

We shall use the word *orthogonality* (Webster defines this as relating to right angles) of vectors as it applies to the dot (or inner) product operation. That is, two vectors are said to be orthogonal to each other if their dot product is zero. This is equivalent to saying that one vector has a zero component in the direction of the other; they have nothing in common. A set of vectors is called an orthogonal set of vectors if each member of the set is orthogonal to every other member of the set. An often used set of orthogonal vectors in three-dimensional Euclidean space is that set composed of the three rectangular unit vectors,

$$\bar{a}_x, \bar{a}_y, \bar{a}_z$$

The bar above a quantity is the notation which we have adopted for a vector. The three unit vectors shown above form an orthogonal set since

$$\bar{a}_x \cdot \bar{a}_y = \bar{a}_x \cdot \bar{a}_z = \bar{a}_y \cdot \bar{a}_z = 0 \tag{1.1}$$

Recall that in three-dimensional space, the dot product of two vectors is given by the product of their lengths multiplied by the cosine of the angle between them.

We know that any three-dimensional vector can be written as a sum of its three rectangular components. This is equivalent to saying that it can be written as a "linear combination" of the three orthogonal unit vectors presented above. If $\bar{A}$ is a general three-dimensional vector, then

$$\bar{A} = c_1\bar{a}_x + c_2\bar{a}_y + c_3\bar{a}_z \tag{1.2}$$

where c_1, c_2, and c_3 are scalar constants. Suppose we take the dot product of each side of Eq. (1.2) with $\bar{a}_x$. Since each side of this equation is certainly a vector, we are justified in doing this.

$$\bar{A} \cdot \bar{a}_x = c_1 \bar{a}_x \cdot \bar{a}_x + c_2 \bar{a}_y \cdot \bar{a}_x + c_3 \bar{a}_z \cdot \bar{a}_x \qquad (1.3)$$

By virtue of the orthogonality of the unit vectors and the fact that their lengths are unity, we have

$$\bar{a}_x \cdot \bar{a}_x = 1 \qquad \bar{a}_y \cdot \bar{a}_x = 0 \qquad \bar{a}_z \cdot \bar{a}_x = 0 \qquad (1.4)$$

and the right side of Eq. (1.3) reduces to just c_1. Similarly, we could have taken the dot product of both sides of Eq. (1.2) with $\bar{a}_y$ or $\bar{a}_z$ in order to find c_2 or c_3. The results would be

$$c_1 = \bar{A} \cdot \bar{a}_x \qquad c_2 = \bar{A} \cdot \bar{a}_y \qquad c_3 = \bar{A} \cdot \bar{a}_z \qquad (1.5)$$

Thus, each weighting constant, c_n, corresponds to the component of the vector, $\bar{A}$, in the direction of the associated unit vector. Note how simple it was to find the values of the c_n's. This would have been far more complicated if the component unit vectors had not been orthogonal to each other.

In the case of a three-dimensional vector, it is obvious that if we break it into its components, we need three component vectors in most cases. Suppose that we were not endowed with such intuitive powers, and wished to express a general three-dimensional vector, $\bar{A}$, as a linear combination of x and y unit vectors only.

$$\bar{A} \approx c_1 \bar{a}_x + c_2 \bar{a}_y \qquad (1.6)$$

The "squiggly" equals sign in Eq. (1.6) means "approximately equal to" and is placed here since one cannot be sure that the vector $\bar{A}$ can be expressed exactly as the sum shown in Eq. (1.6). Indeed, the reader with intuition knows that, in general, this cannot be made into an exact equality. For example, suppose that $\bar{A}$ had a non-zero component in the z-direction. Equation (1.6) could then never be an exact equality.

In many cases of interest, we either cannot use as many approximating vectors as we know are required, or we are not even sure of how many are necessary. A significant question would therefore be, "How does one choose c_1 and c_2 in order to make this approximation as 'good' as is possible?" (whatever "good" means). The standard way of defining goodness of an approximation is by defining an error term. The next step is to try to minimize this error term.

We shall define the error as the difference between the actual $\bar{A}$ and the approximation to $\bar{A}$. This difference will be a vector, and since defining error as a vector is not too satisfying (i.e., What does "minimum" mean for a vector?), we shall use the magnitude, or length of this vector as the definition of error. Since the magnitude of a vector is a term which is virtually impossible to work with mathematically, we shall modify this one more time by working with the square of this magnitude. Therefore,

$$e = \text{Error} = |\bar{A} - c_1\bar{a}_x - c_2\bar{a}_y|^2 \tag{1.7}$$

If we minimize the square of the magnitude, we have simultaneously minimized the magnitude. This is true since the magnitude is always positive, and the square increases as the magnitude itself increases (i.e., a monotonic function). Differentiating the error, e, with respect to the variables which we wish to find, c_1 and c_2, and then setting these derivatives equal to zero in order to minimize the quantity, we have[1]

$$e = [\bar{A} - c_1\bar{a}_x - c_2\bar{a}_y] \cdot [\bar{A} - c_1\bar{a}_x - c_2\bar{a}_y]$$

$$\frac{\partial e}{\partial c_1} = -2[\bar{A} - c_1\bar{a}_x - c_2\bar{a}_y] \cdot \bar{a}_x = 0 \tag{1.8}$$

and

$$\frac{\partial e}{\partial c_2} = -2[\bar{A} - c_1\bar{a}_x - c_2\bar{a}_y] \cdot \bar{a}_y = 0 \tag{1.9}$$

from which we get, after expanding and using the orthogonality property,

$$c_1 = \bar{A} \cdot \bar{a}_x \qquad c_2 = \bar{A} \cdot \bar{a}_y \tag{1.10}$$

Equation (1.10) is a very significant result. It says that, even though the orthogonal set used to approximate a general vector does not contain a sufficient number of elements to describe the vector exactly, the weighting coefficients are chosen as if the set were sufficient. In vector terms, even though the unit vector in the z-direction was missing, c_1 was still chosen as the component of $\bar{A}$ in the x-direction, and c_2 as the component of $\bar{A}$ in the y-direction. This indicates a complete independence between the various pairs of components. The z-direction component has nothing to do with that in the x- or y-direction. This could have been expected due to the orthogonality (nothing in common) property of the original set of vectors. It probably appears obvious in this case. However, it will be generalized and used in the following section where it will certainly no longer be obvious.

1.2 ORTHOGONAL FUNCTION SPACES

In communication systems, we deal primarily with time functions. In order to apply the above results to functions, the concept of vectors will now be generalized with an appropriate modification of the dot product operation. We will consider a set of real time functions, $g_n(t)$, for n between 1 and N. That is, we have N functions of time. This set will be defined to be an orthogonal set of functions if each member is orthogonal to every other member

---

[1]Since the error, as defined, is quadratic, we are assured of a nice minimum. One can easily be convinced that the zero-derivative point corresponds to a minimum and not a maximum of the error.

of the set. Two real time functions are defined to be orthogonal over the interval between t_1 and t_2 if the integral of their product over this interval is zero. This integral therefore corresponds to the inner, or dot, product operation. That is, $f(t)$ is orthogonal to $h(t)$ if

$$\int_{t_1}^{t_2} f(t)h(t)\,dt = 0 \tag{1.11}$$

Therefore, the set of time functions, $g_n(t)$, forms an orthogonal set over the interval between t_1 and t_2 if

$$\int_{t_1}^{t_2} g_j(t)g_k(t)\,dt = 0 \qquad (j \neq k) \tag{1.12}$$

The restriction $j \neq k$ assures that we integrate the product of two different members of the set.

Suppose that we now wish to *approximate* any time function, $f(t)$, by a linear combination of the $g_n(t)$ in the interval between t_1 and t_2, where the $g_n(t)$ are orthogonal over this interval. Notice that we use the word "approximate" since there is no assurance that an exact equality can be obtained no matter what values are chosen for the weighting coefficients. The number of g_n's necessary to attain an exact equality is not at all obvious. Compare this with the three-dimensional vector case where we knew that three unit vectors were sufficient to express any general vector.

The approximation of $f(t)$ is then of the form

$$f(t) \approx c_1 g_1(t) + c_2 g_2(t) + \cdots + c_N g_N(t) \tag{1.13}$$

or in more compact form,

$$f(t) \approx \sum_{n=1}^{N} c_n g_n(t) \tag{1.14}$$

Recalling our previous result [Eq. (1.10)], we conjecture that the c_n's are chosen as if Eq. (1.13) were indeed an exact equality. If it were an equality, we would find the c_n's as we did in Eq. (1.3). That is, take the inner product of both sides of the equation with one member of the orthogonal set. In this case, multiply both sides of Eq. (1.14) by one of the $g_n(t)$, say $g_{17}(t)$ [assuming, of course, the $N \geq 17$ in Eq. (1.13)] and integrate both sides between t_1 and t_2. By virtue of the orthogonality of the $g_n(t)$, each term on the right side would integrate to zero except for the 17th term. That is,

$$\int_{t_1}^{t_2} f(t)g_{17}(t)\,dt = c_1 \int_{t_1}^{t_2} g_1(t)g_{17}(t)\,dt + \cdots$$
$$+ c_{17} \int_{t_1}^{t_2} g_{17}^2(t)\,dt + \cdots + c_N \int_{t_1}^{t_2} g_N(t)g_{17}(t)\,dt$$

Identifying the terms which are zero yields

$$\int_{t_1}^{t_2} f(t)g_{17}(t)\,dt = c_{17} \int_{t_1}^{t_2} g_{17}^2(t)\,dt \tag{1.15}$$

Solving for c_{17}, we get

$$c_{17} = \frac{\int_{t_1}^{t_2} f(t)g_{17}(t)\, dt}{\int_{t_1}^{t_2} g_{17}^2(t)\, dt}$$

Being extremely clever, we can generalize the above to find any c_n.

$$c_n = \frac{\int_{t_1}^{t_2} f(t)g_n(t)\, dt}{\int_{t_1}^{t_2} g_n^2(t)\, dt} \tag{1.16}$$

Note that the numerator of Eq. (1.16) is essentially the component of $f(t)$ in the $g_n(t)$ "direction" (one must be pretty open minded to believe this at this point). The denominator of Eq. (1.16) normalizes the result since the $g_n(t)$, unlike the unit vectors, do not necessarily have a "length" of unity.

1.3 FOURIER SERIES

We have shown that, over a given interval, a function can be represented by a linear combination of members of an orthogonal set of functions. There are many possible orthogonal sets of functions, just as there are many possible orthogonal sets of three-dimensional vectors (e.g., consider any rotation of the three rectangular unit vectors). One such possible set of functions is the set of harmonically related sines and cosines. That is, the functions, $\sin \omega_0 t$, $\sin 2\omega_0 t$, $\sin 3\omega_0 t, \ldots$, $\cos \omega_0 t$, $\cos 2\omega_0 t$, $\cos 3\omega_0 t, \ldots$, for any ω_0, form an orthogonal set. These functions are orthogonal over the interval between any starting point, t_0, and $t_0 + 2\pi/\omega_0$. That is,

$$\int_{t_0}^{t_0 + 2\pi/\omega_0} f(t)g(t)\, dt = 0 \tag{1.17}$$

where $f(t)$ is any member of the set of functions, and $g(t)$ is any other member. This can be verified by a simple integration using the "cosine of sum" and "cosine of difference" trigonometric identities.

Illustrative Example 1.1

Show that the set made up of the functions $\cos n\omega_0 t$, and $\sin n\omega_0 t$ is an orthogonal set over the interval $t_0 < t \leq t_0 + 2\pi/\omega_0$ for any choice of t_0.

Solution

We must show that

$$\int_{t_0}^{t_0 + 2\pi/\omega_0} f_n(t)f_m(t)\, dt = 0$$

for any two distinct members of the set. There are three cases which must be considered:

(a) Both $f_n(t)$ and $f_m(t)$ are sine waves.

(b) Both $f_n(t)$ and $f_m(t)$ are cosine waves.

(c) One of the two is a sine wave and the other, a cosine wave.

Considering each of these cases, we have

Case (a)

$$\int_{t_0}^{t_0+2\pi/\omega_0} \sin n\omega_0 t \sin m\omega_0 t \, dt = \frac{1}{2}\int_{t_0}^{t_0+2\pi/\omega_0} \cos(n-m)\omega_0 t \, dt \\ -\frac{1}{2}\int_{t_0}^{t_0+2\pi/\omega_0} \cos(n+m)\omega_0 t \, dt \tag{1.18}$$

If $n \neq m$, both $n-m$ and $n+m$ are non-zero integers. We note that, for the function $\cos k\omega_0 t$, the interval, $t_0 < t \leq t_0 + 2\pi/\omega_0$ represents exactly k periods. The integral of a cosine function over any whole number of periods is zero, so we have completed this case. Note that it is important that $n \neq m$ since, if $n = m$, we have

$$\frac{1}{2}\int_{t_0}^{t_0+2\pi/\omega_0} \cos(n-m)\omega_0 t \, dt = \frac{1}{2}\int_{t_0}^{t_0+2\pi/\omega_0} 1 \, dt = \pi/\omega_0 \neq 0$$

Case (b)

$$\int_{t_0}^{t_0+2\pi/\omega_0} \cos n\omega_0 t \cos m\omega_0 t \, dt = \frac{1}{2}\int_{t_0}^{t_0+2\pi/\omega_0} \cos(n-m)\omega_0 t \, dt \\ +\frac{1}{2}\int_{t_0}^{t_0+2\pi/\omega_0} \cos(n+m)\omega_0 t \, dt \tag{1.19}$$

This is equal to zero by the same reasoning applied to Case (a).

Case (c)

$$\int_{t_0}^{t_0+2\pi/\omega_0} \sin n\omega_0 t \cos m\omega_0 t \, dt = \frac{1}{2}\int_{t_0}^{t_0+2\pi/\omega_0} \sin(n+m)\omega_0 t \, dt \\ +\frac{1}{2}\int_{t_0}^{t_0+2\pi/\omega_0} \sin(n-m)\omega_0 t \, dt \tag{1.20}$$

To verify this case, we note that

$$\int_{t_0}^{t_0+2\pi/\omega_0} \sin k\omega_0 t \, dt = 0$$

for all integer values of k. This is true since the integral of a sine function over any whole number of periods is zero. (There is no difference between a sine function and a cosine function other than a shift.) Each term present in Case (c) is therefore equal to zero.

The given set is therefore an orthogonal set of time functions over the interval $t_0 < t \leq t_0 + 2\pi/\omega_0$.

We have talked about *approximating* a time function with a linear combination of members of an orthogonal set of time functions. We now present a definition borrowed from mathematics. An orthogonal set of time functions is said to be a *complete* set if the approximation of Eq. (1.14) can be made into an equality (with the word "equality" being interpreted in some special sense) by properly choosing the c_n weighting factors, and for $f(t)$ being any member of a certain class of functions. The three rectangular unit vectors form a complete orthogonal set in three-dimensional space, while the vectors $\bar{a}_x$ and $\bar{a}_y$ by themselves form an orthogonal set which is not complete. We state without proof that the set of harmonic time functions,

$$\cos n\omega_0 t \qquad \sin n\omega_0 t$$

where n can take on any integer value between zero and infinity, is an orthogonal complete set in the space of time functions in the interval between t_0 and $t_0 + 2\pi/\omega_0$. Therefore any time function[2] can be expressed, in the interval between t_0 and $t_0 + 2\pi/\omega_0$, by a linear combination of sines and cosines. In this case, the word equality is interpreted not as a pointwise equality, but in the sense that the "distance" between $f(t)$ and the series representation, given by

$$\int_0^{2\pi/\omega_0} \left[f(t) - \sum_{n=1}^{N} c_n g_n(t) \right]^2 dt$$

approaches zero as more and more terms are included in the sum. This is what we will mean when we talk of equality of two time functions. This type of equality will be sufficient for all of our applications.

For convenience, we define,

$$T \triangleq \frac{2\pi}{\omega_0}$$

Therefore, any time function, $f(t)$, can be written as

$$f(t) = a_0 \cos(0) + \sum_{n=1}^{\infty} [a_n \cos n\omega_0 t + b_n \sin n\omega_0 t] \qquad (1.21)$$

for

$$t_0 < t \le t_0 + T$$

An expansion of this type is known as a *Fourier Series*. We note that the first term in Eq. (1.21) is simply "a_0" since $\cos(0) = 1$. The proper choice of the constants, a_n and b_n is indicated by Eq. (1.16).

$$a_0 = \frac{\int_{t_0}^{t_0+T} f(t)\, dt}{\int_{t_0}^{t_0+T} 1^2\, dt} = \frac{1}{T} \int_{t_0}^{t_0+T} f(t)\, dt \qquad (1.22)$$

---

[2]In the case of Fourier Series, the class of time functions is restricted to be that class which has a finite number of discontinuities and a finite number of maxima and minima in any one period. Also the integral of the magnitude of the function over one period must exist (i.e., not be infinite).

and for $n \neq 0$,

$$a_n = \frac{\int_{t_0}^{t_0+T} f(t) \cos n\omega_0 t \, dt}{\int_{t_0}^{t_0+T} \cos^2 n\omega_0 t \, dt} = \frac{2}{T} \int_{t_0}^{t_0+T} f(t) \cos n\omega_0 t \, dt \qquad (1.23)$$

$$b_n = \frac{\int_{t_0}^{t_0+T} f(t) \sin n\omega_0 t \, dt}{\int_{t_0}^{t_0+T} \sin^2 n\omega_0 t \, dt} = \frac{2}{T} \int_{t_0}^{t_0+T} f(t) \sin n\omega_0 t \, dt \qquad (1.24)$$

Note that a_0 is the average of the time function, $f(t)$. It is reasonable to expect this term to appear by itself in Eq. (1.21) since the average value of the sines or cosines is zero. In any equality such as Eq. (1.21), the time average of the left side must equal the time average of the right side.

A more compact form of the Fourier Series described above is obtained if one considers the orthogonal, complete set of complex harmonic exponentials. That is, the set made up of the time functions,

$$e^{jn\omega_0 t}$$

where n is any integer, positive or negative.

The student should recall that the complex exponential can be viewed as (actually, it is *defined* as) a vector of length "1" and angle, "$n\omega_0 t$" in the complex two-dimensional plane. Thus,

$$e^{jn\omega_0 t} \triangleq \cos n\omega_0 t + j \sin n\omega_0 t$$

As before, the series expansion will apply in the time interval between t_0 and $t_0 + 2\pi/\omega_0$. Therefore, any time function, $f(t)$, can be expressed as a linear combination of these exponentials in the interval between t_0 and $t_0 + T$ ($T = 2\pi/\omega_0$ as before).

$$f(t) = \sum_{n=-\infty}^{\infty} c_n e^{jn\omega_0 t} \qquad (1.25)$$

The c_n are given by

$$c_n = \frac{\int_{t_0}^{t_0+T} f(t) e^{-jn\omega_0 t} \, dt}{\int_{t_0}^{t_0+T} e^{jn\omega_0 t} e^{-jn\omega_0 t} \, dt} = \frac{1}{T} \int_{t_0}^{t_0+T} f(t) e^{-jn\omega_0 t} \, dt \qquad (1.26)$$

This is easily verified by multiplying both sides of Eq. (1.25) by $e^{-jn\omega_0 t}$ and integrating both sides. The observant reader, who may have tried to prove that this set of exponentials is orthogonal, has noticed that a different definition of dot product must be used here. Shouldn't we really have multiplied both sides of Eq. (1.25) by $e^{+jn\omega_0 t}$ and integrated in order to find c_n? The answer is that, in the case of sets of complex time functions, the definition of the inner product is modified. Two complex time functions, $f(t)$ and $g(t)$, are orthogonal to each other if

$$\int_{t_0}^{t_0+T} f(t)g^*(t)\, dt = \int_{t_0}^{t_0+T} f^*(t)g(t)\, dt = 0$$

where the asterisk (*) indicates complex conjugate. In vague terms, this is done since we desire some quantities to be real. That is, Eq. (1.15) contains the square of the time function, $g_n^2(t)$. We would prefer to replace this with the magnitude squared of the time function, which equals the product of the function with its complex conjugate. This is analogous to the length in the vector case, and length should certainly be a real quantity. This argument is an admittedly loose attempt at presenting mathematical results without getting into the details.

The basic results are summed up in Eqs. (1.21) and (1.25). Any time function can be expressed by a weighted sum of sines and cosines or a weighted sum of complex exponentials in an interval. The rules for finding the weighting factors are given in Eqs. (1.22-1.24) and (1.26).

Equations (1.21) through (1.26) could have formed the starting point of our discussion of communication theory. For some students, this might still be the best place to begin.

Examination of the right side of Eq. (1.21) discloses that it is a periodic function outside of the interval $t_0 < t \le t_0 + T$. In fact, its period is T. Therefore, if $f(t)$ happened to be periodic with period T, even though the equality in Eq. (1.21) was written to apply only within the interval $t_0 < t \le t_0 + T$, it does indeed apply for all time! (Please give this statement some thought!)

In other words, if $f(t)$ is periodic, and we write a Fourier Series for $f(t)$ that applies over one complete period of $f(t)$, the series is equivalent to $f(t)$ for all time.

Illustrative Example 1.2

Evaluate the trigonometric Fourier Series expansion of $f(t)$ as shown in Fig. 1.1. This series should apply in the interval $-\pi/2 < t \le \pi/2$.

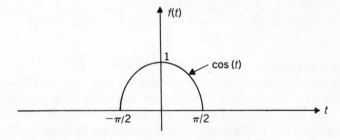

Figure 1.1 $f(t)$ for Illustrative Example 1.2.

Solution

Using the trigonometric Fourier Series form, $T = \pi$, and $\omega_0 = 2\pi/T = 2$. The series is therefore of the form

$$f(t) = a_0 + \sum_{n=1}^{\infty} [a_n \cos 2nt + b_n \sin 2nt] \tag{1.27}$$

where

$$a_0 = \frac{1}{T} \int_{-T/2}^{T/2} f(t)\, dt = \frac{1}{\pi} \int_{-\pi/2}^{\pi/2} \cos t\, dt = \frac{1}{\pi} \sin t \bigg|_{-\pi/2}^{\pi/2} \tag{1.28}$$

$$a_0 = 2/\pi$$

and

$$a_n = \frac{2}{T} \int_{-T/2}^{T/2} f(t) \cos 2nt\, dt = \frac{2}{\pi} \int_{-\pi/2}^{\pi/2} \cos t \cos 2nt\, dt$$

$$a_n = \frac{1}{\pi} \int_{-\pi/2}^{\pi/2} [\cos (2n-1)t + \cos (2n+1)t]\, dt$$

$$= \frac{1}{\pi} \frac{\sin (2n-1)t}{(2n-1)} \bigg|_{-\pi/2}^{\pi/2} + \frac{\sin (2n+1)t}{(2n+1)} \bigg|_{-\pi/2}^{\pi/2} \tag{1.29}$$

We note that

$$\sin \left\{ (2n-1)\left[\frac{\pi}{2} \right] \right\} = (-1)^{n+1}$$

$$\sin \left\{ (2n+1)\left[\frac{\pi}{2} \right] \right\} = (-1)^{n}$$

Therefore,

$$a_n = \frac{2}{\pi} \left[\frac{(-1)^{n+1}}{2n-1} + \frac{(-1)^{n}}{2n+1} \right] \tag{1.30}$$

Proceeding now to the evaluation of b_n, we find

$$b_n = \frac{2}{T} \int_{-T/2}^{T/2} f(t) \sin 2nt\, dt \tag{1.31}$$

Since $f(t)$ is an even function of time, $f(t) \sin 2nt$ is an odd function, and the integral from $-T/2$ to $+ T/2$ will be zero.

$$b_n = 0$$

Finally,

$$f(t) = \frac{2}{\pi} + \sum_{n=1}^{\infty} \left\{ \frac{2}{\pi} \left[\frac{(-1)^{n+1}}{2n-1} + \frac{(-1)^{n}}{2n+1} \right] \cos 2nt \right\} \tag{1.32}$$

Writing out the first few terms of this series, we have

$$f(t) = \frac{2}{\pi} \left[1 + \frac{2}{3} \cos 2t - \frac{2}{15} \cos 4t + \frac{2}{35} \cos 6t - \cdots \right] \tag{1.33}$$

We note that the Fourier Series in Eq. (1.33) is also the expansion of the periodic function, $f_p(t)$ shown in Fig. 1.2. That is, $f_p(t)$ is a periodic function which is equal to $f(t)$ over one period.

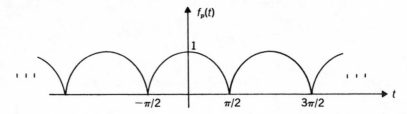

Figure 1.2 $f_p(t)$ representing Eq. (1.33).

Suppose we now calculated the Fourier Series of $g(t)$ shown in Fig. 1.3, to apply in the interval $-2 < t \leq +2$.

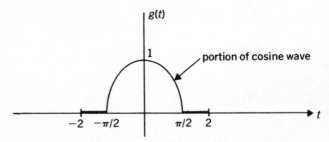

Figure 1.3 $g(t)$ similar to $f(t)$ of Illustrative Example 1.2.

The result will clearly be different from that of Example 1.2. One is readily convinced of this difference once he realizes that the frequencies of the various sines and cosines will be different from those of Example 1.2. However, for t between $-\pi/2$ and $+\pi/2$, both series represent the same function. Both series do not, however, represent $f_p(t)$ of Fig. 1.2. The periodic function corresponding to $g(t)$, denoted as $g_p(t)$, is sketched in Fig. 1.4.

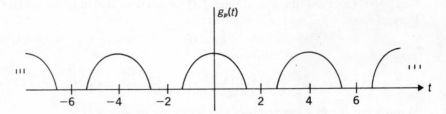

Figure 1.4 Periodic repetition of $g(t)$ of Fig. 1.3.

The series expansion of a function in a finite interval is therefore *not* unique. This should not upset anybody. In fact, there are situations in which one actually takes advantage of this fact in order to choose the type series which simplifies the results. (Solution of partial differential equations by separation of variables is one example.)

1.4   COMPLEX FOURIER SPECTRUM (SOMETIMES CALLED "LINE SPECTRUM")

To each value of n in the complex Fourier Series representation of a time function, we have assigned a complex weighting factor, c_n. We can plot these c_n as a function of "n." Note that this really requires two graphs since the c_n are, in general, complex numbers. That is, we can have one plot of the magnitude of c_n and one plot of the phase. Alternatively, we could graph real and imaginary parts. We further note that this graph would be discrete. That is, it only has non-zero value for discrete values of the abscissa. (e.g., $c_{1/2}$ has no meaning.)

A more meaningful quantity to plot as the abscissa would be n times ω_0, a quantity corresponding to the frequency of the complex exponential for which c_n is a weighting coefficient. This plot of c_n vs. $n\omega_0$ is called the "Complex Fourier Spectrum."

Illustrative Example 1.3

Find the Complex Fourier Spectrum of a full wave rectified (absolute value of) cosine wave,

$$f(t) = |\cos t|$$

as shown in Fig. 1.5.

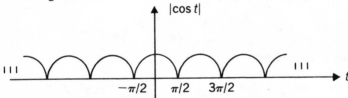

Figure 1.5 $f(t)$ for Illustrative Example 1.3.

Solution

In order to find the Fourier Spectrum, we must first find the exponential (complex) Fourier Series expansion of this waveform.

As in Illustrative Example 1.2 $\omega_0 = 2$. Instead of evaluating the c_n from

$$c_n = \frac{1}{T} \int_{-T/2}^{T/2} f(t) e^{-j2nt}\, dt$$

we can use the results of Example 1.2 to find the Complex Fourier Series directly. Recall that the trigonometric Fourier Series expansion was found to be

$$f(t) = \frac{2}{\pi} + \sum_{n=1}^{\infty} a_n \cos 2nt$$

where

$$a_n = \frac{2}{\pi}\left[\frac{(-1)^{n+1}}{2n-1} + \frac{(-1)^n}{2n+1}\right]$$

We note that Euler's identity for the cosine,

$$\cos 2nt = \frac{e^{j2nt} + e^{-j2nt}}{2} \tag{1.34}$$

can be used to go from the trigonometric to the exponential form of the expansion,

$$\begin{aligned}
f(t) &= \frac{2}{\pi} + \sum_{n=1}^{\infty}\frac{a_n}{2}e^{j2nt} + \sum_{n=1}^{\infty}\frac{a_n}{2}e^{-j2nt} \\
&= \frac{2}{\pi} + \sum_{n=1}^{\infty}\frac{a_n}{2}e^{j2nt} + \sum_{n=-\infty}^{-1}\frac{a_{-n}}{2}e^{+j2nt}
\end{aligned} \tag{1.35}$$

where, in the second line, we have simply made a change of variables. Comparing this to Eq. (1.25), we see that

$$c_0 = \frac{2}{\pi}$$

$$c_n = \frac{a_n}{2} \qquad (n > 0)$$

$$c_n = \frac{a_{-n}}{2} \qquad (n < 0)$$

Writing several terms, we have

$$c_1 = c_{-1} = \tfrac{1}{2}a_1 = \tfrac{2}{3}\pi$$
$$c_2 = c_{-2} = \tfrac{1}{2}a_2 = -\tfrac{2}{15}\pi$$

The series is of the form

$$f(t) = \frac{2}{\pi} + \frac{2}{3\pi}e^{j2t} + \frac{2}{3\pi}e^{-j2t} - \frac{2}{15\pi}e^{j4t} - \frac{2}{15\pi}e^{-j4t}, \ldots \tag{1.36}$$

The Fourier Spectrum is simply a sketch of c_n vs. $n\omega_0$, as shown in Fig. 1.6

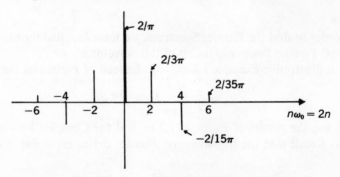

Figure 1.6 Fourier Spectrum for Illustrative Example 1.3.

for this example. Note that only one plot is necessary since, in this particular example, the c_n were all *real* numbers.

Illustrative Example 1.4

It is desired to approximate the time function,

$$f(t) = |\cos t|$$

by a constant. This constant is to be chosen so as to minimize the error. The error is defined as the average of the square difference between $f(t)$ and the approximating constant. Find the constant.

Solution

$$f(t) = |\cos t|$$

Approximation to $f(t) \triangleq C$

Square error, $e^2(t) = [|\cos t| - C]^2$

Find the average square error by integrating,

$$\{e^2(t)\}_{\text{avg}} = \lim_{T \to \infty} \frac{1}{T} \int_{-T/2}^{T/2} [|\cos t| - C]^2 \, dt \tag{1.37}$$

We could evaluate this integral and differentiate with respect to C in order to minimize the quantity, or we can be more devious and borrow a result from orthogonal vector spaces. Recall that, "Even though the orthogonal set used to approximate a general vector does not contain a sufficient number of elements to describe the vector exactly, the weighting coefficients are chosen as if the set were sufficient!"

The above problem is simply just such a case where we must approximate $f(t)$ by the first term in its Fourier Series expansion. The best value to choose for the constant, C, is the a_0, or constant term in the Fourier Series expansion. This expansion was found in Illustrative Example 1.2, where the value of a_0 was $2/\pi$. The function, $f(t)$, and its approximation are shown in Fig. 1.7.

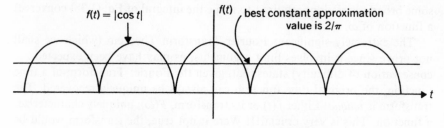

Figure 1.7   Result of Illustrative Example 1.4.

In summary, if we wish to approximate $|\cos t|$ by a constant so as to minimize the mean square error, the best value of the constant to choose is $2/\pi$.

1.5 THE FOURIER TRANSFORM

The vast majority of interesting signals extend for all time and are non-periodic. One would certainly not go through any great effort to transmit a periodic wave since all of the information is contained in one period. He could either transmit the signal over one period only, or transmit the values of the Fourier Series coefficients in the form of a list. The question therefore arises as to whether or not we can write a "Fourier-type" series for a non-periodic signal.

A non-periodic signal can be viewed as a limiting case of a periodic signal, where the period of the signal approaches infinity. If the period, T, approaches infinity, the Fourier Series summation representation of $f(t)$ becomes an integral. In this manner, we could develop the Fourier Integral theory.

In order to avoid the limiting processes required to go from Fourier Series to Fourier Integral, this text will take an axiomatic approach. That is, we will *define* the Fourier Transform, and then show that it is extremely useful. There need be no loss in motivation by approaching the transform in this "pull out of a hat" manner, since its extreme versatility will become rapidly obvious.

You will recall that a common everyday *function* is actually a set of rules which substitutes one number for another number. That is, $f(t)$ is a set of rules which assigns a number, $f(t)$, to any number, t, in the domain. In a similar manner, a *transform* is a set of rules which substitutes one function for another function. We *define* one particular transform as follows,

$$F(\omega) \triangleq \int_{-\infty}^{\infty} f(t)e^{-j\omega t}\, dt \qquad (1.38)$$

Since "t" is a dummy variable of integration, the result of the integral evaluation (after the limits are plugged in) is not a function of t, but only a function of ω. We have therefore given a rule that assigns to every function of t (with some broad restrictions required to make the integral of Eq. (1.38) converge) a function of ω.

The extremely significant Fourier Transform Theorem (which we shall not prove since it involves limit arguments, as may have been expected . . . conservation of difficulty) states that, given the Fourier Transform of a time function, the original time function can always be uniquely recovered. The transform is *unique*! Either $f(t)$ or its transform, $F(\omega)$, uniquely characterizes a function. This is very crucial!!! Were it not true, the transform would be useless.

An example of a useless transform (the Roden Transform) follows:

To every function, $f(t)$, assign the function
$$R(\omega) = \omega^2 + 1.3$$

This transform defines a function of "ω" for every function of "t." The reason it has not become famous is that, among other factors, it is not unique. Given that the Roden Transform of a time function is $\omega^2 + 1.3$, one hasn't got a prayer of finding the $f(t)$ which led to that transform.

Actually, the Fourier Transform Theorem goes one step further than stating uniqueness. It gives the rule for recovering $f(t)$ from its Fourier Transform. The rule exhibits itself as an integral, and is almost of the same form as the original transform rule. That is, given $F(\omega)$, one can recover $f(t)$ by evaluating the following integral,

$$f(t) = \frac{1}{2\pi} \int_{-\infty}^{\infty} F(\omega)e^{j\omega t}\, d\omega \tag{1.39}$$

Equation (1.39) is sometimes referred to as "the inverse transform of $F(\omega)$." It follows that this is also unique.

We will usually use the same letter for the time function and its corresponding transform, the capital version being used for the transform. That is, if we have a time function called $g(t)$, we shall call its transform, $G(\omega)$. In cases where this is not possible, we find it necessary to adopt some alternative notational forms to associate a time function with its transform. The script capital "$\mathcal{F}$" and "$\mathcal{F}^{-1}$" are often used to denote taking the transform, or the inverse transform respectively. Thus, if $F(\omega)$ is the transform of $f(t)$, we can write

$$\mathcal{F}[f(t)] = F(\omega)$$
$$\mathcal{F}^{-1}[F(\omega)] = f(t)$$

A double ended arrow is also often used to relate a time function to its transform, sometimes known as a transform pair. Thus we would write

$$f(t) \leftrightarrow F(\omega)$$

or

$$F(\omega) \leftrightarrow f(t)$$

There are infinitely many unique transforms.[3] Why then has the Fourier Transform achieved such widespread fame and use? Certainly it must possess properties which make it far more useful than any other transform.

Indeed, we shall presently discover that the Fourier Transform is useful in a way which is analogous to the common logarithm. (Remember them from high school?) You may recall that, in order to multiply two numbers together, one often found the logarithm of each of the numbers, added the logs, and then found the number corresponding to the resulting logarithm. One goes through all of this trouble in order to avoid multiplication (a frightening prospect to high school students).

---

[3] As two examples, consider either time scaling or multiplication by a constant. That is, define $F_1(\omega) = f(2\omega)$ or $F_2(\omega) = 2f(\omega)$. For example, if $f(t) = \sin t$, $F_1(\omega) = \sin 2\omega$ and $F_2(\omega) = 2 \sin \omega$. The extension to an infinity of possible transform rules should be obvious.

$$a \quad \times \quad b \quad = \quad c$$
$$\downarrow \qquad \downarrow \qquad \downarrow \qquad \qquad \uparrow$$
$$\log(a) + \log(b) = \log(c)$$

There is an operation which must often be performed between two time functions which we call *convolution*. It is enough to scare even those few who are not frightened by multiplication. It will now be shown that, if one first takes the Fourier Transform of each of the two time functions, he can perform an operation on the transforms which corresponds to convolution of the time functions, but is considerably simpler. That operation, which corresponds to convolution of the two time functions, is multiplication of the two transforms. (Multiplication is no longer difficult once one graduates from high school.) Thus, we will multiply the two transforms together, and then find the time function corresponding to the resulting transform. It is now time for us to determine what this horrible operation of convolution involves.

1.6 CONVOLUTION

The *convolution* of two time functions, $f(t)$ and $g(t)$, is defined by the following integral operation,

$$f(t) * g(t) \triangleq \int_{-\infty}^{\infty} f(\tau)g(t - \tau)\, d\tau \tag{1.40}$$

The asterisk notation is conventional, and is read, "$f(t)$ convolved with $g(t)$."

The reason that one would ever want to perform such an operation will be deferred to Chapter 2. Suffice it to say that this operation is basic to almost any system we will study.

Note that the convolution of two functions of "t" is, itself, a function of "t" since "τ" is a dummy variable which is integrated out. The integral of Eq. (1.40) is, in general, very difficult to evaluate in closed form. An apparently simple example will back up this statement.

Illustrative Example 1.5

Evaluate the convolution of $f(t)$ with $g(t)$, where $f(t)$ and $g(t)$ are the square pulses shown in Fig. 1.8.

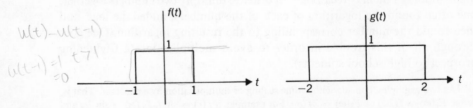

Figure 1.8 $f(t)$ and $g(t)$ for Illustrative Example 1.5.

Solution

We note that the functions can be written in the form

$$f(t) = U(t + 1) - U(t - 1)$$
$$g(t) = U(t + 2) - U(t - 2)$$

where $U(t)$ is the unit step function defined by

$$U(t) = \begin{cases} 1 & t > 0 \\ 0 & t < 0 \end{cases}$$

The convolution is defined by

$$f(t) * g(t) \triangleq \int_{-\infty}^{\infty} f(\tau)g(t - \tau) \, d\tau$$

We see that

$$f(\tau) = U(\tau + 1) - U(\tau - 1) \tag{1.41}$$

and

$$g(t - \tau) = U(t - \tau + 2) - U(t - \tau - 2) \tag{1.42}$$
$$f(\tau)g(t - \tau) = U(\tau + 1)U(t - \tau + 2) - U(\tau + 1)U(t - \tau - 2)$$
$$-U(\tau - 1)U(t - \tau + 2) + U(\tau - 1)U(t - \tau - 2)$$

Therefore, breaking the integral into parts,

$$f(t) * g(t) = \int_{-\infty}^{\infty} U(\tau + 1)U(t - \tau + 2) \, d\tau$$

$$- \int_{-\infty}^{\infty} U(\tau + 1)U(t - \tau - 2) \, d\tau$$

$$- \int_{-\infty}^{\infty} U(\tau - 1)U(t - \tau + 2) \, d\tau$$

$$+ \int_{-\infty}^{\infty} U(\tau - 1)U(t - \tau - 2) \, d\tau \tag{1.43}$$

We now note that $U(\tau + 1)$ is equal to zero for $\tau < -1$, and $U(\tau - 1)$ is zero for $\tau < 1$. Taking this into account, the limits of integration in Eq. (1.43) can be reduced to yield

$$f(t) * g(t) = \int_{-1}^{\infty} U(t - \tau + 2) \, d\tau - \int_{-1}^{\infty} U(t - \tau - 2) \, d\tau$$

$$- \int_{1}^{\infty} U(t - \tau + 2) \, d\tau + \int_{1}^{\infty} U(t - \tau - 2) \, d\tau \tag{1.44}$$

To derive this, we have replaced one of the step functions by its value, unity, in the range in which this applies. We now try to evaluate each integral separately. We note that

$$U(t - \tau + 2) = 0 \qquad (\tau > t + 2)$$

and

$$U(t - \tau - 2) = 0 \qquad (\tau > t - 2)$$

Using these facts, we have

$$\int_{-1}^{\infty} U(t - \tau + 2)\, d\tau = \int_{-1}^{t+2} d\tau = t + 3 \tag{1.45}$$

provided that $t + 2 > -1$, or equivalently, $t > -3$. Otherwise, the integral evaluates to zero. Likewise, if $t - 2 > -1$, that is, $t > 1$, we have

$$\int_{-1}^{\infty} U(t - \tau - 2)\, d\tau = \int_{-1}^{t-2} d\tau = t - 1 \tag{1.46}$$

If $t + 2 > 1$, that is, $t > -1$,

$$\int_{1}^{\infty} U(t - \tau + 2)\, d\tau = \int_{1}^{t+2} d\tau = t + 1 \tag{1.47}$$

If $t - 2 > 1$, that is, $t > 3$, we have

$$\int_{1}^{\infty} U(t - \tau - 2)\, d\tau = \int_{1}^{t-2} d\tau = t - 3 \tag{1.48}$$

Using these four results, Eq. (1.44) becomes

$$f(t) * g(t) = (t + 3)U(t + 3) - (t - 1)U(t - 1) - (t + 1)U(t + 1) \\ + (t - 3)U(t - 3) \tag{1.49}$$

These four terms, together with their sum, are sketched in Fig. 1.9. From this modest example, we can see that, if either $f(t)$ or $g(t)$ contains step functions, the evaluation of the convolution becomes quite involved.

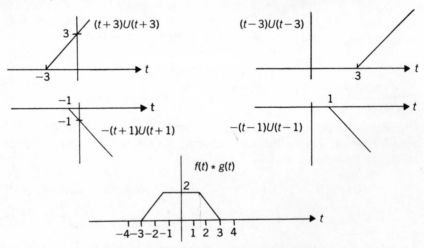

Figure 1.9   Result of Illustrative Example 1.5.

Graphical Convolution

The claim is made that, for simple (what we mean by simple should be clear at the end of this section) $f(t)$ and $g(t)$, the result of the convolution can be obtained *almost* by inspection. Even in cases where $f(t)$ and $g(t)$ are quite complex, or even not precisely known, certain observations can be made about their convolution without actually performing the detailed calculations. In many communications' applications, these general observations will be sufficient, and the exact convolution will not be required.

This inspection procedure is known as *graphical convolution*. We shall arrive at the technique by examining the definition of convolution.

$$f(t) * g(t) \triangleq \int_{-\infty}^{\infty} f(\tau)g(t - \tau)\, d\tau$$

One of the original functions is $f(\tau)$, where the independent variable is now called "τ."

The mirror image of $g(\tau)$ is represented by $g(-\tau)$, that is, $g(\tau)$ flipped around the *y*-axis.

The definition now tells us that for a given "t," we form $g(t - \tau)$, which represents the function $g(-\tau)$ shifted to the right by "t." We then take the product,

$$f(\tau)g(t - \tau)$$

and integrate this product (i.e., find the area under it) in order to find the value of the convolution for that particular value of t. This procedure is illustrated for 10 different values of t, and the functions $f(t)$ and $g(t)$ from Illustrative Example 1.5 (*see* Fig. 1.10). The fact that $g(\tau)$ was flipped is not obvious in this case since $g(t)$ was an even function to begin with.

Viewing Fig. 1.10, we can interpolate between each pair of values of t, and plot the area as a function of t. This yields the convolution as sketched in Fig. 1.11.

With practice, the calculation at many points will become unnecessary, and the answer could truly be arrived at by inspection.

Note that the result is identical to that found in Illustrative Example 1.5.

Illustrative Example 1.6

We wish to find $f(t) * f(t)$, where the only information we have about $f(t)$ is that it is zero for $|t| > 1$. That is, $f(t)$ is limited to the range between $t = -1$ and $t = +1$. A typical, but not exclusive, $f(t)$ is sketched in Fig. 1.12.

Solution

Not knowing $f(t)$ exactly, we certainly cannot find $f(t) * f(t)$ exactly. To see how much information we can obtain concerning this convolution, we attempt graphical convolution. Figure 1.13 is a sketch of $f(\tau)$ and $f(t - \tau)$.

We note that, as "t" increases from zero, the two functions illustrated have less and less of the τ-axis in common. When "t" reaches $+2$, the two functions separate. That is, at $t = 2$, we have

$$f(\tau)f(2 - \tau) = 0$$

as sketched in Fig. 1.14.

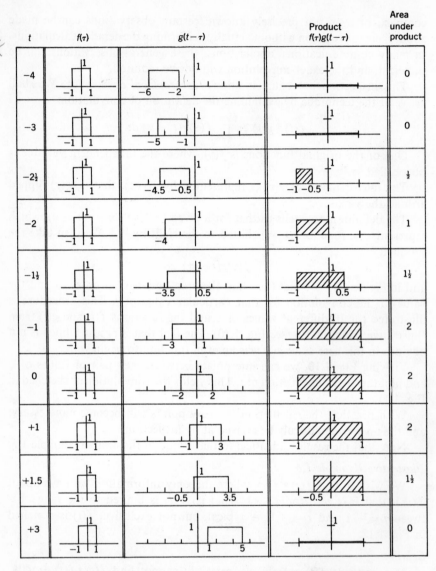

Figure 1.10 Illustration of graphical convolution of Illustrative Example 1.5.

The two functions continue to have nothing in common for all $t > 2$. Likewise, for negative "t" one can see that the product of the two functions is zero as long as $t < -2$. That is, $f(-\tau)$ can move to the left by a distance of 2 units before it separates from $f(\tau)$.

Therefore, although we don't know $f(t) * f(t)$ exactly, we do know what range it occupies along the t-axis. We find that it is zero for $|t| > 2$. A

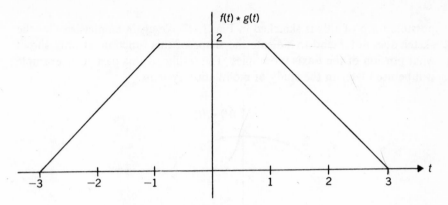

Figure 1.11 Convolution result from Fig. 1.10.

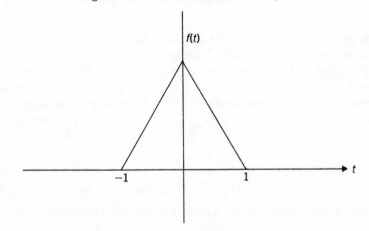

Figure 1.12 Typical $f(t)$ for Illustrative Example 1.6.

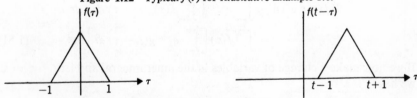

Figure 1.13 $f(\tau)$ and $f(t - \tau)$ for representative $f(t)$ of Fig. 1.12.

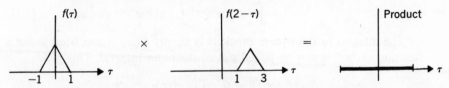

Figure 1.14 Convolution product of Illustrative Example 1.6 goes to zero at $t = 2$.

23

possible form of this is sketched in Fig. 1.15. We again emphasize that the sketch does not intend to indicate the shape of this function. It only shows what portion of the t-axis it occupies. The results of this particular example will be used later in the study of modulation systems.

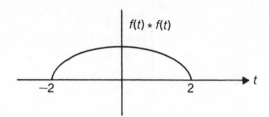

Figure 1.15   Typical result of the convolution in Illustrative Example 1.6.

Now that we have a feel for the operation known as convolution, let us return to our study of the Fourier Transform.

As promised, the *convolution theorem* states that the Fourier Transform of a time function which is the convolution of two time functions is equal to the product of the two corresponding Fourier Transforms. That is, if

$$f(t) \longleftrightarrow F(\omega)$$

and

$$g(t) \longleftrightarrow G(\omega)$$

then

$$f(t) * g(t) \longleftrightarrow F(\omega)G(\omega) \qquad (1.50)$$

The proof of this is rather direct. Simply evaluate the Fourier Transform of $f(t) * g(t)$.

$$\mathfrak{F}[f(t) * g(t)] \triangleq \int_{-\infty}^{\infty} e^{-j\omega t} \left[\int_{-\infty}^{\infty} f(\tau)g(t - \tau)\, d\tau \right] dt$$

$$= \int_{-\infty}^{\infty} f(\tau) \left[\int_{-\infty}^{\infty} e^{-j\omega t}g(t - \tau)\, dt \right] d\tau \qquad (1.51)$$

If we now make a change of variables in the inner integral and let $t - \tau = k$, we have

$$\mathfrak{F}[f(t) * g(t)] = \int_{-\infty}^{\infty} f(\tau) \left[\int_{-\infty}^{\infty} e^{-j\omega\tau}g(k)e^{-j\omega k}\, dk \right] d\tau$$

$$= \int_{-\infty}^{\infty} f(\tau)e^{-j\omega\tau} \left[\int_{-\infty}^{\infty} g(k)e^{-j\omega k}\, dk \right] d\tau \qquad (1.52)$$

The integral in the square brackets is simply $G(\omega)$. Since $G(\omega)$ is not a function of "τ," it can be pulled out of the outer integral. This yields

$$\mathfrak{F}[f(t) * g(t)] = G(\omega) \int_{-\infty}^{\infty} f(\tau)e^{-j\omega\tau}\, d\tau$$

$$= G(\omega)F(\omega) \qquad [\text{q.e.d.}] \qquad (1.53)$$

Convolution is an operation performed between two functions. They need not be functions of the independent variable "t." We could just as easily have convolved two Fourier Transforms together to get a third function of "ω."

$$H(\omega) = F(\omega) * G(\omega) = \int_{-\infty}^{\infty} F(k)G(\omega - k) \, dk \qquad (1.54)$$

Since the integral defining the Fourier Transform and that yielding the inverse transform are quite similar, one might guess that convolution of two transforms corresponds to multiplication of the two corresponding time functions. Indeed, one can prove, exactly in an analogous way to the above proof, that

$$F(\omega) * G(\omega) \longleftrightarrow 2\pi f(t)g(t) \qquad (1.55)$$

In order to prove this, simply calculate the inverse Fourier Transform of $F(\omega) * G(\omega)$.

Equation (1.50) is sometimes called the "time convolution theorem" and Eq. (1.55), the "frequency convolution theorem."

Illustrative Example 1.7

Use the convolution theorem in order to evaluate the following integral,

$$\int_{-\infty}^{\infty} \frac{\sin 3\tau}{\tau} \frac{\sin (t - \tau)}{t - \tau} \, d\tau$$

Solution

We recognize that the above integral represents the convolution of the following two time functions,

$$\frac{\sin 3t}{t} * \frac{\sin t}{t}$$

The transform of the integral is therefore the product of the transforms of the above two time functions. These two transforms may be found in Appendix II. The appropriate transforms and their product are sketched in Fig. 1.16.

The time function corresponding to the convolution is simply the inverse transform of this product. This is easily seen to be

$$\frac{\pi \sin t}{t}$$

Note that when $(\sin t)/t$ is convolved with $(\sin 3t)/t$, the only change that takes place is the addition of a scale factor, π. In fact, if $(\sin t)/t$ had been convolved with $(\sin 3t)/\pi t$, it would not have changed at all. This surprising result is no accident. We will see, in the next chapter, that there are entire classes of functions which remain unchanged after convolution with $(\sin t)/\pi t$. If this were not true, many of the most basic communication systems could never function.

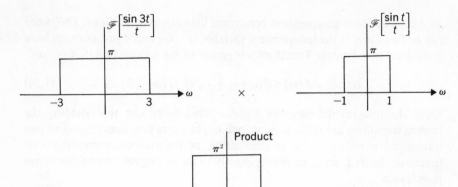

Figure 1.16 Transforms and product for Illustrative Example 1.7.

Parseval's Theorem

There is little similarity between a function and its Fourier Transform. That is, the waveshapes of the two functions are, in general, completely different. Certain relationships do exist, however, between the "energy" of a time function and the "energy" of its transform. Thus, if we knew the transform of a time function, and only wished to know the energy of the time function, such relationships would eliminate the need to evaluate the inverse transform.

Parseval's Theorem is such a relationship. It is easily derived from the frequency convolution theorem. Starting with that theorem, we have

$$f(t)g(t) \longleftrightarrow \frac{1}{2\pi} F(\omega) * G(\omega)$$

$$\mathcal{F}[f(t)g(t)] = \int_{-\infty}^{\infty} f(t)g(t)e^{-j\omega t}\, dt$$

$$= \frac{1}{2\pi} \int_{-\infty}^{\infty} F(k)G(\omega - k)\, dk \qquad (1.56)$$

Since the above equality holds for all values of ω, we can take the special case of $\omega = 0$. For this value of ω, Eq. (1.56) becomes

$$\int_{-\infty}^{\infty} f(t)g(t)\, dt = \frac{1}{2\pi} \int_{-\infty}^{\infty} F(k)G(-k)\, dk \qquad (1.57)$$

Equation (1.57) is one form of Parseval's formula. It can be made to relate to energy by further taking the special case of

$$g(t) = f^*(t)$$

where the asterisk (*) represents complex conjugate.

We can show that, if $g(t) = f^*(t)$, then

$$G(\omega) = F^*(-\omega)$$

We illustrate this last relationship by finding the transform of $f(t)$, evaluating it at "$-\omega$," and taking the conjugate,

$$F(\omega) \triangleq \int_{-\infty}^{\infty} f(t)e^{-j\omega t}\, dt$$

$$F(-\omega) = \int_{-\infty}^{\infty} f(t)e^{j\omega t}\, dt$$

$$F^*(-\omega) = \int_{-\infty}^{\infty} f^*(t)e^{-j\omega t}\, dt \triangleq \mathcal{F}[f^*(t)] \tag{1.58}$$

Plugging these relationships into Parseval's formula (Eq. 1.57), we get

$$\int_{-\infty}^{\infty} |f^2(t)|\, dt = \frac{1}{2\pi} \int_{-\infty}^{\infty} |F^2(\omega)|\, d\omega \tag{1.59}$$

We have used the fact that the product of a function with its complex conjugate is equal to the magnitude of the function, squared.

Equation (1.59) shows that the energy under the time function is equal to the energy under its Fourier Transform, with a factor of 2π thrown in.[4] The integral of the magnitude squared is often called the energy of the signal since this would be the amount of energy, in watt-seconds, dissipated in a 1-ohm resistor if the signal represented the voltage across the resistor (or the current through it).

1.7 PROPERTIES OF THE FOURIER TRANSFORM

We shall illustrate some of the more important properties of the Fourier Transform by means of examples.

Illustrative Example 1.8

Evaluate the Fourier Transform of $f(t)$, where

$$f(t) = \begin{cases} 1 & \text{for } |t| < 1 \\ 0 & \text{otherwise} \end{cases}$$

as shown in Fig. 1.17.

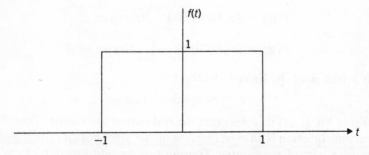

Figure 1.17 $f(t)$ for Illustrative Example 1.8.

[4]The factor of 2π which appears in many of our relationships can be eliminated if one redevelops the various formulae using f for frequency instead of ω. Recall that $\omega = 2\pi f$.

Solution

From the definition of the Fourier Transform,

$$F(\omega) \triangleq \int_{-\infty}^{\infty} f(t)e^{-j\omega t}\, dt$$

$$= \int_{-1}^{1} e^{-j\omega t}\, dt$$

$$= \frac{-e^{-j\omega t}}{j\omega}\Big|_{-1}^{+1} = \frac{e^{j\omega} - e^{-j\omega}}{j\omega} = 2\frac{\sin \omega}{\omega} \qquad (1.60)$$

This transform is sketched in Fig. 1.18.

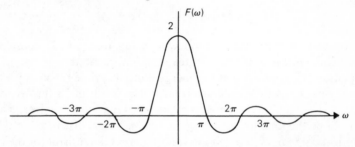

Figure 1.18   Transform of $f(t)$ of Illustrative Example 1.8.

We note several surprising things about this result. Although the Fourier Transform is, in general, a complex function of "ω," this particular transform turned out to be a purely real function. It also is an even function of "ω." Is there any reason that we might have expected these properties?

Suppose, first, that $f(t)$ is a real function of time, as will most often be the case. The real and imaginary parts of $F(\omega)$ can be then identified as follows,

$$F(\omega) \triangleq R(\omega) + jX(\omega) = \int_{-\infty}^{\infty} f(t)e^{-j\omega t}\, dt$$

where

$$R(\omega) = Re\{F(\omega)\} = \int_{-\infty}^{\infty} f(t) \cos \omega t\, dt$$

$$X(\omega) = Im\{F(\omega)\} = -\int_{-\infty}^{\infty} f(t) \sin \omega t\, dt \qquad (1.61)$$

This is true since, by Euler's identity,

$$e^{-j\omega t} = \cos \omega t - j \sin \omega t$$

From Eq. (1.61) it is clear that the real part of the Fourier Transform of a real time function is an even function of "ω" since $\cos \omega t$ is itself even. The imaginary part of the Fourier Transform is an odd function of "ω" since $\sin \omega t$ is odd. We can also see that, since the limits of integration in Eq. (1.61) are symmetrical, $X(\omega) = 0$ for even $f(t)$, and $R(\omega) = 0$ for odd $f(t)$. That is, if $f(t)$ is real and even, $f(t) \sin \omega t$ is odd, and integrates to zero. Likewise, if $f(t)$ is odd, $f(t) \cos \omega t$ is odd, and integrates to zero.

In summary, if $f(t)$ is real and even, $F(\omega)$ is real and even. If $f(t)$ is real and odd, $F(\omega)$ is imaginary and odd.

Illustrative Example 1.9

Find the Fourier Transform of $f(t)$,

$$f(t) = \begin{cases} 1 & \text{for } 0 < t \leq 2 \\ 0 & \text{otherwise} \end{cases}$$

as sketched in Fig. 1.19.

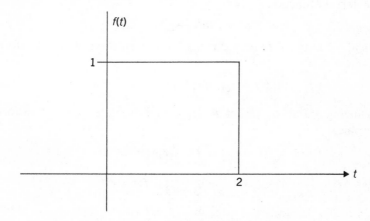

Figure 1.19 $f(t)$ for Illustrative Example 1.9.

Solution

From the definition of the Fourier Transform,

$$
\begin{aligned}
F(\omega) &\triangleq \int_{-\infty}^{\infty} f(t) e^{-j\omega t} \, dt \\
&= \int_{0}^{2} e^{-j\omega t} \, dt \\
&= \frac{-e^{-j\omega t}}{j\omega} \Big|_{0}^{2} = \frac{1}{j\omega} [1 - e^{-2j\omega}] \\
&= \frac{e^{-j\omega}}{j\omega} [e^{j\omega} - e^{-j\omega}] \\
F(\omega) &= e^{-j\omega} \left[2 \frac{\sin \omega}{\omega} \right]
\end{aligned}
$$

We first note that this $F(\omega)$ is complex as expected, since $f(t)$ is neither even nor odd.

Is it not somewhat intriguing that the Fourier Transform of the $f(t)$ of Illustrative Example 1.9 is almost the same as that of Illustrative Example

1.8, except for multiplication by $e^{-j\omega}$? Maybe we could have saved ourselves
work and somehow derived this second transform from the first. Indeed, this
is the case. We shall invoke the so-called "time shift theorem." This theorem
is certainly not basic nor earthshaking, but it does possess the potential of
saving computation time, so we shall spend a paragraph on it.

Time Shift Theorem

Suppose that we know the transform of $f(t)$ is $F(\omega)$, and we wish to find
the transform of $f(t - t_0)$, that is, the transform of the time function formed
by shifting the original $f(t)$ by t_0 in time. We shall prove that this transform
of the delayed function is given by

$$e^{-j\omega t_0}F(\omega) \leftrightarrow f(t - t_0)$$

Proof. The proof follows directly from evaluation of the transform
of $f(t - t_0)$.

$$\mathfrak{F}[f(t - t_0)] \triangleq \int_{-\infty}^{\infty} f(t - t_0)e^{-j\omega t}\, dt \tag{1.62}$$

Changing variables, we let $t - t_0 = \tau$. For finite t_0, the transform
becomes

$$\mathfrak{F}[f(t - t_0)] = \int_{-\infty}^{\infty} f(\tau)e^{-j\omega(\tau+t_0)}\, d\tau$$

$$= e^{-j\omega t_0} \int_{-\infty}^{\infty} f(\tau)e^{-j\omega\tau}\, d\tau$$

$$= e^{-j\omega t_0}F(\omega) \tag{1.63}$$

Using this theorem, we could have immediately seen that the $f(t)$ of Illustrative Example 1.9 was that of Illustrative Example 1.8 shifted by $t_0 = 1$.
Therefore its transform is the solution to Example 1.8 multiplied by $e^{-j\omega}$,

$$e^{-j\omega}\left[2\frac{\sin\omega}{\omega}\right]$$

which we previously calculated in a much more roundabout manner. In an
exactly analogous way, we can prove a dual theorem.

Frequency Shift Theorem

Since the Fourier Transform integral and the Inverse Fourier Transform
integral are so similar, one could expect a dual of the time shift theorem to
exist. Indeed, suppose that we know that the transform of $f(t)$ is $F(\omega)$. Then
the function which has $F(\omega - \omega_0)$ as its transform, that is, $F(\omega)$ shifted by
ω_0, is $e^{j\omega_0 t}f(t)$.

$$F(\omega - \omega_0) \leftrightarrow e^{j\omega_0 t}f(t) \tag{1.64}$$

Proof. The proof follows directly from evaluation of the inverse transform of $F(\omega - \omega_0)$.

$$\mathfrak{F}^{-1}[F(\omega - \omega_0)] = \frac{1}{2\pi} \int_{-\infty}^{\infty} F(\omega - \omega_0)e^{j\omega t}\, d\omega \qquad (1.65)$$

Changing variables, we let

$$\omega - \omega_0 = k$$

$$\begin{aligned}
\mathfrak{F}^{-1}[F(\omega - \omega_0)] &= \frac{1}{2\pi} \int_{-\infty}^{\infty} F(k)e^{jt(k+\omega_0)}\, dk \\
&= \frac{1}{2\pi}e^{j\omega_0 t} \int_{-\infty}^{\infty} F(k)e^{jkt}\, dk \\
&= e^{j\omega_0 t}f(t)
\end{aligned} \qquad (1.66)$$

Illustrative Example 1.10

Find the Fourier Transform of $f(t)$, where

$$f(t) = \begin{cases} e^{jt} & |t| < 1 \\ 0 & \text{otherwise} \end{cases}$$

Solution

This $f(t)$ is the same as that of Illustrative Example 1.8, except for a multiplying factor of e^{jt}. The frequency shift theorem indicates that if a function of time is multiplied by e^{jt}, its transform is shifted by 1. We therefore take the transform found in Illustrative Example 1.8 and substitute $(\omega - 1)$ for (ω).

$$F(\omega) = 2\frac{\sin(\omega - 1)}{\omega - 1}$$

This transform is sketched in Fig. 1.20.

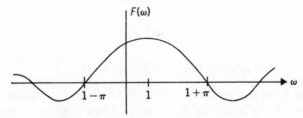

Figure 1.20 $F(\omega)$ of Illustrative Example 1.10.

Note that in this case the time function is neither even nor odd. However, the Fourier Transform turned out to be a real function. Such a situation can arise only because the time function is not real.

The final property that we wish to exploit is undoubtedly the most important. We have saved it for last since it is probably the most obvious, and we can both use a breathing spell before getting to the next section.

Illustrative Example 1.11

Find the Fourier Transform of $f(t)$, where

$$f(t) = \begin{cases} 1 & -1 < t \le 0 \\ 2 & 0 < t \le 1 \\ 1 & 1 < t \le 2 \\ 0 & \text{otherwise} \end{cases}$$

as sketched in Fig. 1.21.

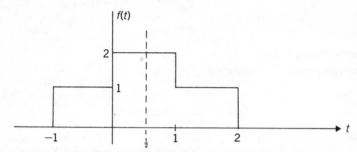

Figure 1.21 $f(t)$ for Illustrative Example 1.11.

Solution

We can evaluate this transform directly, and with much labor, or we can take advantage of the fact that the Fourier Transform is a *linear* transform. That is, the Fourier Transform of a sum of time functions is equal to the sum of their individual Fourier Transforms. More formally, if

$$f_1(t) \leftrightarrow F_1(\omega)$$

and

$$f_2(t) \leftrightarrow F_2(\omega)$$

then

$$af_1(t) + bf_2(t) \leftrightarrow aF_1(\omega) + bF_2(\omega) \tag{1.67}$$

where a and b are any two constants. The proof of this follows in one line directly from the definition of the Fourier Transform, and should be performed by the student.

Using the linearity property, we observe that $f(t)$ given above is simply the sum of the time function of Illustrative Example 1.8 with that of Illustrative Example 1.9. Therefore, the transform is given by the sum of the two transforms.

$$F(\omega) = 2\frac{\sin \omega}{\omega}[1 + e^{-j\omega}] \tag{1.68}$$

This transform is complete in the above form. However, we shall perform a slight manipulation to illustrate a point.

$$F(\omega) = 2\frac{\sin \omega}{\omega}e^{-j\omega/2}[e^{j\omega/2} + e^{-j\omega/2}]$$

$$= 4\frac{\sin \omega \cos (\omega/2)}{\omega}[e^{-j\omega/2}]$$

(1.69)

We now recognize that $F(\omega)$ is in the form of a real, even function of "ω" multiplied by $e^{-j\omega/2}$. This indicates that the corresponding time function must be the result of shifting a real, even time function by $t_0 = \frac{1}{2}$. This is indeed the case as can be easily observed from Fig. 1.21.

1.8 SINGULARITY FUNCTIONS

Suppose that we wished to evaluate the transform of a function, $g(t)$, which was equal to 1 for $|t| < a$, that is, a pulse of width $2a$. We would proceed exactly as we did in Illustrative Example 1.8, where "a" was given as "1." We would find that

$$G(\omega) = 2\frac{\sin a\omega}{\omega}$$

(1.70)

As a check, for $a = 1$, this reduces to the $F(\omega)$ found in Example 1.8. The functions $g(t)$ and $G(\omega)$ are sketched in Fig. 1.22.

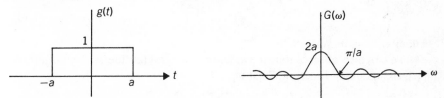

Figure 1.22   A pulse and its transform.

Suppose that we now wanted to find the Fourier Transform of a constant, say $f(t) = 1$ for all t. We could consider this as the limit of the above $g(t)$ as "a" approaches infinity. We attempt this roundabout approach since the straightforward technique fails in this case. That is, plugging $f(t) = 1$ into the defining integral for the Fourier Transform yields

$$F(\omega) = \int_{-\infty}^{\infty} e^{-j\omega t}\, dt$$

(1.71)

This integral does not converge (i.e., it oscillates). If we proceed via the limit argument, we find that unfortunately, the limit of the transform in Eq. (1.70) as "a" approaches infinity does not exist. We can make the general observations that, as "a" increases, the height of the main peak of $G(\omega)$ increases while its width decreases. Thus, we can only timidly suggest that, in the limit, the height goes to infinity and the width to zero. This sounds like a pretty ridiculous function, and, indeed, it is not a function at all since it is not defined at $\omega = 0$. If we insist upon saying anything about the Fourier Transform of a constant, we must now restructure our way of thinking.

This restructuring is begun by defining a new function which is not really a function at all. This new "function" shall be given the name, *impulse*. We will get around the fact that it is not really a function by also *defining* its behavior in every conceivable situation. The symbol for this function is the Greek letter *delta* (δ), and the function is denoted as $\delta(t)$.

The usual, though non-rigorous, definition of the impulse function is formed by making three simple observations, two of which have already been hinted at. That is,

$$\begin{aligned} \delta(t) &= 0 \qquad (t \neq 0) \\ \delta(t) &= \infty \qquad (t = 0) \end{aligned} \tag{1.72}$$

The third property is that the total area under the impulse function is equal to unity.

$$\int_{-\infty}^{\infty} \delta(t)\, dt = 1 \tag{1.73}$$

Since all of the area of $\delta(t)$ is concentrated at one point, the limits on the above integral can be moved in toward $t = 0$ without changing the value of the definite integral. Thus,

$$\int_{b}^{a} \delta(t)\, dt = 1 \tag{1.74}$$

as long as $a < 0$ and $b > 0$.

One can also observe that the indefinite integral (antiderivative) of $\delta(t)$ is $U(t)$, the unit step function. That is,

$$\int_{-\infty}^{t} \delta(\tau)\, d\tau = \begin{cases} 1 & t > 0 \\ 0 & t < 0 \end{cases} \triangleq U(t) \tag{1.75}$$

We called the definition comprised of the above three observations "non-rigorous." Some elementary study of singularity functions shows that we have not uniquely defined a delta function. That is, there are other functions in addtion to the impulse which satisfy the above three conditions. However these three conditions can be used to indicate (not prove) a fourth. This fourth property serves as a unique definition of the delta function, and will represent the only property of the impulse which we ever make use of. We shall integrate the product of an arbitrary time function with $\delta(t)$.

$$\int_{-\infty}^{\infty} f(t)\delta(t)\, dt = \int_{-\infty}^{\infty} f(0)\delta(t)\, dt \tag{1.76}$$

In Eq. (1.76) we have claimed that we could replace $f(t)$ by a constant function equal to $f(0)$ without changing the value of the integral. This requires some justification.

Suppose that we have two functions, $g_1(t)$ and $g_2(t)$, and we consider the product of each of these with a third function, $h(t)$. As long as $g_1(t) = g_2(t)$ at all values of "t" for which $h(t)$ is non-zero, then $g_1(t)h(t) = g_2(t)h(t)$. At those values of "t" for which $h(t) = 0$, the values of $g_1(t)$ and $g_2(t)$ have no effect on the product. One possible example is sketched in Fig. 1.23. For $g_1(t)$, $g_2(t)$ and $h(t)$ as shown in Fig. 1.23,

$$g_1(t)h(t) = g_2(t)h(t)$$

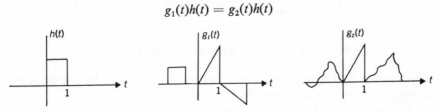

Figure 1.23   Example of $h(t)$, $g_1(t)$, and $g_2(t)$.

Returning to Eq. (1.76), we note that $\delta(t)$ is zero for all $t \neq 0$. Therefore, the product of $\delta(t)$ with any time function only depends upon the value of the time function at $t = 0$. We illustrate several possible functions which will have the same product with $\delta(t)$ as does $f(t)$. (*See* Fig. 1.24.)

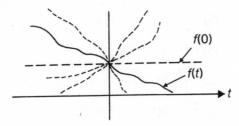

Figure 1.24   Several possible functions all having the same product with $\delta(t)$.

Out of the infinity of possibilities, the constant time function is a wise choice, since we can now factor it out of the integral to get

$$\int_{-\infty}^{\infty} f(t)\delta(t)\,dt = f(0) \int_{-\infty}^{\infty} \delta(t)\,dt = f(0) \tag{1.77}$$

This is a very significant result, and is sometimes appropriately given the name of "the sifting or sampling property of the delta (impulse) function," since the impulse sifts out the value of $f(t)$ at the time, $t = 0$. Note that a great deal of information has been lost since the result depends only upon the value of $f(t)$ at one point.

In a similar manner, a change of variables yields a shifted impulse with the analogous sifting property.

$$\int_{-\infty}^{\infty} f(t)\delta(t - t_0)\,dt = \int_{-\infty}^{\infty} f(k + t_0)\delta(k)\,dk \tag{1.78}$$
$$= f(k + t_0)|_{k=0} = f(t_0)$$

Figure 1.25 shows a sketch of $\delta(t)$ and $\delta(t - t_0)$. The upward pointing arrow is a generally accepted technique to indicate an actual value of infinity.

Figure 1.25   Pictorial representation of delta function.

The number next to the arrow indicates the total area under the impulse, sometimes known as its "strength."

Equations (1.77) and (1.78) are the only things that one must know about the impulse function. Indeed, either of these can be treated as the definition of the impulse. Note that $f(t)$ can be any time function for which $f(t_0)$ can be defined.

Illustrative Example 1.12

Evaluate the following integrals:

(a) $\displaystyle\int_{-\infty}^{\infty} \delta(t)[t^2 + 1]\, dt$

(b) $\displaystyle\int_{-1}^{2} \delta(t - 1)[t^2 + 1]\, dt$

(c) $\displaystyle\int_{3}^{5} \delta(t - 1)[t^3 + 4t + 2]\, dt$

(d) $\displaystyle\int_{-\infty}^{\infty} \delta(1 - t)[t^4 + 2]\, dt$

Solution

(a) Straightforward application of the sifting property yields

$$\int_{-\infty}^{\infty} \delta(t)[t^2 + 1]\, dt = 0^2 + 1 = 1 \tag{1.79}$$

(b) Since the impulse falls within the range of integration,

$$\int_{-1}^{2} \delta(t - 1)[t^2 + 1]\, dt = 1^2 + 1 = 2 \tag{1.80}$$

(c) The impulse occurs at $t = 1$, which is outside the range of integration. Therefore,

$$\int_{3}^{5} \delta(t - 1)[t^3 + 4t + 2]\, dt = 0 \tag{1.81}$$

(d) $\delta(1 - t)$ falls at $t = 1$ since this is the value of "t" which makes the argument equal to zero. Therefore,[5]

$$\delta(1 - t) = \delta(t - 1)$$

---

[5] Many arguments can be advanced for regarding the impulse as an even function. For one thing, it can be considered as the derivative of an odd function. For another, we will see that its transform is real and even.

and

$$\int_{-\infty}^{\infty} \delta(1 - t)[t^4 + 2] \, dt = 1^4 + 2 = 3 \qquad (1.82)$$

From Eq. (1.77) it is a simple matter to find the Fourier Transform of the impulse function.

$$\delta(t) \longleftrightarrow \int_{-\infty}^{\infty} \delta(t)e^{-j\omega t} \, dt = e^{-j\omega 0} = 1 \qquad (1.83)$$

This is indeed a very nice Fourier Transform for a function to have. One can guess that it may prove significant since it is the unity, or identity, multiplicative element. That is, anything multiplied by 1 is left unchanged.

One manifestation of this occurs when one convolves an arbitrary time function with $\delta(t)$.

$$\delta(t) * f(t) \triangleq \int_{-\infty}^{\infty} \delta(\tau)f(t - \tau) \, d\tau = f(t - 0) = f(t) \qquad (1.84)$$

That is, any function convolved with an impulse remains unchanged. The proof of this is equally trivial in the frequency domain. Recall that convolution in the time domain corresponds to multiplication in the frequency domain. Since the transform of $\delta(t)$ is unity, the product of this transform with the transform of $f(t)$ is simply the transform of $f(t)$.

$$f(t) * \delta(t) \longleftrightarrow F(\omega)\mathcal{F}[\delta(t)] = F(\omega) \longleftrightarrow f(t) \qquad (1.85)$$

Thus, convolution with $\delta(t)$ does not change the function.

If we convolve $f(t)$ with the shifted impulse, $\delta(t - t_0)$, we find

$$\delta(t - t_0) * f(t) = \int_{-\infty}^{\infty} \delta(\tau - t_0)f(t - \tau) \, d\tau = f(t - t_0) \qquad (1.86)$$

The same derivation in the frequency domain involves the time shift theorem. The transform of the shifted impulse is $e^{-j\omega t_0}$, by virtue of the time shift theorem.

$$\delta(t - t_0) \longleftrightarrow e^{-j\omega t_0}$$

Therefore, by the time convolution theorem, the transform of the convolution of $f(t)$ with $\delta(t - t_0)$ is the product of the two transforms.

$$\delta(t - t_0) * f(t) \longleftrightarrow e^{-j\omega t_0}F(\omega) \qquad (1.87)$$

$F(\omega)e^{-j\omega t_0}$ is recognized as the transform of $f(t - t_0)$ by virtue of the time shift theorem.

In summation, convolution of $f(t)$ with an impulse function does not change the functional form of $f(t)$. It may only cause a time shift in $f(t)$ if the impulse does not occur at $t = 0$.

We are almost ready to apply all of the theory we have been laboriously developing to a practical problem. Before doing that, we need simply evaluate several transforms which involve impulse functions.

Let us now return to the evaluation of the transform of a constant, $f(t) = A$. If we applied the definition and tried to evaluate the integral, we would find that it does not converge as already stated,

$$\int_{-\infty}^{\infty} Ae^{-j\omega t} \, dt \longleftrightarrow f(t) = A \tag{1.88}$$

For $\omega \neq 0$, the above integral is bounded by $2A/\omega$ but does not converge. A function need not approach infinity if it does not converge, but may oscillate instead. This case is analogous to finding the limit of $\cos t$ as "t" approaches infinity. For $\omega = 0$, the above integral clearly blows up. If we are to find any representation of the transform, we must obviously take a different approach. As before, limit arguments are often invoked. Also as before, we shall avoid them.

Since the integral defining the Fourier Transform and that used to evaluate the inverse transform are quite similar (a change of sign and factor of 2π differentiate them), one might guess that the transform of a constant is an impulse. That is, since an impulse transforms to a constant, a constant should transform to an impulse. In the hope that this reasoning is valid, let us find the inverse transform of an impulse,

$$\delta(\omega) \longleftrightarrow \frac{1}{2\pi} \int_{-\infty}^{\infty} \delta(\omega)e^{j\omega t} \, d\omega = \frac{1}{2\pi} \tag{1.89}$$

Our guess was correct! The inverse transform of $\delta(\omega)$ is a constant. Therefore, by a simple scaling factor, we have

$$A \longleftrightarrow 2\pi A\delta(\omega) \tag{1.90}$$

A straightforward application of the frequency shift theorem yields another useful transform pair,

$$Ae^{j\omega_0 t} \longleftrightarrow 2\pi A\delta(\omega - \omega_0) \tag{1.91}$$

The above "guess and check" technique deserves some comment. We stressed earlier that the uniqueness of the Fourier Transform is extremely significant. That is, given $F(\omega)$, $f(t)$ can be uniquely recovered. Therefore, the guess technique is perfectly rigorous to use in order to find the Fourier Transform of a time function. If we can somehow guess at an $F(\omega)$ which yields $f(t)$ when plugged into the inversion integral, we have found the one and only transform of $f(t)$. As in the above example, this technique is very useful if the transform integral cannot be easily evaluated, while the inverse transform integral evaluation is an obvious one.

Illustrative Example 1.13
Find the Fourier Transform of $f(t) = \cos \omega_0 t$.

Solution
We make use of Euler's identity to express the cosine function in the following form,

$$\cos \omega_0 t = \frac{e^{j\omega_0 t} + e^{-j\omega_0 t}}{2}$$

By the linearity property of the Fourier Transform, we have,

$$F(\omega) = \mathfrak{F}[\tfrac{1}{2}e^{j\omega_0 t}] + \mathfrak{F}[\tfrac{1}{2}e^{-j\omega_0 t}]$$
$$= \pi\delta(\omega - \omega_0) + \pi\delta(\omega + \omega_0) \tag{1.92}$$

This transform is sketched in Fig. 1.26.

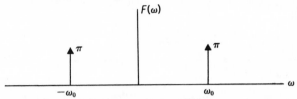

Figure 1.26   Transform of cos $\omega_0 t$, Illustrative Example 1.13.

We can now reveal something somewhat underhanded that has been attempted in this text. Although the definition of the Fourier Transform is a strictly mathematical definition, and "ω" is just a dummy independent functional variable, we have slowly tried to brainwash the reader into thinking of it as a frequency variable. Indeed, the choice of "ω" for the symbol of the independent variable brings the word "frequency" to mind. We know that the Fourier Transform of a real, even time function must be real and even. It can therefore not be identically zero for negative values of "ω". Indeed, for any non-zero, real time function, the Fourier Transform cannot be identically zero for negative "ω." Since negative values of frequency have no meaning, we could never be completely correct in calling "ω" a frequency variable.

Let us view the positive ω-axis only. For this region, the transform of cos $\omega_0 t$ is non-zero only at the point $\omega = \omega_0$. The only case for which we have previously experienced the definition of frequency is that in which the time function is a pure sinusoid. Since for the pure sinusoid, the positive ω-axis appears to have meaning when interpreted as frequency, we shall consider ourselves justified in calling "ω" a frequency variable. We shall continue to do this until such time that a contradiction arises (as it turns out, none will, since for other than a pure sinusoid, we will *define* frequency in terms of the Fourier Transform).

Another transform pair that will be needed in our later work is that of a unit step function and its transform. Here again, if one simply plugs this into the defining integral, he finds that the resulting integral does not converge. We could again attempt the guess technique, but due in part to the discontinuity of the step function, the technique is somewhat hopeless. The transform is relatively easy to evaluate once one realizes that

$$U(t) = \frac{1 + sgn(t)}{2} \tag{1.93}$$

where the "sign" function is defined by

$$sgn(t) = \begin{cases} +1 & t > 0 \\ -1 & t < 0 \end{cases}$$

(See Fig. 1.27.)

Figure 1.27 Representation of $U(t)$ as $\frac{1}{2}[1 + sgn(t)]$.

The transform of $\frac{1}{2}$ is $\pi\delta(\omega)$, and that of $sgn(t)$ can be evaluated by a simple limiting process (see Fig. 1.28).

$$sgn(t) = \lim_{a \to 0} [e^{-a|t|}sgn(t)]$$

Assuming that the order of limiting and taking the transform can be interchanged (as engineers, we first do it, and if the result converges, we assume that it was a permissible thing to do), we have

$$\mathfrak{F}[sgn(t)] = \lim_{a \to 0} \mathfrak{F}[e^{-a|t|}sgn(t)]$$

$$= \lim_{a \to 0} \left[\frac{1}{j\omega + a} + \frac{1}{j\omega - a} \right]$$

$$\mathfrak{F}[sgn(t)] = \lim_{a \to 0} \left[\frac{-2j\omega}{\omega^2 + a^2} \right] = \frac{2}{j\omega} \qquad (1.94)$$

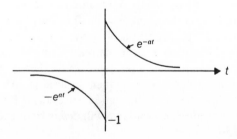

Figure 1.28 $sgn(t)$ as a limit.

From Eqs. (1.93) and (1.94) we see that the transform of the unit step is given by

$$U(t) \longleftrightarrow \frac{1}{j\omega} + \pi\delta(\omega) \qquad (1.95)$$

The student may recall that the one-sided LaPlace Transform of the unit step is $1/s$. At first glance, it appears that the Fourier Transform of any function which is zero for negative "t" should be the same as the one-sided LaPlace Transform, with "s" replaced by "$j\omega$." However, we see that in the case of $f(t) = U(t)$, the two differ by a very important factor, $\pi\delta(\omega)$. The explanation of this apparent discrepancy requires a study of convergence in the complex s-plane. Since this is not critical to the applications of communication theory, we shall omit it in this text.

1.9 PERIODIC TIME FUNCTIONS

In the Illustrative Example 1.13, we found the Fourier Transform of the cosine function. It was composed of two impulses occurring at the frequency of the cosine, and at the negative of this frequency. It will now be shown that the Fourier Transform of any periodic function of time is a discrete function of frequency. That is, the transform is non-zero only at discrete points along the ω-axis. The proof follows from Fourier Series expansions and the linearity of the Fourier Transform.

Suppose that we wish to find the Fourier Transform of a time function, $f(t)$, which happens to be periodic with period T. We choose to express the function in terms of the complex Fourier Series representation.

$$f(t) = \sum_{n=-\infty}^{\infty} c_n e^{jn\omega_0 t}$$

where

$$\omega_0 = \frac{2\pi}{T}$$

Recall the transform pair, Eq. (1.91),

$$Ae^{j\omega_0 t} \longleftrightarrow 2\pi A\delta(\omega - \omega_0)$$

From this transform pair, and the linearity of the transform, we have

$$\mathcal{F}[f(t)] = \sum_{n=-\infty}^{\infty} c_n \mathcal{F}[e^{jn\omega_0 t}]$$
$$= 2\pi \sum_{n=-\infty}^{\infty} c_n \delta(\omega - n\omega_0) \tag{1.96}$$

This transform is sketched in Fig. 1.29 for a representative $f(t)$. Thus, the Fourier Transform of a periodic time function consists of a train of equally

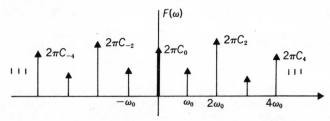

Figure 1.29 Transform of a periodic $f(t)$.

spaced delta functions, each of whose strength (area) is the corresponding c_n of the exponential Fourier Series representation of the time function. Note that the c_n may turn out to be complex numbers. Therefore, the area of the delta functions may also be complex. This does not contradict anything which we said previously, so it should not bother anybody. The c_n will be real if the time function is real and even.[6]

Illustrative Example 1.14

Find the Fourier Transform of the periodic function made up of unit impulses, as shown in Fig. 1.30.

$$f(t) = \sum_{n=-\infty}^{\infty} \delta(t - nT)$$

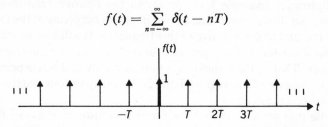

Figure 1.30 Periodic pulse train $f(t)$ for Illustrative Example 1.14.

Solution

Using Eq. (1.96), we have

$$F(\omega) = 2\pi \sum_{n=-\infty}^{\infty} c_n \delta(\omega - n\omega_0)$$

where

$$\omega_0 = \frac{2\pi}{T}$$

and

$$c_n = \frac{1}{T} \int_{-T/2}^{T/2} f(t)e^{-jn\omega_0 t}\, dt$$

Within the range of integration, $-T/2$ to $+T/2$, the only contribution of $f(t)$ is that due to the impulse at the origin. Therefore,

$$c_n = \frac{1}{T} \int_{-T/2}^{T/2} \delta(t)e^{-jn\omega_0 t}\, dt = \frac{1}{T} \tag{1.97}$$

This is an interesting Fourier Series expansion. All of the coefficients are equal. Each frequency component possesses the same amplitude as every other component. Is this analogous to the observation that the Fourier Transform of a single delta function is a constant?

---

[6]From this point forth, unless otherwise stated, all time functions will be assumed to be real. This certainly conforms to real life, where one would never talk about a complex voltage or current.

Proceeding, we have

$$F(\omega) = \frac{2\pi}{T} \sum_{n=-\infty}^{\infty} \delta(\omega - n\omega_0) \qquad (1.98)$$

with

$$\omega_0 = \frac{2\pi}{T}$$

This transform is sketched in Fig. 1.31.

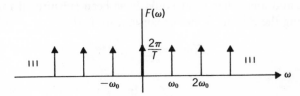

Figure 1.31   Transform of impulse train of Illustrative Example 1.14.

The Fourier Transform of a periodic train of impulses is itself a periodic train of impulses! This is the only case we will encounter in which the Fourier Transform and the original time function resemble each other. This result will be used as a basis for an intuitive proof of the sampling theorem, a *remarkable* result which we shall now present.

1.10   THE SAMPLING THEOREM

The *sampling theorem* (sometimes called Shannon's theorem, or, if in a Russian text, Kotelnikov's theorem) states that, if the Fourier Transform of a time function is zero for $|\omega| > \omega_m$, and the values of the time function are known for $t = nT_s$ (all integer values of n), where $T_s < \pi/\omega_m$, then the time function is exactly known for ALL values of "t." Stated in a different way, $f(t)$ can be uniquely determined from its values at a sequence of equidistant points in time. Think about this. Knowing the values of the function at discrete points, we know its value *exactly* at all points between these discrete points. The upper limit of T_s, π/ω_m, is known as the *Nyquist sampling rate*.

Before going any further, we should observe that the spacing between these "sample points" is inversely related to the maximum frequency, ω_m. This is intuitively satisfying since the higher ω_m is, the faster we would expect the function to vary. The faster the function varies, the closer together one could expect the sample points to be in order to reproduce the function.

The theorem is so significant that we will present two different proofs of it. The first is straightforward, but, as with many straightforward proofs, supplies little insight.

Proof 1. Since $F(\omega)$ is only non-zero along a finite portion of the ω-axis (by assumption), we can expand it in a Fourier Series in the interval

$$-\omega_m < \omega \leq \omega_m$$

In expanding $F(\omega)$ in a Fourier Series, one must be a little careful not to let the change in notation confuse the issue. The "t" used in the Fourier Series previously presented is a dummy functional independent variable, and any other letter could have been substituted for it. Performing the Fourier Series expansion, we find

$$F(\omega) = \sum_{n=-\infty}^{\infty} c_n e^{jn\omega_0\omega} \tag{1.99}$$

where

$$\omega_0 = \frac{2\pi}{2\omega_m}$$

Perhaps we should use "t_0" in place of "ω_0" since the quantity does indeed have the units of time. However, we shall continue with "ω_0" so as to require the least possible change in notation from that used previously.

The c_n in the above expansion are given by

$$c_n = \frac{1}{2\omega_m} \int_{-\omega_m}^{\omega_m} F(\omega)e^{-jn\omega_0\omega}\, d\omega \tag{1.100}$$

However, the Fourier inversion integral tells us that

$$f(t) = \frac{1}{2\pi} \int_{-\infty}^{\infty} F(\omega)e^{j\omega t}\, d\omega = \frac{1}{2\pi} \int_{-\omega_m}^{\omega_m} F(\omega)e^{j\omega t}\, d\omega \tag{1.101}$$

In Eq. (1.101), the limits were reduced since $F(\omega)$ is equal to zero outside of the interval between $-\omega_m$ and $+\omega_m$. Upon comparing Eq. (1.100) with Eq. (1.101), we can easily see that

$$c_n = \frac{\pi}{\omega_m} f(-n\omega_0) = \frac{\pi}{\omega_m} f\left[-\frac{n\pi}{\omega_m}\right] \tag{1.102}$$

This says that the c_n are known once $f(t)$ is known at the points $t = n\pi/\omega_m$. Plugging these values of c_n into the series expansion for $F(\omega)$ (Eq. 1.99), we find

$$F(\omega) = \sum_{n=-\infty}^{\infty} c_n e^{jn\omega_0\omega} = \frac{\pi}{\omega_m} \sum_{n=-\infty}^{\infty} f\left[-\frac{n\pi}{\omega_m}\right] e^{j\omega n\pi/\omega_m} \tag{1.103}$$

Equation (1.103) is essentially a mathematical statement of the sampling theorem. It states that $F(\omega)$ is completely known and determined by the sample values, $f(-n\pi/\omega_m)$. These are the values of $f(t)$ at equally spaced points in time. We can plug this $F(\omega)$ given by Eq. (1.103) into the

Fourier inversion integral to actually find $f(t)$ in terms of its samples. This will yield the following complex appearing result.

$$f(t) = \frac{1}{2\pi} \sum_{n=-\infty}^{\infty} \frac{\pi}{\omega_m} \int_{-\omega_m}^{\omega_m} f\left[-\frac{n\pi}{\omega_m}\right] e^{jn\omega\pi/\omega_m} e^{j\omega t} \, d\omega$$

$$= \sum_{n=-\infty}^{\infty} f\left[-\frac{n\pi}{\omega_m}\right]\left[\frac{\sin(\omega_m t + n\pi)}{\omega_m t + n\pi}\right]$$

(1.104)

Equation (1.104) tells us how to find $f(t)$ for any "t" from those values of $f(t)$ at the points $t = n\pi/\omega_m$. The student could never truly appreciate Eq. (1.104) unless he sketches a few terms of this sum for a representative $f(t)$.

The sampling theorem is critical to any form of pulse communications (we are getting way ahead of the game). If somebody wanted to send a time signal, $f(t)$, from one point to another, all he would have to do is transmit a list of numbers which represented the values of $f(t)$ at the sampling instants. This is certainly easier than transmitting the value of $f(t)$ at every single instant of time, since there are an infinite number of instants between any two points in time.

Proof 2. Given that $f(t)$ has a transform, $F(\omega)$, which is zero outside of the interval $-\omega_m < \omega < \omega_m$, we will use the sketch of $F(\omega)$ as shown in Fig. 1.32 to represent a general $F(\omega)$ with this "band limited" prop-

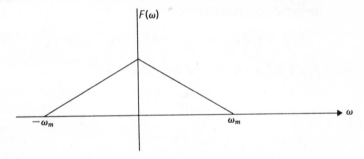

Figure 1.32 A representative band limited $F(\omega)$.

erty. We do not mean to restrict $F(\omega)$ to be of this exact shape. We now consider the product of $f(t)$ with a periodic train of delta function, $f_\delta(t)$, which we shall call a sampling, or gating function.

$$f_\delta(t) = \sum_{n=-\infty}^{\infty} \delta(t - nT_s)$$

(1.105)

The sifting property of the impulse function tells us that

$$f(t)\delta(t - t_0) = f(t_0)\delta(t - t_0)$$

(1.106)

Using this property, we see that the sampled version of $f(t)$ (call it $f_s(t)$) is given by

$$f_s(t) = f(t)f_\delta(t) = \sum_{n=-\infty}^{\infty} f(t)\delta(t - nT_s)$$

$$= \sum_{n=-\infty}^{\infty} f(nT_s)\delta(t - nT_s) \qquad (1.107)$$

A typical $f(t)$, a sampling function, and the sampled version of $f(t)$ are sketched in Fig. 1.33. $f_s(t)$ is truly a sampled version of $f(t)$ since it only

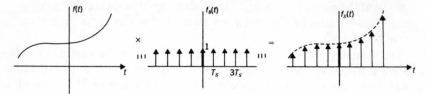

Figure 1.33 The sampling operation.

depends upon the values of $f(t)$ at the sample points in time. We wish to show that, under certain conditions, $f(t)$ can be recovered from $f_s(t)$. We do this by finding the transform of $f_s(t)$, $F_s(\omega)$. By the uniqueness of the Fourier Transform, if we can show that the transform of $f(t)$, $F(\omega)$, can be recovered from $F_s(\omega)$, then $f(t)$ can be recovered from $f_s(t)$. From the frequency convolution theorem, we have

$$f_s(t) \triangleq f(t)f_\delta(t) \leftrightarrow \frac{1}{2\pi}F(\omega) * F_\delta(\omega) = F_s(\omega) \qquad (1.108)$$

We found $F_\delta(\omega)$ in Illustrative Example 1.14

$$F_\delta(\omega) = \frac{2\pi}{T_s} \sum_{n=-\infty}^{\infty} \delta(\omega - n\omega_0)$$

where

$$\omega_0 = 2\pi/T_s$$

Recall that convolution with a delta function simply shifts the original function. Therefore,

$$F(\omega) * \delta(\omega - n\omega_0) = F(\omega - n\omega_0)$$

and

$$F_s(\omega) = \frac{1}{2\pi}F(\omega) * \frac{2\pi}{T_s} \sum_{n=-\infty}^{\infty} \delta(\omega - n\omega_0)$$

$$= \frac{1}{T_s} \sum_{n=-\infty}^{\infty} F(\omega - n\omega_0) \qquad (1.109)$$

This $F_s(\omega)$ is sketched in Fig. 1.34, where we have let $1/T_s = k$. From the sketch of $F_s(\omega)$, it is clear that $F(\omega)$ can be identified. In practice, we need only multiply $F_s(\omega)$ by a gating function, $H(\omega)$, as shown in Fig. 1.35. Devices

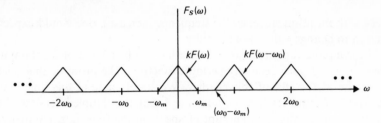

Figure 1.34 $F_s(\omega)$, the transform of the sampled wave.

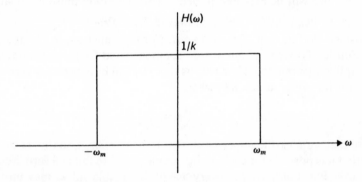

Figure 1.35 $H(\omega)$ to recover $F(\omega)$ from $F_s(\omega)$.

which perform this multiplication in the frequency domain (frequency separation) are called filters, and are analyzed in detail in the next chapter.

One assumption which was made in the sketching of $F_s(\omega)$ will now be examined. The assumption is that the humps in the figure do not overlap. That is, the point labelled "$\omega_0 - \omega_m$" must fall to the right of that labelled "ω_m." If this were not true, the center blob (centered about $\omega = 0$) would result from an interaction of several shifted versions of $F(\omega)$, and would therefore not be an accurate reproduction of $F(\omega)$. In this case, $f(t)$ could not be uniquely recovered. This restriction can be rewritten as follows:

$$\omega_0 - \omega_m > \omega_m \Rightarrow \omega_0 > 2\omega_m \qquad (1.110)$$

Since

$$\omega_0 = \frac{2\pi}{T_s}$$

the restriction finally becomes

$$T_s < \frac{\pi}{\omega_m} \qquad (1.111)$$

as was stated in the condition of the sampling theorem. This restriction says that the samples cannot be spaced more widely apart than π/ω_m. They can certainly be as close together, consistent with the π/ω_m upper limit, as desired. We again note that as ω_m increases, the samples must be closer together. This agrees with intuition since, as the frequency increases, one would expect the function to change value more rapidly.

The sampling function, $f_\delta(t)$, did not have to be a train of ideal impulses. Indeed, it could have been any periodic function. The proof of this is left as an exercise at the end of this chapter. We simply note at this time that the ideal train of impulses has the advantage that the sampled version of $f(t)$, $f_s(t)$, takes up a minimum amount of space on the time axis. (Actually, for ideal impulses, the time required for each sample is zero.) This will prove significant, and will be explored in detail when we study pulse modulation.

Illustrative Example 1.15   (*This is a Classic Trick Problem.*)

Consider the function, $f(t) = \sin \omega_0 t$. For this function, $\omega_m = \omega_0$. That is, the Fourier Transform contains no components at frequencies above ω_m. The sampling theorem therefore tells us that $f(t)$ can be completely recovered if we know its values at $t = nT_s$ where

$$T_s < \frac{\pi}{\omega_m} = \frac{\pi}{\omega_0}$$

Suppose that we pick the upper limit, that is $T_s = \pi/\omega_0$. Looking at $\sin \omega_0 t$, we see that it is possible for each of the sample points to fall at a zero crossing of $f(t)$. (*See* Fig. 1.36.) That is, every one of the sample values may turn out to be zero. Obviously, $f(t)$ could not be recovered if this were the case.

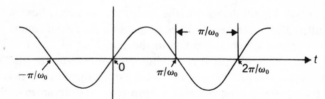

Figure 1.36   Sampling at zero points of sinusoid.

Conclusion

Looking back at the sketch of $F_s(\omega)$ (Fig. 1.34) we said that the humps must not overlap. However, if $T_s = \pi/\omega_m$, while they do not overlap, each pair does meet at one point, $\omega = n\omega_m$. While the value of $F(\omega)$ at one single point in frequency usually cannot affect the value of the time function, $f(t)$ [an integral operation is involved in going from $F(\omega)$ to $f(t)$], if $F(\omega)$ happens to be an impulse at this one point, $f(t)$ certainly can be changed. Since the transform of $\sin \omega_0 t$ does indeed possess an impulse at $\omega = \omega_m$, this "point overlap" of the blobs cannot be tolerated. In the above example, the two impulses that fall at $\omega = \omega_m$ were cancelling each other out. To eliminate this possibility, the statement of the sampling theorem must carefully say

$$T_s < \frac{\pi}{\omega_m} \tag{1.112}$$

where the inequality must mean "strictly less than" and not "less than or equal to." In the above example, if T_s were a little bit less than π/ω_0, the sample values could not all equal zero, and $f(t)$ could be recovered from $f_s(t)$.

We can make another important observation from this same example. Assume that we sample at twice the minimum rate. That is, let the sampling period, $T_s = \pi/2\omega_0$. We can certainly recover $f(t)$ from the samples. Figure 1.37 shows the sampled version of $f(t) = \sin \omega_0 t$. Note that every other sample falls at a zero crossing.

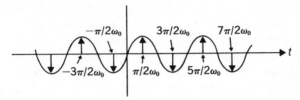

Figure 1.37   Sin $\omega_0 t$ sampled at twice the minimum sampling rate.

One may ask how we can be sure that the samples of Fig. 1.37 correspond to $f(t) = \sin \omega_0 t$, and not, for example, the sawtooth type waveform of Fig. 1.38.

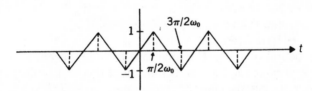

Figure 1.38   Sawtooth waveform with sample values of Fig. 1.37.

The answer lies in the observation that the waveform of Fig. 1.38 is not frequency band limited to ω_m (in fact, it possesses infinite frequencies). Therefore, the conditions of the sampling theorem would have been violated.

In summation, we can uniquely recover a waveform from its sample values *provided* that we are sure the conditions of the sampling theorem ($T_s < \pi/\omega_m$) were not violated in producing the sampled waveform.

We conclude this section with the idea that the restriction on $F(\omega)$ imposed by the sampling theorem is not very severe in practice. All signals of interest in real life do possess Fourier Transforms which are zero (approximately) above some frequency. No physical device can transmit infinitely high frequencies since all wires contain inductance, and all circuits contain parasitic capacitances. A famous man once said that this upper frequency cutoff of any device can be proven by the "flashlight test." Shine a flashlight on the input terminals of any device, and see if you get light at the output. If not, the device obviously has some upper frequency above which it cannot respond. This anecdote, while ridiculous (can you give a reason for it being irrelevant, indeed, untrue?), helps to point out that, for practical reasons, all signals of interest have Fourier Transforms which are essentially zero above some cutoff frequency.

1.11 THE DISCRETE FOURIER TRANSFORM

We have seen that a continuous, band limited time signal can be represented by time samples. Practical system limitations often make it desirable to work with sample values rather than with continuous functions of time. As a trivial example, suppose a weather station were asked to monitor temperature. It could plot the temperature as a function of time, or it could report the temperature at periodic sampling times (e.g., every hour). The latter technique might be used if the weather station were responding to a request from a television network news bureau.

It is therefore often useful to be able to analyze sampled time functions. The advantages of using transform analysis were previously observed for continuous time functions. It would therefore seem reasonable to look toward transform techniques that would be applicable to sampled systems. We shall explore two types of transform analysis. The first is the *Discrete Fourier Transform* (*DFT*), including a variation known as the *Fast Fourier Transform* (*FFT*). The second is known as the *z-Transform*.

Suppose we have a series of N samples of a time function spaced apart by a sampling period of T sec. The total time record is therefore NT sec long. We wish to calculate the Fourier Transform of this time limited function. If the integral in the definition of the continuous Fourier Transform is replaced by a summation, the Discrete Fourier Transform results. Thus, instead of

$$F(\omega) = \int_0^{NT} f(t)e^{-j\omega t}\, dt$$

we have

$$F(\omega) = \sum_0^{N-1} f(nT)e^{-j\omega nT}T \tag{1.113}$$

Two additional modifications are now made. First, we sample the frequency at multiples of Ω, and second, we scale the transform by dividing by NT, the record length. (*Note:* We can cancel the effects of any constant scale factor when we get around to defining the equation for the inverse transform.) Doing this, we get

$$F(m\Omega) = \frac{1}{N} \sum_0^{N-1} f(nT)e^{-jmn\Omega T} \tag{1.114}$$

To simplify Eq. (1.114) further, we now assume that the limited time record of $f(t)$ represents one period of a periodic function. If the period is NT sec, the fundamental frequency of the periodic function would be $2\pi/NT$, and the Fourier Transform would have non-zero value only at multiples of this fundamental frequency. It would therefore make sense to choose Ω to be equal to this fundamental frequency. Using this value, ΩT becomes $2\pi/N$, and Eq. (1.114) can be rewritten as

$$F(m\Omega) = \frac{1}{N} \sum_{0}^{N-1} f(nT) e^{-jmn2\pi/N} \tag{1.115}$$

Finally, defining W as the Nth complex root of unity, that is, $W = e^{j2\pi/N}$, Eq. (1.115) becomes the simplified form

$$F(m\Omega) = \frac{1}{N} \sum_{0}^{N-1} f(nT) W^{-mn} \tag{1.116}$$

This is the usual form of the Discrete Fourier Transform. The student should verify that this transform is periodic in m with period N. That is, $F[(N + m)W] = F(mW)$. For this reason, it is only necessary to calculate the transform for N frequency points.

Illustrative Example 1.16

Find the DFT of the following sampled time function:

t	0	T	$2T$	$3T$	$4T$	$5T$	$6T$	$7T$	$8T$
$f(t)$	1	1	1	1	1	1	0	0	0

This is a sampled step as shown in Fig. 1.39.

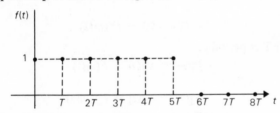

Figure 1.39   Sampled step of Example 1.16.

Solution

The number of sample points $N = 8$. Therefore,

$$F(m\Omega) = \frac{1}{8} \sum_{0}^{7} f(nT) W^{-mn}$$
$$= \frac{1}{8}(1 + W^{-m} + W^{-2m} + W^{-3m} + W^{-4m} + W^{-5m})$$

where

$$W = e^{j2\pi/8} = \frac{1+j}{\sqrt{2}}$$

and

$$\Omega = \frac{1}{8T}$$

Therefore,

$$F(\Omega) = \frac{1}{8}(1 + e^{-j\pi/4} + e^{-2j\pi/4} + e^{-3j\pi/4} + e^{-4j\pi/4} + e^{-5j\pi/4})$$

$$= \frac{1}{8}\left(-j - \frac{j}{\sqrt{2}} - \frac{1}{\sqrt{2}}\right)$$

$$F(2\Omega) = \frac{1}{8}(1 - j)$$

$$F(3\Omega) = \frac{1}{8}\left(j + \frac{1}{\sqrt{2}} - \frac{j}{\sqrt{2}}\right)$$

$$F(4\Omega) = 0$$

$$F(5\Omega) = \frac{1}{8}\left(-j + \frac{1}{\sqrt{2}} + \frac{j}{\sqrt{2}}\right)$$

$$F(6\Omega) = \frac{1}{8}(1 + j)$$

$$F(7\Omega) = \frac{1}{8}\left(j + \frac{j}{\sqrt{2}} - \frac{1}{\sqrt{2}}\right)$$

Due to the periodicity of the DFT, there is no need to calculate any additional samples of the transform. We note from the above solution that the real part of the DFT is symmetric about the midpoint and that the imaginary part of the DFT is antisymmetric about the midpoint. This is true in general and is derived from the property of continuous Fourier Transforms; that is, $F(-\omega) = F*(\omega)$. One can easily see from the DFT defining equation (Eq. (1.115)) that

$$F(-m\Omega) = F^*(m\Omega)$$

Since the DFT is periodic,

$$F[(m + N)\Omega] = F(m\Omega)$$

we have

$$F[(N - m)\Omega] = F^*(m\Omega)$$

which is a mathematical statement of the above observation. The DFT can be shown to have other properties similar to those of the continuous Fourier Transform. Some of these properties are explored in the problems at the end of this chapter.

The Discrete Fourier Transform is a convenient way to approximate the Fourier Transform when sampled waves are involved, and it lends itself readily to computation on a digital computer. Computation of the DFT as in Eq. (1.116) would involve summing N products (complex) for each of N frequency sample points. If N is large, this can become quite tedious.

The Fast Fourier Transform (FFT) is an algorithm which greatly simplifies and shortens the calculations which must be done to find the DFT. We begin by rewriting Eq. (1-116) in matrix form:

$$\overline{F(m\Omega)} = [W]\overline{f(nT)}$$

where $\overline{F(m\Omega)}$ is a column vector,

$$\overline{F(m\Omega)} = \begin{bmatrix} F(0) \\ F(1\Omega) \\ F(2\Omega) \\ F(3\Omega) \\ \cdot \\ \cdot \\ F\{(N-1)\Omega\} \end{bmatrix}$$

and $\overline{f(nT)}$ is a column vector,

$$\overline{f(nT)} = \begin{bmatrix} f(0) \\ f(1t) \\ f(2t) \\ f(3t) \\ \cdot \\ \cdot \\ f\{(N-1)t\} \end{bmatrix}$$

$[W]$ is a matrix of the form

$$[W] = \begin{bmatrix} W^0 & W^0 & W^0 & W^0 & \cdots & W^0 \\ W^0 & W^{-1} & W^{-2} & W^{-3} & \cdots & W^{-N} \\ W^0 & W^{-2} & W^{-4} & W^{-6} & \cdots & W^{-2N} \\ W^0 & W^{-3} & W^{-6} & W^{-8} & \cdots & W^{-3N} \\ \cdot & \cdot & \cdot & \cdot & \cdots & \cdot \\ W^0 & W^{-1N} & W^{-2N} & W^{-3N} & \cdots & W^{-N^2} \end{bmatrix}$$

The Fast Fourier Transform algorithm is derived by factoring the matrix $[W]$ into a product of matrices. This can be done in such a way that each of the new matrices contains many zeros. The calculation is therefore greatly simplified. The interested reader is referred to the references at the end of this chapter for detailed information on the algorithm.

1.12 THE z-TRANSFORM

As in the case of the DFT, the *z-Transform* is well suited to the analysis of discrete time signals and systems. Although the *z*-Transform is related to the Fourier Transform, it does not follow from it so directly as does the DFT. Calculation of the DFT often presents computational problems which make the use of a digital computer (and the FFT) almost necessary. This is not the case with the *z*-Transform. Finally, the time signal need not be time limited (i.e., periodic) to permit analysis using the *z*-Transform.

Given a sequence of numbers, $f(n)$, the z-Transform is defined by

$$F(z) = \sum_{n=0}^{\infty} f(n)z^{-n} \qquad (1.117)$$

where z is a complex variable. $f(n)$ can be a sequence of numbers, or it can represent time samples of a continuous time function, $f(t)$.

Illustrative Example 1.17

Find the z-Transform of $f(n) = 1$ for all positive n. This is a unit step, often denoted by $u(n)$.

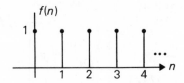

Solution

From Eq. (1.117), $F(z)$ is given by

$$F(z) = \sum_{n=0}^{\infty} 1z^{-n}$$
$$= 1 + z^{-1} + z^{-2} + \cdots$$
$$= \frac{1}{1 - z^{-1}} = \frac{z}{z - 1}$$

This solution converges for all $|z| > 1$.

Illustrative Example 1.18

Find the z-Transform of

$$f(n) = \begin{cases} 1 & n = 0 \\ 0 & \text{otherwise} \end{cases}$$

This is a unit impulse, δ_n, sometimes known as the *Kronecker delta*.

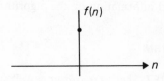

Solution

By direct substitution into Eq. (1.117), we find

$$F(z) = 1$$

Illustrative Example 1.19

Find the z-Transform of $f(n) = k^n$.

Solution

Again, we use the defining relationship, Eq. (1.117):

$$F(z) = \sum_{n=0}^{\infty} k^n z^{-n}$$
$$= \sum_{n=0}^{\infty} \left(\frac{z}{k}\right)^{-n}$$
$$= \frac{z}{z-k}$$

The z-Transform has properties which are similar to those of the Fourier Transform. Derivation of some of these properties is left to the exercises at the end of this chapter. One property which we will need here is the *time shift property*. Given that $F(z)$ is the z-Transform of $f(n)$, we wish to find the z-Transform of $f(n - m)$. This is a shifted sequence. If $f(n)$ were a sequence of time samples of a continuous signal, $f(n - m)$ would represent samples of the signal shifted by mT to the right, where T is the sampling period.

To derive the relationship between the transform of the original sequence and that of the shifted sequence, we proceed to calculate the z-Transform of $f(n - m)$, using the definition

$$\sum_{n=0}^{\infty} f(n - m)z^{-n} = \sum_{n=0}^{\infty} f(n - m)z^{-(n-m)}z^{-m}$$
$$= \sum_{k=-m}^{\infty} f(k)z^{-k}z^{-m}$$
$$= z^{-m}F(z)$$

In the last step of the above derivation, we assumed that $f(k)$ is equal to zero for k less than zero. Indeed, this is a necessary assumption since the z-Transform of $f(n)$ is independent of the values of $f(n)$ for negative argument. That is, using the z-Transform (as we defined it), negative arguments are ignored, and the transform is unique only if we make some uniform assumption about these values (e.g., assume all sequences are zero for negative argument).

Now that we know what the z-Transform is and have found some representative z-Transforms, we arrive at the next logical question, Who cares? Why was the z-Transform invented in the first place?

Just as the *convolution* operation justified the existence of the Fourier Transform, the same is true of the z-Transform. The only difference is that we must now consider a discrete form of the convolution operator. We shall find this discrete convolution operation to be critical to the analysis of discrete linear systems, as presented in Chapter 2.

Given two sequences, $x(n)$ and $h(n)$, the discrete convolution of these two sequences results in a third sequence, $y(n)$, according to

$$y(n) = x(n) * h(n)$$
$$= \sum_{m=0}^{\infty} x(m)h(n - m) \tag{1.118}$$

The summation starts at zero in order to be consistent with our previous assumption about all sequences—that they are zero for negative argument.

Equation (1.118) requires an infinite summation for every value of n. This is clearly a tedious computation to perform. However, just as we found in the case of continuous convolutions, the transform will greatly simplify the problem. In the case of continuous convolution, we found that the Fourier Transform of the result is equal to the product of the transforms of each of the input functions. The exact same relationship is true of discrete convolution and the z-Transform. That is, the z-Transform of the convolution of two sequences is the product of the z-Transforms of each of the two sequences. To prove this relationship, we evaluate the z-Transform of $y(n)$ in Eq. (1.118):

$$\sum_{n=0}^{\infty} y(n)z^{-n} = \sum_{n=0}^{\infty} \sum_{m=0}^{\infty} x(m)h(n-m)z^{-n} \tag{1.119}$$

We now use the time shift property derived previously to find

$$\sum_{n=0}^{\infty} h(n-m)z^{-n} = z^{-m}H(z) \tag{1.120}$$

Substituting Eq. (1.120) into Eq. (1.119) yields

$$\sum_{n=0}^{\infty} y(n)z^{-n} = \sum_{m=0}^{\infty} z^{-m}x(m)H(z)$$
$$= X(z)H(z) \tag{1.121}$$

Equation (1.121) is the desired result. That is, the z-Transform of the convolution of two sequences is the product of the z-Transforms of each of the sequences.

PROBLEMS

1.1. Find the x and y components of a unit vector which is orthogonal to $\bar{x}$, where

$$\bar{x} = 2\bar{a}_x + 3\bar{a}_y$$

Note that there are two possible answers to this problem.

1.2. Find the squared error that exists when we approximate the vector

$$\bar{x} = 2\bar{a}_x + 3\bar{a}_y + 4\bar{a}_z$$

by

$$\bar{\bar{x}} = k_1\bar{a}_x + k_2\bar{a}_y$$

where k_1 and k_2 are chosen to minimize this error.

1.3. Show that the two functions

$$\cos t \quad \text{and} \quad \cos 2t$$

are *not* orthogonal over the interval $-\pi/2 < t \leq \pi/2$.

. **1.4.** Evaluate the Fourier Series Expansion of each of the following periodic functions. Use either the complex exponential or the trigonometric form of the series.

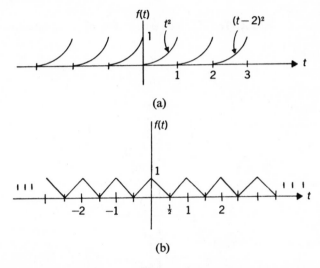

(a)

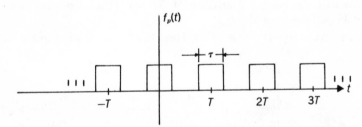

(b)

1.5. Find the Fourier Series representation of the periodic function $g(t)$.

$$g(t) = 2 \sin t + 3 \sin 2t$$

* **1.6.** Find the Fourier Series expansion of a periodic function, $f_p(t)$, as shown.

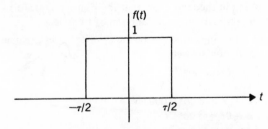

1.7. Find a Fourier Series expansion of $f(t)$ which applies for $-\tau/2 < t < \tau/2$.

Does your answer surprise you? Does this seem to contradict the answer to Problem 1.6? (Don't just answer "yes" or "no.")

1.8. Find the complex Fourier Series representation of $f(t) = t^2$ which applies in the interval $0 < t < 1$. How does this compare with your answer to 1.4(b)?

1.9. Find a trigonometric Fourier Series representation of the function $f(t) = \cos t$ in the interval $0 < t < 2\pi$.

1.10. Evaluate the Fourier Transform of each of the following time functions.

(a) $f(t) = \dfrac{\sin at}{\pi t}$

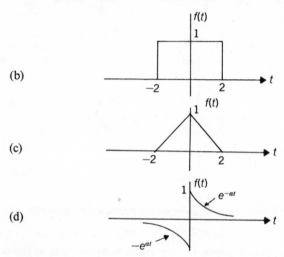

(b)

(c)

(d)

1.11. Evaluate the Fourier Transform of $f(t) = e^{-at}U(t)$. Compare this to the LaPlace Transform of $f(t)$.

1.12. Convolve $e^{at}U(-t)$ with $e^{-at}U(t)$. These two functions are sketched below.

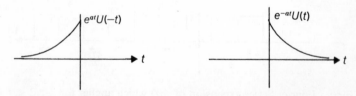

1.13. Without looking in the text, show that the Fourier Transform of a real, even time function is a real and even function of frequency.

1.14. Which of the following could not be the Fourier Series expansion of a periodic signal? Explain your answer.

(a) $f(t) = 2 \cos t + 3 \cos 3t$
(b) $f(t) = 2 \cos \frac{1}{2}t + 3 \cos 3.5t$
(c) $f(t) = 2 \cos \frac{1}{4}t + 3 \cos 0.00054t$
(d) $f(t) = 2 \cos \pi t + 3 \cos 2t$
(e) $f(t) = 2 \cos \pi t + 3 \cos 2\pi t$

1.15. The Fourier Transform of a time function, $f(t)$, is $F(\omega)$ as shown. Find the time function, $f(t)$.

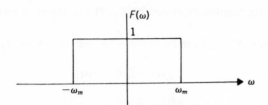

1.16. Find $\int_{-\infty}^{\infty} e^{-2t} U(t)\, dt$ by using Parseval's Theorem.

(Hint: Use the fact that $e^{-2t}U(t) = |e^{-t}U(t)|^2$.)

1.17. Given that the Fourier Transform of $f(t)$ is $F(\omega)$:

(a) What is the Fourier Transform of df/dt in terms of $F(\omega)$?

(b) What is the Fourier Transform of $\int_{-\infty}^{t} f(\sigma)\, d\sigma$ in terms of $F(\omega)$?

(Hint: Write the Fourier inversion integral and differentiate or integrate both sides of the expression.)

1.18. Using the frequency shift theorem, show that

$$f(t)\cos \omega_0 t \longleftrightarrow \frac{F(\omega + \omega_0) + F(\omega - \omega_0)}{2}$$

This is sometimes called the "modulation theorem." We will make extensive use of it later.

1.19. Use the time shift theorem to find the Fourier Transform of

$$\frac{f(t + T) - f(t)}{T}$$

From this result find the Fourier Transform of df/dt and compare this with your answer to 1.17(a).

1.20. Show that if $F(\omega) = 0$ for $|\omega| > \sigma$, then

$$f(t) * \frac{\sin at}{\pi t} = f(t)$$

provided that $a > \sigma$. Note that the Fourier Transform of $(\sin at)/\pi t$ is as shown in the accompanying diagram.

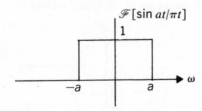

1.21. (a) Write the convolution of $f(t)$ with $U(t)$, $f(t) * U(t)$, in integral form. See if you can identify this as the integral (anti-derivative) of $f(t)$.

(b) Using the convolution theorem, what is the transform of $f(t) * U(t)$?

(c) Compare your answer to (a) and (b) with that which you found in 1.17(b).

1.22. Prove the relationship expressed by Eq. (1.55), the frequency convolution theorem.

1.23. Any arbitrary function can be expressed as the sum of an even and an odd function.

$$f(t) = f_e(t) + f_0(t)$$

where

$$f_e(t) = \tfrac{1}{2}[f(t) + f(-t)]$$
$$f_0(t) = \tfrac{1}{2}[f(t) - f(-t)]$$

(a) Show that $f_e(t)$ is indeed even and that $f_0(t)$ is odd.
   (Hint: Use the definition of an even and an odd function.)
(b) Show that $f(t) = f_e(t) + f_0(t)$.
(c) Find $f_e(t)$ and $f_0(t)$ for $f(t) = U(t)$, the unit step.
(d) Find $f_e(t)$ and $f_0(t)$ for $f(t) = \cos 10t$.

1.24. Given a function $f(t)$ which is zero for negative t, [i.e., $f(t) = f(t)U(t)$] find a relationship between $f_e(t)$ and $f_0(t)$. Can this relationship be used to find a relationship between the real and imaginary parts of the Fourier Transform of $f(t)$?

1.25. A time signal, $f(t)$, is put through a gate and truncated in time. The gate is closed for $1 < t < 2$. Therefore

$$\hat{f}(t) = \begin{cases} f(t) & 1 < t < 2 \\ 0 & \text{otherwise} \end{cases}$$

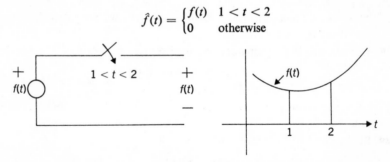

Find the Fourier Transform of $\hat{f}(t)$ in terms of $\mathcal{F}[f(t)]$.
(Hint: Can you express $f(t)$ as the product of two time functions and use the frequency convolution theorem?)

1.26. (a) Find the time derivative of the function shown.

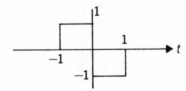

(b) Find the Fourier Transform of this derivative.
(c) Using the formula for the transform of the integral of a time function [Problems 1.17(b) or 1.21(a)] find the transform of the time function in part (a) using the answer to part (b).

1.27. Find the Fourier Transform of $f(t) = \sin \omega_0 t$ from the Fourier Transform of $\cos \omega_0 t$ and the time shift theorem.

1.28. Find the Fourier Transform of $\cos^2 \omega_0 t$ using the frequency convolution theorem and the transform of $\cos \omega_0 t$. Check your answer by expanding $\cos^2 \omega_0 t$ by trigonometric identites.

1.29. Evaluate the following integrals:

(a) $\displaystyle\int_{-\infty}^{\infty} \frac{\sin 3\tau}{\tau} \delta(t - \tau) \, d\tau$

(b) $\displaystyle\int_{-\infty}^{\infty} \frac{\sin 3(\tau - 3)}{(\tau - 3)} \frac{\sin 5(t - \tau)}{(t - \tau)} \, d\tau$

1.30. Evaluate the following integral,

$$\int_{-\infty}^{\infty} \delta(1 - t)(t^3 + 4) \, dt$$

1.31. In Eq. (1.104) (the final statement of the sampling theorem) verify that $f(t)$ given in this form coincides with the sample values at the sample points $(t = n\pi/\omega_m)$. Sketch the first few terms in the summation for a typical $f(t)$.

1.32. The $f_p(t)$ of Problem 1.6 with $T = 2\pi$ and $\tau = \pi$ multiplies a time function $g(t)$, with $G(\omega)$ as shown below.

Part (a)　　　　　　　　　　　Part (c)

(a) Sketch the Fourier Transform of $g_s(t) = g(t)f_p(t)$.
(b) Can $g(t)$ be recovered from $g_s(t)$?
(c) If $G(\omega)$ is as sketched in Part (c), can $g(t)$ be recovered from $g_s(t)$? Explain your answer.

1.33. You are given a function of time,

$$f(t) = \frac{\sin t}{t}$$

This function is sampled by being multiplied by $f_\delta(t)$ as shown below.

$$f_s(t) = f(t)f_\delta(t)$$

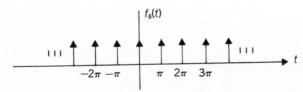

(a) What is the Fourier Transform of the sampled function, $F_s(\omega)$?

(b) From your answer to part (a), what is the actual form of $f_s(t)$?

(c) Should your answer to part (b) have been a train of impulses? Did it turn out that way? Explain any apparent discrepancies.

(d) Describe a method of recovering $f(t)$ from $f_s(t)$ and show that it works in this specific case.

1.34. A signal, $f(t)$, with $F(\omega)$ as shown, is sampled by two different sampling functions, $f_{\delta 1}(t)$ and $f_{\delta 2}(t)$, where

$$f_{\delta 2}(t) = f_{\delta 1}\left(t - \frac{T}{2}\right) \quad \text{and} \quad T = \frac{\pi}{\omega_m}$$

Find the Fourier Transform of the sampled waveforms,

$$f_{s1}(t) = f_{\delta 1}(t)f(t)$$
$$f_{s2}(t) = f_{\delta 2}(t)f(t)$$

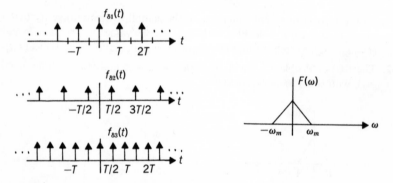

Now consider $f(t)$ sampled by $f_{\delta 3}(t)$, a train of impulses spaced $T/2$ apart. Note that this new sampling function is equal to $f_{\delta 1}(t) + f_{\delta 2}(t)$. Show that the transform of $f_{s3}(t)$ is equal to the sum of the transforms of $f_{s2}(t)$ and $f_{s1}(t)$. (Hint: Don't forget the phase of $f_{\delta 2}(t)$. You should find that when $F_{\delta 1}(\omega)$ is added to $F_{\delta 2}(\omega)$, every second impulse is cancelled out.)

1.35. You are given a time function, $f(t)$, with z-Transform, $F(z)$. That is, the z-Transform of a sampled version of $f(t)$ is $F(z)$. $f(t)$ is now multiplied by e^{-at}. What is the effect of this multiplication upon the z-Transform?

1.36. Find the DFT of the following sampled function.

1.37. Show that (a) the DFT of an even function is real and (b) the DFT of an odd function is imaginary.

1.38. What is the effect upon the DFT of a function, $f(t)$, if the function is delayed by n sampling periods? That is, find the DFT of $f(t - nT_s)$ in terms of the DFT of $f(t)$.

1.39. Find the DFT of the convolution $f(t) * g(t)$ in terms of the DFT of $f(t)$ and the DFT of $g(t)$.

REFERENCES

CHILDERS, D. and DURLINE, A., *Digital Filtering and Signal Processing*, West Publishing, New York, 1975.

PAPOULIS, A., *The Fourier Integral and its Applications*, McGraw-Hill, New York, 1962.

RABINER, L. R., and RADER, C. M., eds., *Digital Signal Processing*, IEEE Press, New York, 1972. (*Note:* This volume contains 32 excellent papers covering all aspects of the FFT.)

STANLEY, W. D., *Digital Signal Processing*, Reston Publishing, Reston, Va., 1975.

LINEAR
SYSTEMS
2

In Chapter 1 of this text, the basic mathematical tools required for waveform analysis were developed. We shall now apply these techniques to the study of linear systems in order to determine system capabilities and features. We will then be in a position to pick those particular characteristics of linear systems which are desirable in communication systems.

It is first necessary to define some common terms. A *system* is defined as a set of rules that associates an "output" time function to every "input" time function. This is sometimes shown in block diagram form as Fig. 2.1.

The input, or source signal, is $f(t)$; and $g(t)$ is the output, or response signal due to this input. The actual physical structure of the system determines the exact relationship between $f(t)$ and $g(t)$.

For example, if the system under study is an electric circuit, $f(t)$ can be a voltage or current source, and $g(t)$ can be a voltage or current anywhere in the circuit. Don't be disturbed by the fact that two wires are needed to describe a voltage. The lines in Fig. 2.1 are not restricted to representing wires.

In the special case of a two terminal electrical network, $f(t)$ can be a sinusoidal voltage across the two terminals, and $g(t)$ can be the current flowing into one of the terminals and out of the other due to this impressed

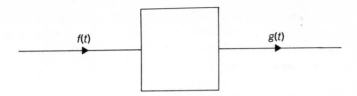

Figure 2.1 Block diagram of a system.

voltage. In this case, the relationship between $f(t)$ and $g(t)$ is the familiar complex impedance between the two terminals of the network.

Any system can be described by specifying the response, $g(t)$, associated with every possible input, $f(t)$. This is an obviously exhaustive process. We would certainly hope to find a much simpler way of describing the system.

Before exploring alternate techniques of characterizing systems, we need some additional basic definitions.

A system is said to obey *superposition* if the output due to a sum of inputs is equal to the sum of the corresponding individual outputs. That is, given that the response (output) due to an excitation (input) of $f_1(t)$ is $g_1(t)$, and that the response due to $f_2(t)$ is $g_2(t)$, then the output due to $f_1(t) + f_2(t)$ is $g_1(t) + g_2(t)$.

A single ended arrow is often used as a shorthand method of relating an input to its resulting output. That is,

$$f(t) \rightarrow g(t) \tag{2.1}$$

and is read, "an input, $f(t)$, causes an output, $g(t)$."

In terms of this notation, the definition of superposition is as follows. If $f_1(t) \rightarrow g_1(t)$ and $f_2(t) \rightarrow g_2(t)$, then

$$f_1(t) + f_2(t) \rightarrow g_1(t) + g_2(t) \tag{2.2}$$

Some thought will convince one that in order for a system to obey super-position, the source free, or transient response (response due to initial conditions) must be zero. In practice, one often replaces a circuit having non-zero initial conditions with one containing zero initial conditions. Additional sources are added to replace the initial condition contributions.

A concept closely related to superposition is *linearity*. Assume again that $f_1(t) \rightarrow g_1(t)$ and $f_2(t) \rightarrow g_2(t)$. The system is said to be linear if the following relationship holds true for all values of the constants a and b:

$$af_1(t) + bf_2(t) \rightarrow ag_1(t) + bg_2(t)$$

In the remainder of this text, we will use the words *linearity* and *superposition* interchangeably.

A system is said to be *time invariant* if the response due to an input is not dependent upon the actual time of occurrence of the input. That is, a time shift in input signal causes an equal time shift in the output waveform. In symbolic form, if $f(t) \rightarrow g(t)$, then $f(t - t_0) \rightarrow g(t - t_0)$ for all real t_0.

It should be clear that a sufficient condition for an electrical network to be time invariant is that its component element values do not change with time (assuming constant initial conditions). That is, resistances, capacitances, and inductances remain constant.

2.1 THE SYSTEM FUNCTION

Returning to the task of characterizing a system, we shall see that for a time invariant system which obeys superposition, a very simple description is possible. That is, instead of requiring that we know the response due to *every* possible input, it will turn out that we need only know it for one "test" input. The derivation of this highly significant result follows.

Recall that any time function convolved with an impulse function remains unchanged. That is, we can write any general input, $f(t)$, in the following form,

$$f(t) = f(t) * \delta(t),$$
$$= \int_{-\infty}^{\infty} f(\tau)\delta(t - \tau) \, d\tau \tag{2.3}$$

This integral can be considered as a limiting case of a weighted sum of delayed delta functions. That is,

$$f(t) = \lim_{\Delta t \to 0} \sum_{-\infty}^{\infty} f(n\Delta t)\delta(t - n\Delta t) \, \Delta t \tag{2.4}$$

Suppose that we know the system response due to $\delta(t)$. Suppose also that the system is time invariant and obeys superposition. If we call this impulse response, $h(t)$, then the response due to

$$f(t) = a\delta(t - t_1) + b\delta(t - t_2)$$

will be

$$g(t) = ah(t - t_1) + bh(t - t_2)$$

That is, if

$$\delta(t) \longrightarrow h(t)$$

then

$$a\delta(t - t_1) + b\delta(t - t_2) \longrightarrow ah(t - t_1) + bh(t - t_2) \tag{2.5}$$

Generalizing this to the infinite sum (integral) input, we see that the output due to $f(t)$ in Eq. (2.4), call it $g(t)$, is given by

$$f(t) \longrightarrow g(t) = \lim_{\Delta t \to 0} \sum_{-\infty}^{\infty} f(n\Delta t)h(t - n\Delta t) \, \Delta t$$
$$= \int_{-\infty}^{\infty} f(\tau)h(t - \tau) \, d\tau \tag{2.6}$$
$$= f(t) * h(t)$$

In words, the output due to any general input is found by convolving that input with the system's response to an impulse. That is, if the impulse response is known, then the response to any other input is automatically also known. Equation (2.6) is sometimes called the superposition integral equation. It is probably the most significant result of system theory.

Recall from Chapter 1 that the Fourier Transform of $\delta(t)$ is unity. There-fore, in some sense, $\delta(t)$ contains all frequencies to an equal degree. Perhaps this observation hints at the delta function's suitability as a test function for system behavior.

Taking the Fourier Transform of both sides of Eq. (2.6), we have the corresponding equation in the frequency domain,

$$G(\omega) = F(\omega)H(\omega) \tag{2.7}$$

The Fourier Transform of the output due to $f(t)$ as an input to a linear (we are being sloppy about linearity as promised) system is equal to the product of the Fourier Transform of the input with the Fourier Transform of the impulse response. $H(\omega)$ is sometimes given the name *system function* since it truly characterizes the linear system. It is also sometimes given the name, *transfer function* since it is the ratio of the output Fourier Transform to the input Fourier Transform.

2.2 COMPLEX TRANSFER FUNCTION

In the sinusoidal steady state analysis of an electrical network, a complex transfer function is defined. It represents the ratio of the output phasor (complex number giving amplitude and relative phase of a sinusoid) to the input phasor. This ratio is a complex function of frequency. In the special case in which the input is a current flowing between two terminals and the output is the voltage due to this current, the transfer function becomes the familiar complex impedance between these two terminals.

For example, in the circuit of Fig. 2.2, if $i_1(t)$ is the input, and $v(t)$ is the output, the transfer function is

$$H(j\omega) = \frac{2j\omega}{1 + 2j\omega} \tag{2.8}$$

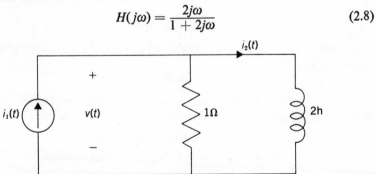

Figure 2.2 Circuit with Eqs. (2.8) and (2.9) as two possible transfer functions.

Alternatively, if $i_2(t)$ is regarded as the output and $i_1(t)$ as input, then the transfer function is

$$H(j\omega) = \frac{1}{1 + 2j\omega} \tag{2.9}$$

It is the intention of this section to show that this $H(j\omega)$, derived for the simple sinusoidal steady state case, is actually the system function, $H(\omega)$ (i.e., the transform of the impulse response). That is, we will show[1]

$$H(\omega) = H(j\omega)$$

For example, in the previous circuit the complex transfer function between $i_1(t)$ and $i_2(t)$ was found as

$$H(j\omega) = \frac{1}{1 + 2j\omega}$$

We now claim that the transform of the impulse response is

$$H(\omega) = \frac{1}{1 + 2j\omega}$$

That is, if $i_1(t) = \delta(t)$, then the transform of $i_2(t)$ is

$$I_2(\omega) = \frac{1}{1 + 2j\omega}$$

If this is indeed the case, the output, $i_2(t)$, due to $i_1(t) = \delta(t)$ would be given by the Fourier inversion integral,

$$i_2(t) = \frac{1}{2\pi} \int_{-\infty}^{\infty} \frac{1}{1 + 2j\omega} e^{j\omega t} \, d\omega \tag{2.10}$$

The general proof of this result follows.

Suppose that the input to a circuit is

$$f(t) = e^{j\omega_0 t}$$

Then the output due to this input is found by sinusoidal steady state techniques to be

$$g(t) = H(j\omega_0)e^{j\omega_0 t} \tag{2.11}$$

The complex transfer function evaluated at ω_0 is $H(j\omega_0)$. This gives us the input-output pair,

$$e^{j\omega_0 t} \longrightarrow H(j\omega_0)e^{j\omega_0 t}$$

Taking the transform of the input and of the corresponding output (noting that $H(j\omega_0)$ is a constant) yields

$$
\begin{aligned}
e^{j\omega_0 t} &\longleftrightarrow 2\pi\delta(\omega - \omega_0) = F(\omega) \\
H(j\omega_0)e^{j\omega_0 t} &\longleftrightarrow 2\pi H(j\omega_0)\delta(\omega - \omega_0) = G(\omega)
\end{aligned} \tag{2.12}
$$

---

[1]It is unfortunate that the same functional letter, H, has gained acceptance to indicate both forms. One should *not* look at this equation and say "aha! $H(\omega) = H(j\omega)$. $H(\omega)$ is therefore of the form $1 + \omega^4 + \omega^8 + \cdots$." An example will clarify this.

Recalling that, for a linear system, $F(\omega)H(\omega) = G(\omega)$ (Eq. 2.7), where $H(\omega)$ is the *system* function, we have

$$2\pi\delta(\omega - \omega_0)H(\omega) = 2\pi H(j\omega_0)\delta(\omega - \omega_0) \tag{2.13}$$

If we invoke the "sifting" property of the impulse function,

$$f(t)\delta(t - t_0) = f(t_0)\delta(t - t_0)$$

Eq. (2.13) becomes

$$2\pi\delta(\omega - \omega_0)H(\omega_0) = 2\pi H(j\omega_0)\delta(\omega - \omega_0) \tag{2.14}$$

and

$$H(\omega_0) = H(j\omega_0)$$

This holds for any value of ω_0, and thus, our proof is complete.

Illustrative Example 2.1

In the circuit of Fig. 2.3, the capacitor is initially uncharged. If $v(t) = \delta(t)$, find $i(t)$.

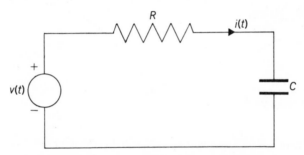

Figure 2.3 Circuit for Illustrative Example 2.1.

Solution

The complex transfer function relating $i(t)$ to $v(t)$ is

$$H(j\omega) = \frac{1}{R + 1/j\omega C} = \frac{j\omega C}{1 + j\omega CR} \tag{2.15}$$

The problem asks for $i(t)$ due to $v(t) = \delta(t)$. This is simply the impulse response, whose transform is the system function, $H(\omega)$. Since $H(\omega) = H(j\omega)$, we have

$$i(t) = \frac{1}{2\pi} \int_{-\infty}^{\infty} \frac{j\omega C e^{j\omega t}}{1 + j\omega CR} \, d\omega \tag{2.16}$$

At this point, it should be emphasized that the evaluation of integrals cannot be considered as a significant part of communication theory. One can look up the answer in a table or as a last resort approximate the evaluation on a computer. Therefore, in this text we will only evaluate integrals in those cases when the result is of instructional significance. Otherwise, the final solutions will be left in integral form.

In this particular example, we need not evaluate the integral at all since the inversion can be performed by inspection. Those familiar with LaPlace Transform will notice that this is analogous to partial fraction expansions. We note that $H(\omega)$ can be rewritten as follows,

$$H(\omega) = \frac{j\omega C}{1 + j\omega CR} = \frac{1}{R} - \frac{1/R}{1 + j\omega CR}$$

$$i(t) = \frac{1}{R}\,\delta(t) - \frac{1}{R^2 C}c^{-t/RC}U(t)$$

(2.17)

This current waveform is sketched as Fig. 2.4.

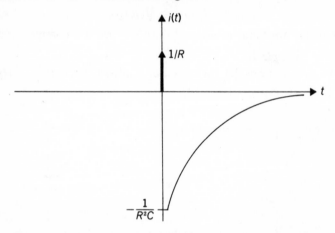

Figure 2.4 $i(t)$ for Illustrative Example 2.1 (circuit of Fig. 2.3).

Note that the impulse in $i(t)$ has appeared automatically. This is noteworthy since so-called classical (traditional) techniques handle impulses with a great deal of difficulty or by resorting to physical reasoning.

It should be realized that the above solution is only correct for zero initial charge on the capacitor. Otherwise, superposition is violated (prove it) and the output is not the convolution of $h(t)$ with the input.

This apparent shortcoming of the Fourier Transform analysis of systems with non-zero initial conditions is easily circumvented as mentioned earlier (page 65). However, the consideration of initial conditions is not critical to communication theory as we will use it, and we will therefore always assume zero initial conditions.

2.3 FILTERS

The word "filter" refers to the removal of the undesired parts of something. In linear systems theory, it was probably originally applied to systems which eliminate undesired frequency components from a time waveform. It has been modified to include systems which simply weigh the various frequency components of a function.

In other words, a linear filter is any (usually passive) linear system which was intended to have the characteristics that it does have. Not as dramatic as you had hoped, perhaps, but extremely useful.

We shall be using the term "ideal distortionless filter" many times in the work to follow. It would therefore be appropriate to define *distortion* at this time.

We define a distorted time signal to be one whose basic shape has been altered. That is, $f(t)$ can be multiplied by a constant and shifted in time without distorting the actual waveform. In household communications reception, we account for the constant multiplier with a volume control. We don't really care about the absolute incoming level as long as we can adjust the volume control to a satisfactory level. We take the time shift into account by not really caring whether the eleven o'clock news comes at exactly 11, or at 50 microseconds after 11 o'clock. If one were listening to anything other than the news, he might not even be concerned with a time shift of several minutes.

In mathematical terms, we consider $Af(t - t_0)$ to be an undistorted version of $f(t)$, where A and t_0 are any real constants (and, of course, $A \neq 0$). The transform of $Af(t - t_0)$ can be found from the time shift theorem.

$$Af(t - t_0) \leftrightarrow Ae^{-j\omega t_0}F(\omega) \qquad (2.18)$$

where

$$f(t) \leftrightarrow F(\omega)$$

We can consider this as the output of a linear system with input $f(t)$ and system function, $H(\omega) = Ae^{-j\omega t_0}$, as shown in Fig. 2.5. Here, since $H(\omega)$ is complex, both the magnitude and phase of $H(\omega)$ have been plotted. The real and imaginary parts would have also sufficed, but would not have been as instructive.

Viewing Fig. 2.5, it seems reasonable that the magnitude function turned out to be a constant. This indicates that all frequencies of $f(t)$ are multiplied by the same factor. Why did the phase turn out to be a linear function of frequency? Why aren't all frequencies shifted by the same "amount"? The answer is clear from a simple example. Suppose that we wished to shift the

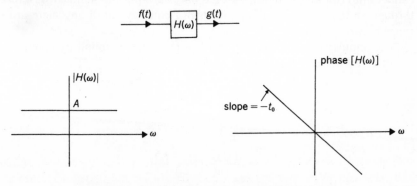

Figure 2.5 System function of a distortionless system.

function, $f(t) = \cos 2\pi t$ by $\frac{1}{4}$ second in time. This would represent a shift of $\pi/2$ radians, or 90° in the argument of the cosine. If we now wished to shift a function of twice the frequency, $g(t) = \cos 4\pi t$ by the same $\frac{1}{4}$ second, it represents a shift of π radians, or 180°. The extension to a continuous representation of frequencies should now be obvious.

Ideal Low Pass Filter

An ideal low pass filter is a linear system which acts like an ideal distortionless system, provided that the input signal contains no frequency components above the "cutoff" frequency of the filter. Frequency components above this cutoff are completely blocked, and do not appear in the output.

Denoting the maximum output frequency (cutoff of the filter) as ω_m, (the "m" denotes maximum), we see that the system function is given by

$$H(\omega) = \begin{cases} Ae^{-j\omega t_0} & |\omega| < \omega_m \\ 0 & |\omega| > \omega_m \end{cases} \qquad (2.19)$$

This transfer function is sketched in Fig. 2.6.

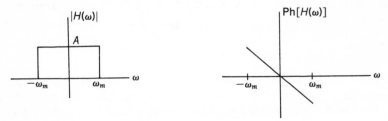

Figure 2.6 Ideal low pass filter characteristic.

We now digress for a moment to our discussion of the sampling theorem. In that case, when we wanted to recover $f(t)$ from its sampled version we should have used a low pass filter. Recall that the transform of the sampled wave contained $F(\omega)$ plus an infinity of shifted versions of $F(\omega)$. In order to extract $F(\omega)$ (the center blob), we need only have put the sampled waveform through an ideal low pass filter with high frequency cutoff anywhere in the range between ω_m and $(2\pi/T_s - \omega_m)$.

The impulse response of the ideal low pass filter is found by computing the inverse Fourier Transform of $H(\omega)$.

$$h(t) = \frac{1}{2\pi} \int_{-\infty}^{\infty} H(\omega)e^{j\omega t}\,d\omega$$

$$= \frac{1}{2\pi} \int_{-\omega_m}^{\omega_m} Ae^{-j\omega t_0}e^{j\omega t}\,d\omega$$

$$= \frac{A \sin \omega_m(t - t_0)}{\pi(t - t_0)} \qquad (2.20)$$

This $h(t)$ is shown in Fig. 2.7.

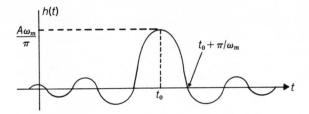

Figure 2.7 Impulse response of ideal low pass filter.

Ideal Band Pass Filter

Rather than transmit (pass) frequencies between zero and ω_m as the low pass filter does, one might desire a system which passes frequencies between two other limits, say ω_1 and ω_2.

An ideal band pass filter is such a linear system. It acts like an ideal distortionless system provided that the input signal contains no frequency components outside of the filter's "pass band."

The system function of an ideal band pass filter with pass band $\omega_1 < |\omega| < \omega_2$ is given by

$$H(\omega) = \begin{cases} Ae^{-j\omega t_0} & \text{for } \omega_1 < |\omega| < \omega_2 \\ 0 & \text{otherwise} \end{cases} \tag{2.21}$$

This $H(\omega)$ is shown in Fig. 2.8.

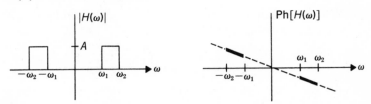

Figure 2.8 Ideal band pass filter characteristic.

The impulse response of the band pass filter, $h(t)$, can be found from the result for the low pass filter and the frequency shifting theorem. That is, given $H_{lp}(\omega)$ as shown in Fig. 2.9, $H(\omega)$ is given by,

$$H(\omega) = H_{lp}\left[\omega - \frac{\omega_1 + \omega_2}{2}\right] + H_{lp}\left[\omega + \frac{\omega_1 + \omega_2}{2}\right]. \tag{2.22}$$

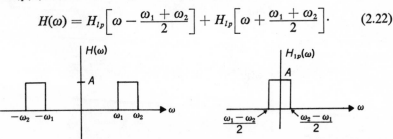

Figure 2.9 Band pass characteristic and corresponding low pass characteristic.

Note that for convenience, we have chosen $t_0 = 0$. This makes the phase equal to zero which simplifies the derivation. We can always reinsert a time shift into the final answer for $h(t)$. Therefore, from Eq. (2.22) we have

$$h(t) = h_{l_p}(t)e^{jxt} + h_{l_p}(t)e^{-jxt} \tag{2.23}$$

where, for notational convenience,

$$x \triangleq \tfrac{1}{2}(\omega_1 + \omega_2)$$

The result from Eq. (2.23) follows from the frequency shift theorem. Continuing,

$$\begin{aligned}
h(t) &= h_{l_p}(t)[e^{jxt} + e^{-jxt}] \\
&= 2h_{l_p}(t)\cos xt \\
&= 2h_{l_p}(t)\cos \tfrac{1}{2}(\omega_1 + \omega_2)t
\end{aligned} \tag{2.24}$$

$h_{l_p}(t)$ is the impulse response of the low pass filter with $t_0 = 0$. From Eq. (2.20), we have

$$h_{l_p}(t) = \frac{A}{\pi t}\sin \tfrac{1}{2}(\omega_2 - \omega_1)t$$

Finally, for the band pass filter, the impulse response is given by

$$h(t) = \frac{2A}{\pi t}\sin\left[\frac{\omega_2 - \omega_1}{2}t\right]\cos\left[\frac{\omega_1 + \omega_2}{2}t\right] \tag{2.25}$$

If the phase factor is included, it can be simply accounted for by a time shift. For the ideal band pass filter, we therefore have

$$h(t) = \frac{2A}{\pi(t - t_0)}\sin\left[\frac{\omega_2 - \omega_1}{2}(t - t_0)\right]\cos\left[\frac{\omega_1 + \omega_2}{2}(t - t_0)\right]$$

This impulse response is sketched in Fig. 2.10. The outline of this sketch resembles the impulse response of the low pass filter. We note that as the two limiting frequencies become high compared to the difference between them, the impulse response starts resembling a shaded-in version of the low pass impulse response (and its mirror image). In more precise terms, this happens when the center frequency of the band pass filter becomes large compared to

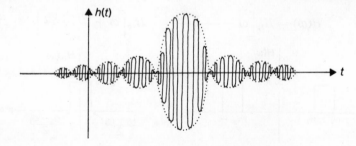

Figure 2.10   Impulse response of ideal band pass filter.

the width of its pass band. This observation will prove highly significant in our later studies of amplitude modulation.

2.4 CAUSALITY

Things are going much too well. We must now darken the picture by showing that it is impossible to build ideal filters. We do this by introducing the idea of *causality*. The concept of causality refers to the "cause and effect" relationship. In loose terms, it states that the effect, or response, due to a cause, or input, cannot anticipate the input in a causal system. That is, a causal system's output at any particular time depends only upon the input prior to that time, and not upon any future values of the input.

$$g(t_0) \text{ depends only upon } f(t) \qquad (t \leq t_0)$$

In order for a linear system to be causal, it is necessary and sufficient that the impulse response, $h(t)$, be zero for $t < 0$. It follows that the response due to a general $f(t)$ input depends only upon past values of $f(t)$. To prove this, we write the expression for the output of a linear system in terms of its input and impulse response.

$$g(t) = \int_{-\infty}^{\infty} h(\tau)f(t - \tau) \, d\tau$$

If $h(t) = 0$ for $t < 0$, this becomes

$$g(t) = \int_{0}^{\infty} h(\tau)f(t - \tau) \, d\tau \qquad (2.26)$$

If we now change variables, letting $t - \tau = k$, this becomes

$$g(t) = \int_{-\infty}^{t} f(k)h(t - k) \, dk \qquad (2.27)$$

which, as stated, shows that $g(t)$ is known in terms of the values of $f(k)$ for $k \leq t$, that is all past values of the input. This proves sufficiency. In order to prove necessity, we note that $\delta(t) = 0$ for all $t < 0$. Therefore, in a causal system, the inputs $f(t) = 0$ and $f(t) = \delta(t)$ must yield the same outputs at least until time $t = 0$. That is, if the output cannot anticipate the input, it has no way of telling the difference between "0" and $\delta(t)$ prior to time $t = 0$. Therefore, $h(t)$ must equal zero for $t < 0$, and the necessity part of the theorem is proven.

The study of causality is important since, in general, causal systems are physically realizable, and non-causal systems are not (much to the dismay of astrologers, fortune tellers, earthquake predictors, and gamblers).

The above criterion is easy to apply if $h(t)$ is explicitly known. That is, one need simply examine $h(t)$ and see if it is zero for negative "t." Since the system function, $H(\omega)$, is often known, and the integral evaluation of $h(t)$ from $H(\omega)$ is sometimes difficult to perform, it would be nice to translate the "$h(t) = 0$

for $t < 0$" requirement into a corresponding restriction on $H(\omega)$. Then, if causality were the only thing we were looking for, it would not be necessary to find the time function corresponding to $H(\omega)$. The *Paley-Wiener criterion* does just this. If

$$\int_{-\infty}^{\infty} \frac{|\ln H(\omega)|}{1 + \omega^2} \, d\omega < \infty$$

and   (2.28)

$$\int_{-\infty}^{\infty} |H(\omega)|^2 \, d\omega < \infty$$

then, for an appropriate choice of phase function for $H(\omega)$, $h(t) = 0$ for $t < 0$.

We note that Eq. (2.28) does not take the phase of $H(\omega)$ into account at all. The actual form of $h(t)$ certainly does depend upon this phase. Indeed, if some $H(\omega)$ corresponds to a causal system, we can form a non-causal system by shifting the original impulse response to the left in time. This time shift corresponds to a linear change in the phase of $H(\omega)$, which does not affect Eq. (2.28). Thus, the Paley-Wiener criterion really tells us whether it is possible for $H(\omega)$ to be the transform of a causal time function[2] (i.e., a necessary but not sufficient condition). Assuming that Eq. (2.28) were satisfied, the phase of $H(\omega)$ would have to be examined before final determination about the causality of the system could be made. Because of all these complications, we will only use the Paley-Wiener condition to make one simple, but significant observation.

Since $\ln (0) = -\infty$, one can observe that if $H(\omega) = 0$ for any range along the ω-axis, the first integral of Eq. (2.28) does not converge, and the system cannot be physically realizable. In particular, all of our ideal low pass and band pass filters must be non-causal. We already knew this from the actual observation of their impulse responses, $h(t)$, which could be easily evaluated for the ideal filters. For this reason, practical filters must be, at best, approximations to their ideal versions.

The previous observation about $H(\omega)$ is a special case of a much more general theorem. Suppose $h(t)$ is such that

$$\int_{-\infty}^{\infty} |h(t)|^2 \, dt < \infty$$ (2.29)

With this restriction, if $h(t)$ is identically zero for any finite range along the t-axis, then its Fourier Transform cannot be zero for any finite range along the ω-axis, and vice versa. This is sometimes stated in loose terms as "a function which is time limited cannot be band limited."

---

[2]Since the condition for causality is that $h(t) = 0$ for negative time, the word "causality" is often extended to apply to all time functions obeying this property. That is, a time function is causal if it equals zero for negative "t."

This leads to an unfortunate fact of life. Any time function which does not exist for all time cannot be band limited. Nothing in real life does exist for all time since no sources were turned on at the dawn of creation (even if they were, that would not be early enough). Alas, it is time to approximate again.

The next section briefly discusses some of the most common approximations that are used in place of the ideal filters. We emphasize that the remainder of this text will be presented as if the results using ideal filters were exact representations of what happens in real life. The student must bear in mind the fact that, while the mathematics is exact, approximations must be involved in applying it to real life situations.

The next section will show that, provided one has enough electrical elements available for use, he can get arbitrarily close to the ideal characteristics. We prematurely emphasize this fact now since, the student who already believes it, can skip the next section without any loss of continuity. Once one possesses a vague awareness of practical limitations, the design of networks can be divorced from the discipline of communication theory. Communication theorists must leave some work for the network designers to do.

2.5 PRACTICAL FILTERS

We now present actual circuits which approximate the ideal band pass and low pass filters. Throughout this section, we will assume that the closer $H(\omega)$ approaches the system function of an ideal filter, the more the filter will behave in an ideal manner in our applications. This fact is not at all obvious. A very "small" change in $H(\omega)$ can lead to horrible changes in $h(t)$. One can examine the consequences and categorize the effects of deviation from the constant amplitude characteristic or from the linear phase characteristic of the ideal system functions. We will not concern ourselves with this aspect of the analysis.[3]

Before any of the known network synthesis techniques can be applied to filter design, a mathematical expression must be found for $H(\omega)$. It is not sufficient to work from a sketch of $H(\omega)$.

Let us begin by analyzing the low pass filter.

Low Pass Filter

The amplitude characteristic of the ideal low pass filter can be approximated by the function

$$|H_n(\omega)| = \frac{1}{\sqrt{1 + \omega^{2n}}} \tag{2.30}$$

[3]See A. Papoulis, *The Fourier Integral and its Applications*, McGraw-Hill, New York, 1962, Chapter 6.

This is sketched for several values of "n" in Fig. 2.11. We have only illustrated the positive half of the ω-axis since the function is clearly an even function of "ω," as it must be in order to represent the magnitude of the transform of a real impulse response. We have chosen $\omega_m = 1$ for simplicity, but this does not restrict our approach to the synthesis.

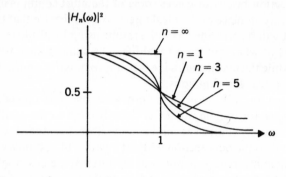

Figure 2.11   Approximation to ideal low pass filter characteristic (Eq. (2.30)).

We note that as "n" gets larger, the amplitude characteristic approaches that of the ideal filter.

Filters which have amplitude characteristics as in Eq. (2.30) are known as *Butterworth filters*. They are used very often since they represent the "best" approximation on the basis of "maximal flatness" in the filter pass band. The concept of maximal flatness relates to the number of derivatives of the magnitude function which are zero at $\omega = 0$. The more derivatives which are zero, the more slowly one would expect the function to turn away from its value at $\omega = 0$. (Think of a Taylor series expansion of the magnitude function.) One can verify that the first $2n - 1$ derivatives of the amplitude function of Eq. (2.30) are zero at $\omega = 0$.

There are other useful approximations to the ideal low pass filter characteristic, most of which optimize with respect to a criterion other than maximal flatness. We will only examine the Butterworth filter since our only intention is to demonstrate realizability of one good approximation to the ideal filters.

Before actually designing the Butterworth filter, we must say something about the phase of $H(\omega)$. The result of Problem 1.24 indicates that, for a causal time function, the odd part of the Fourier Transform can be found from the even part. In a similar way, the phase is determined from the amplitude. Thus, for a causal filter, we are not free to choose the amplitude and phase of $H(\omega)$ independently. Equation (2.30) therefore gives us all of the information that we need to design the filter.

We recall that, for any real time function, $h(t)$,

$$H(\omega) = H^*(-\omega) \tag{2.31}$$

where the asterisk (*) indicates the operation of taking the complex conjugate. We use this fact in factoring $|H(\omega)|^2$.

$$|H(\omega)|^2 = H(\omega)H(-\omega) \tag{2.32}$$

In splitting $|H(\omega)|^2$ into the two factors, we must make sure that $H(\omega)$ is the Fourier Transform of a causal time function. We illustrate this factoring procedure by an example.

Illustrative Example 2.2

In this example, we shall design a third order ($n = 3$) Butterworth filter with $\omega_m = 1$. From Eq. (2.30),

$$|H(\omega)|^2 = \frac{1}{1 + \omega^6} \tag{2.33}$$

If we make a change of variables, $s = j\omega$, we can take advantage of the properties of pole-zero diagrams in the s-plane.

$$|H(s)|^2 = H(s)H(-s) = \frac{1}{1 - s^6} \tag{2.34}$$

The poles of $|H(s)|^2$ are the six roots of unity as sketched in Fig. 2.12. Three of the poles are associated with $H(s)$ and the other three with $H(-s)$. We associate the three poles in the left half plane with $H(s)$ since this will lead to an $h(t)$ which is causal. We are assuming that the reader is familiar with LaPlace Transforms and basic circuit theory. Otherwise he may simply accept this result.

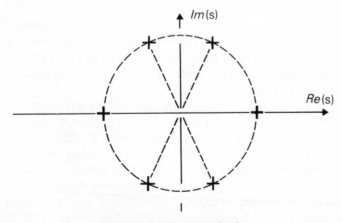

Figure 2.12   Six roots of unity.

Finally, $H(s)$ is found from its poles,

$$H(s) = \frac{1}{(s - p_1)(s - p_2)(s - p_3)} = \frac{1}{s^3 + 2s^2 + 2s + 1} \quad (2.35)$$

There are well known techniques for synthesizing a circuit given $H(s)$. If $v(t)$ is the response and $i(t)$, the source, the above system function corresponds to the circuit of Fig. 2.13.

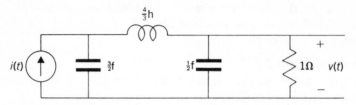

Figure 2.13   Third order Butterworth filter.

Higher order filters (larger n) would lead to other ladder networks with additional elements. That is, additional series inductor, parallel capacitor combinations would appear.

We state without elaboration that, as n increases, the filter approaches an ideal filter in performance. However, as n approaches infinity, the delay between input and output (i.e., slope of the phase characteristic) also approaches infinity.

Band Pass Filter

One can derive a band pass filter corresponding to every low pass filter by simply substituting element combinations for each element in the low pass filter circuit.[4] This has the effect of shifting the frequency term in the filter transfer function. Thus the Butterworth band pass filter has poles lying on a circle centered at the center frequency of the filter. This contrasts to the poles lying on a circle centered at the origin for the low pass filter. A typical pole-zero diagram for a band pass filter is sketched as Fig. 2.14.

The band pass version of the Butterworth filter is what is actually used in household radios as we shall see in Chapter 4.

The synthesis of filters becomes routine with a digital computer. Since computer processing is becoming more and more popular, synthesis of actual filter circuitry is becoming less important. Entire systems are being constructed without the use of a single capacitor, inductor, or resistor. The received signals are modified and fed into a computer. The computer is programmed to simulate the operations of the various electronic devices. The output can then be printed on paper, converted to an analog electrical signal, or even displayed on an oscilloscope screen. The versatility of the com-

[4]See Franklin Kuo, *Network Analysis and Synthesis*, Wiley, New York, 1962.

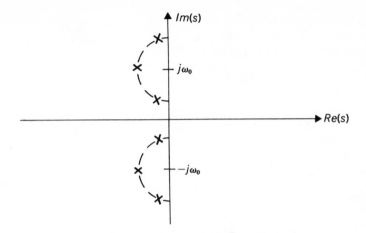

Figure 2.14 Poles of Butterworth band pass filter.

puterized system is quite impressive. One can change the system function of a filter by simply changing one parameter in the program. Indeed, if the input is known for all time, one can even construct a non-causal system in the computer.

2.6 DIGITAL FILTERS

A *digital filter* is a system which converts one sequence of numbers (the input) into another sequence of numbers (the output). Each output number is a linear combination of past input numbers. This leads naturally to the convolution relationship of

$$y(n) = \sum_{m=0}^{n} x(m)h(n-m) \tag{2.36}$$

$$x(n) \longrightarrow \boxed{\begin{array}{c} \text{DIGITAL} \\ \text{FILTER} \end{array}} \xrightarrow{y(n)}$$

Digital filter outputs may also depend upon past output values in addition to past input values. That is,

$$y(n) = \sum_{m=0}^{n} x(m)h(n-m) + \sum_{m=0}^{n-1} y(m)k(n-m) \tag{2.37}$$

The latter type of filter is known as *recursive* since the outputs must be calculated in sequence. That is, for example, before $y(3)$ can be found, $y(0)$, $y(1)$, and $y(2)$ must first be evaluated.

Realization of Digital Filters

We present herein only one form of practical realization of each type of filter.

Illustrative Example 2.3

Derive a block diagram of a non-recursive digital filter with $h(n) = 1, 1, 2, 4$ for $n = 0, 1, 2,$ and 3, respectively. $h(n) = 0$ for n greater than 3.

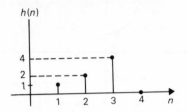

Solution

We expand the summation of Eq. (2.36) to get

$$y(n) = 4x(n - 3) + 2x(n - 2) + x(n - 1) + x(n)$$

This can be realized with the block diagram in Fig. 2.15.

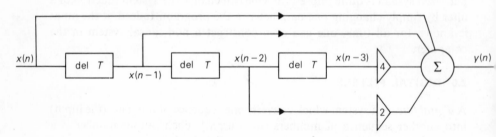

Figure 2.15 Non-recursive filter for Example 2.3.

The "delay T" block is often denoted as follows:

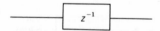

This notation is used since z^{-1} is the z-Transform of a system which delays the input by one period.

Illustrative Example 2.4

Derive the block diagram of a recursive filter with $h(n)$ and $k(n)$ as sketched below.

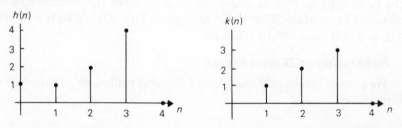

Solution

Expanding the summation in Eq. (2.37), we have

$$y(n) = 4x(n - 3) + 2x(n - 2) + x(n - 1) + x(n) + y(n - 1)$$
$$+ 2y(n - 2) + 3y(n - 3)$$

This equation can be implemented with the block diagram in Fig. 2.16.

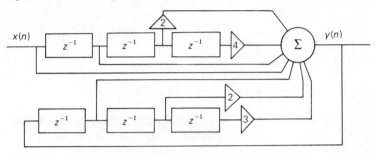

Figure 2.16   Recursive filter for Example 2.4.

The block diagrams of Fig. 2.15 and 2.16 can be implemented using *tapped delay lines* in analog form or by using logic circuits (e.g., registers or, indeed, computers) in the digital form.

2.7 ACTIVE FILTERS

Section 2.5 presented some realizations of filters using inductors, capacitors, and resistors. Such filters are called *passive* since all component parts either absorb or store energy.

A filter is called *active* if there are devices within it which give out energy. Active filters have the advantage of not absorbing part of the desired signal. They are versatile and simple to design, and arbitrary causal transfer functions can be realized. For some applications, such as audio filtering, the passive filter would require impractically large inductors and capacitors.

The basic building block of an active filter is the *operational amplifier* (*OP AMP*). The operational amplifier has characteristics which approach those of an ideal, infinite gain, amplifier. That is, it approximately possesses infinite input resistance, zero output resistance, and infinite voltage gain. Such amplifiers can be purchased as integrated circuits (ICs), and several operational amplifiers are usually included within one IC package. While an ideal OP AMP has infinite bandwidth, practical devices often have highly limited bandwidth.

The OP AMP is almost always used with a large amount of negative feedback. This is true since, as a basic amplifier, the open loop device would have very high gain, but this gain would vary widely with, among other parameters, input frequency.

Assume that an OP AMP is connected as shown in Fig. 2.17. The OP AMP of Fig. 2.17 is being operated as a differential amplifier. It amplifies the difference in voltage between the terminal marked "+" and that marked "−."

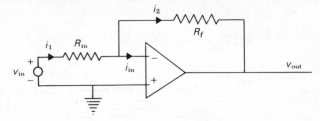

Figure 2.17   OP AMP with feedback.

Since the amplifier has infinite input resistance, we can assume that $i_{in} = 0$. Therefore, $i_1 = i_2$. In a practical system, v_{out} is limited by the supply voltages within the system. Since the amplification is assumed to be infinite, the assumption of finite v_{out} leads to the simplifying conclusion that $v_1 = 0$. Therefore,

$$i_1 = \frac{v_{in}}{R_{in}}$$

and

$$i_2 = -\frac{v_{out}}{R_f}$$

Since $i_1 = i_2$, we have

$$\frac{v_{in}}{R_{in}} = -\frac{v_{out}}{R_f}$$

and

$$\frac{v_{out}}{v_{in}} = -\frac{R_f}{R_{in}} \tag{2.38}$$

Thus, the amplification of the system is a function only of the elements attached to the operational amplifier.

Suppose now that R_{in} is replaced by a capacitor, as in Fig. 2.18. Equation (2.38) now becomes

$$Cv'_{in} = -\frac{v_{out}}{R_f}$$

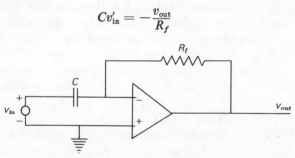

Figure 2.18   OP AMP with capacitor input.

and

$$v_{\text{out}} = -R_f C v'_{\text{in}}$$

where v'_{in} is the time derivative, dv_{in}/dt. Thus, in the configuration of Fig. 2.18, the OP AMP acts as a differentiator. The student should derive relationships for the case in which the capacitor replaces R_f instead of R_{in}. In addition, the effects of using an inductor instead of a capacitor should be easily derivable.

If more complex frequency sensitive circuits replace R_{in} and/or R_f, various filter characteristics result. As one example, a first order low pass filter is shown in Fig. 2.19. Since neither input to the OP AMP is grounded,

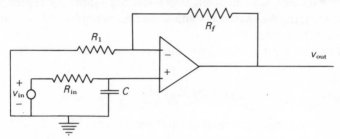

Figure 2.19   First order low pass filter.

we start with the assumption that $v^+ - v^- = 0$. Therefore, $v^+ = v^-$, and from the previous analysis,

$$v_{\text{out}} = -\left(\frac{R_f}{R_1}\right)v^- = -\left(\frac{R_f}{R_1}\right)v^+$$

Therefore, the entire system operates simply as the circuitry attached to the v^+ input. In this case, v_{out} is proportional to the voltage across the capacitor, and the transfer function is basically that of the RC circuit.

The advantage of the active system over the simple RC circuit is that v_{out} does not draw current from the RC circuit. In a passive RC circuit, if the voltage across the capacitor feeds further circuitry, the current drain can seriously affect the relationship between capacitor voltage and input voltage.

The simple first order low pass filter is presented as an example and is not intended to imply that the filter of Fig. 2.19 is often used. The reader is referred to the references for additional information.

2.8   RISE TIME AND PULSE WIDTH

In designing a communication system, an important consideration is that of the bandwidth of the system. The bandwidth is the range of frequencies which the system must be capable of handling.

Bandwidth clearly relates to the Fourier Transform of a time function. It is not directly definable in terms of the time function. That is, one cannot look at a time function and say, "Aha; the bandwidth of that time function is . . . ." He must first calculate the Fourier Transform.

The Fourier Transform does *not* exist anywhere in real life. One cannot directly display it on any device in the laboratory. It is simply a mathematical definition which proves useful in the analysis of real life systems. We would therefore like to relate the concept of bandwidth to properties of signals which *do* exist in real life.

Both the minimum width of a time pulse and the minimum time in which the output of a system can jump from one level to another are truly physical terms. We wish to show that both the minimum width of a time pulse and the minimum transition time between two levels are intimately related to the system bandwidth. We shall start with a specific example, and then attempt to generalize the results.

Returning to the ideal low pass filter response, recall that the impulse response was found to be (we repeat Eq. 2.20)

$$h(t) = \frac{\omega_m}{\pi} \frac{\sin \omega_m(t - t_0)}{\omega_m(t - t_0)} \tag{2.39}$$

Both this $h(t)$ and the corresponding $H(\omega)$ are shown in Fig. 2.20.

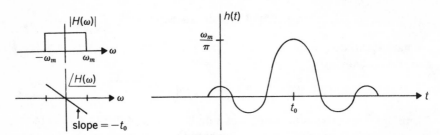

Figure 2.20 $H(\omega)$ and $h(t)$ for ideal low pass filter.

Note that $h(t)$ is symmetrical about $t = t_0$, and has its maximum, ω_m/π, at this point, $t = t_0$.

A general result for time invariant linear systems is that the output due to the integral of a particular signal is the integral of the output due to the signal itself. That is, if $f(t) \rightarrow g(t)$, then

$$\int_{-\infty}^{t} f(\tau)d\tau \rightarrow \int_{-\infty}^{t} g(\tau) \, d\tau \tag{2.40}$$

Although we will not need it here, it is also true that

$$\frac{df}{dt} \rightarrow \frac{dg}{dt} \tag{2.41}$$

This derivative relationship is easily proven since, for a time invariant linear system,

$$\frac{1}{\Delta t}[f(t + \Delta t) - f(t)] \rightarrow \frac{1}{\Delta t}[g(t + \Delta t) - g(t)]$$

Letting Δt approach zero, we get

$$\frac{df}{dt} \longrightarrow \frac{dg}{dt}$$

The integral property can be derived from the derivative relationship by a change of variables (do it!).

Using this result, we can find the response of a low pass filter when its input is a unit step function. Since the unit step is the integral of an impulse function, the step response, sometimes denoted as $a(t)$, is simply the integral of $h(t)$.

$$U(t) \longrightarrow a(t) = \int_{-\infty}^{t} h(\tau) \, d\tau \qquad (2.42)$$

The step response of a low pass filter is sketched in Fig. 2.21.

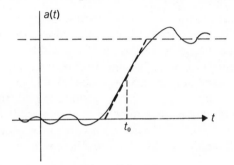

Figure 2.21   Step response of ideal low pass filter.

Since the slope of $a(t)$ is equal to $h(t)$, the maximum slope, ω_m/π, occurs at the point $t = t_0$.

There are several common definitions of rise time. They all try to mathematically define the length of time it takes the output to respond to a change, or jump, in the input. In practice, it is difficult to actually define the point in time when the output has finished responding to the input jump. We present one particular definition which happens to work well when applied to the ideal low pass filter example.

The rise time may be defined as the time required for a signal to go from the initial to the final value (steady state value) along a ramp with constant slope equal to the maximum slope of the actual function. That is,

$$\text{Rise time} = t_r \triangleq \frac{f_{\text{final}} - f_{\text{initial}}}{df/dt \text{ maximum}} \qquad (2.43)$$

For the low pass filter example, the rise time is therefore given by

$$t_r = \frac{1}{\omega_m/\pi} = \frac{\pi}{\omega_m}$$

The ramp with slope equal to the maximum derivative is shown as a dashed line in Fig. 2.21. We note that since the low pass filter in question passes frequencies between zero and ω_m, its bandwidth is ω_m.[5] Therefore the above result becomes

$$t_r = \frac{\pi}{BW}$$

From this result, we make the bold generalization that, for any band limited linear system (i.e., $H(\omega) = 0$ outside of some interval in the ω-axis), the rise time is inversely proportional to the bandwidth of the system. The bandwidth is defined as the width of the interval occupied by $H(\omega)$ on the positive ω-axis. The actual constant of proportionality depends upon the shape of $H(\omega)$, upon the definition of rise time, and upon the definition of bandwidth. The bandwidth definition is required in cases where $H(\omega)$ does not equal exactly zero outside the interval, but, for example, may asymptotically approach zero.

In a slightly more general way, we would now like to show that an inverse relationship also exists between the width of a pulse in the time domain and the bandwidth of its Fourier Transform (i.e., the width of the corresponding pulse in the frequency domain). The similarity between pulse width and rise time should be somewhat obvious. The student should intuitively justify the statement that the minimum pulse width is related to twice the rise time.

Since the time limited pulse cannot be band limited (*see* "Paley-Wiener condition"), a new definition of bandwidth is required. Actually our definitions to follow will require neither the time function nor its transform to be strictly limited.

We assume a positive pulse, $f(t)$, symmetrical about $t = 0$ (even). We shall choose a definition of pulse width which yields a simple result, even though it is not analogous to our previous definition of rise time. The pulse width, T, will be defined as the width of a rectangular pulse of height $f(0)$, and area the same as that of the original pulse (*see* Fig. 2.22). Thus,

$$T \triangleq \frac{\int_{-\infty}^{\infty} f(t) \, dt}{f(0)} \tag{2.44}$$

Since $f(t)$ is assumed to be real and even, its transform $F(\omega)$, will be a real and even function of ω. We define the bandwidth of $F(\omega)$ in exactly the same way in which we defined the pulse width of $f(t)$. The bandwidth, W, is then given by (*see* Fig. 2.22)

$$W \triangleq \frac{\int_{-\infty}^{\infty} F(\omega) \, d\omega}{F(0)} \tag{2.45}$$

Note that the definition in Eq. (2.45) does not restrict the function to resemble the sketch of Fig. 2.22. That is, the transform of a time pulse does

---

[5]There is some confusion as to whether the bandwidth of the low pass filter is ω_m or $2\omega_m$. This text will define the bandwidth as the difference between the highest and lowest frequencies passed by the filter.

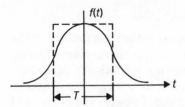
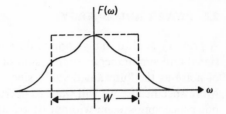

Figure 2.22   Pictorial definition of pulse width and bandwidth.

not necessarily resemble a pulse in frequency. Indeed, it doesn't necessarily have a maximum at $\omega = 0$. (Can you think of an example in which it doesn't?)

We would now like to find a relationship between T and W. From the definition of the Fourier Transform,

$$F(\omega) \triangleq \int_{-\infty}^{\infty} f(t)e^{-j\omega t}\, dt \tag{2.46}$$

we see that

$$F(0) = \int_{-\infty}^{\infty} f(t)\, dt \tag{2.47}$$

From the inverse Fourier Transform Theorem,

$$f(t) = \frac{1}{2\pi} \int_{-\infty}^{\infty} F(\omega)e^{j\omega t}\, d\omega \tag{2.48}$$

we see that

$$f(0) = \frac{1}{2\pi} \int_{-\infty}^{\infty} F(\omega)\, d\omega \tag{2.49}$$

Therefore, Eq. (2.44) becomes

$$T = \frac{F(0)}{f(0)}$$

and Eq. (2.45) becomes

$$W = \frac{2\pi f(0)}{F(0)}$$

Multiplying W by T, we get the desired result,

$$WT = 2\pi$$

or

$$W = \frac{2\pi}{T} \tag{2.50}$$

The bandwidth and pulse width are inversely related, the constant of proportionality again depending upon the actual definition of width.

Therefore, the faster we desire a signal to change from one level to another, the more space on the frequency axis we must allow for it. This fact of life will prove significant in the digital communication schemes to be discussed later.

2.9 POWER AND ENERGY

A primary goal of many communication systems is the enhancement of the signal and simultaneous suppression of the noise (use an intuitive definition of noise as the "unwanted stuff" since we have not yet formally defined it). More specifically, we will want to decrease the noise power at the output of our processing system without decreasing the signal power. The system will increase the so-called *signal to noise ratio*. Before systems can be evaluated with respect to this property, facility must be gained in calculating the power content of a signal and in determining what effect a linear system has on this power content.

We shall associate a number, E_f, with any function of time, $f(t)$.

$$E_f \triangleq \int_{-\infty}^{\infty} |f^2(t)| \, dt \qquad (2.51)$$

E_f is called the *energy* of $f(t)$. This definition makes intuitive sense since, if $f(t)$ were the voltage across, or current through, a 1-ohm resistor, E_f would be the total energy dissipated as heat in the resistor. This number, E_f, is defined by the integral of Eq. (2.51) for any time function, $f(t)$, even if the function does not represent an electrical voltage or current. This distinction between the mathematical definition of energy and the electrical definition is emphasized in Problem 2.12 at the end of this chapter.

E_f is often infinite. We therefore are forced to also define the *average power* of $f(t)$ as the average time derivative of the energy. It is certainly possible for a function to be infinite, while its derivative remains finite (consider a ramp function). Defining average power in this way, we have,

$$P_f \triangleq \text{average power} = \overline{f^2(t)}$$
$$= \lim_{T \to \infty} \frac{1}{2T} \int_{-T}^{T} |f(t)|^2 \, dt \qquad (2.52)$$

The student should convince himself that if E_f is finite, P_f must equal zero, and that if P_f is non-zero, E_f must be infinite.

We will use the average power of $f(t)$, P_f, to classify time functions into one of three groups.

Group I: $P_f = 0$
This group includes all finite energy signals.

Group II: $0 < P_f < \infty$
This group includes, among other things, periodic functions of time.

Group III: $P_f = \infty$
These are messy, unbounded signals. While several common *theoretical* waveforms (i.e., an impulse train) are included in this group, we shall ignore these signals as far as power and energy studies are concerned since we cannot handle them mathematically (this approach is used more often than you think). Fortunately, this class does not occur in the real world.

90

We shall now say as much as we can about each of the first two groups. It is unfortunate that, as in the development of any new concept, several definitions will first have to be presented. An attempt will be made to supply some physical insight into the definitions. This will be meant only as an aid in the memorization since justification of a definition is never necessary (by definition of a definition).

Signals with Finite Energy (Group I)

By invoking Parseval's Theorem, the total energy in a time function, $f(t)$, can be rewritten as

$$E_f = \int_{-\infty}^{\infty} |f^2(t)| \, dt = \frac{1}{2\pi} \int_{-\infty}^{\infty} |F(\omega)|^2 \, d\omega \qquad (2.53)$$

We choose to define the integrand of the right hand term in Eq. (2.53) as a new variable, using the Greek letter, psi (ψ).

$$\psi_f(\omega) \triangleq |F(\omega)|^2 \qquad (2.54)$$

Since the magnitude of the Fourier Transform of a real time function is an even function of "ω." $\psi_f(\omega)$ will necessarily be even and Eq. (2.53) can be rewritten as

$$E_f = \frac{1}{2\pi} \int_{-\infty}^{\infty} \psi_f(\omega) \, d\omega = \frac{1}{\pi} \int_{0}^{\infty} \psi_f(\omega) \, d\omega \qquad (2.55)$$

Does the question, "What part of the total energy of $f(t)$ lies between frequencies $\omega = \omega_1$ and $\omega = \omega_2$?" have any reasonable interpretation?

Suppose that $f(t)$ is the input to an ideal band pass filter which only passes (transmits) input signals with frequencies[6] between ω_1 and ω_2. (*See* Fig. 2.23).

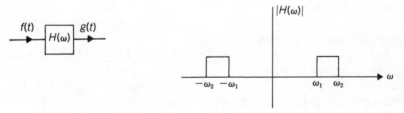

Figure 2.23 Ideal band pass filter.

If $f(t)$ is a finite energy signal (Group I), it is easy to intuitively convince oneself that $g(t)$ is also a finite energy signal. If this were not true, we would have succeeded in solving a significant ecological problem by inventing a

---

[6]We often allow our verbal statements to become a little sloppy. For example, the statement, "input signals with frequencies between . . ." should be read "input time signals whose Fourier Transforms are identically zero outside of the range. . . ."

simple linear system which could replace all of the atomic, fossil fuel, and hydro-electric generating plants in the world.

Thus, the total energy of $g(t)$ is well defined. Since $g(t)$ is, in some sense, that part of $f(t)$ with frequencies between ω_1 and ω_2, we claim that the total energy of $g(t)$ can be thought of as that part of the energy of $f(t)$ which lies between ω_1 and ω_2. This total energy of $g(t)$ is given by

$$
\begin{aligned}
E_g &= \int_{-\infty}^{\infty} |g^2(t)|\, dt \\
&= \frac{1}{\pi} \int_0^{\infty} |G(\omega)|^2\, d\omega
\end{aligned}
\tag{2.56}
$$

From basic linear system theory, we know that

$$
G(\omega) = F(\omega)H(\omega)
$$

Therefore,

$$
\begin{aligned}
E_g &= \frac{1}{\pi} \int_0^{\infty} |H(\omega)F(\omega)|^2\, d\omega = \frac{1}{\pi} \int_0^{\infty} |H(\omega)|^2 |F(\omega)|^2\, d\omega \\
E_g &= \frac{1}{\pi} \int_0^{\infty} |H(\omega)|^2 \psi_f(\omega)\, d\omega = \frac{1}{\pi} \int_{\omega_1}^{\omega_2} \psi_f(\omega)\, d\omega
\end{aligned}
\tag{2.57}
$$

Thus, the energy of $f(t)$ between frequencies ω_1 and ω_2 is found by integrating $\psi_f(\omega)$ between these two frequency limits and dividing the result by π. Because of this property, $\psi_f(\omega)$ is called the *energy spectral density* (or simply energy density) of $f(t)$.

The energy spectral density of the output of a linear system is simply determined in terms of the input density function and the system function, $H(\omega)$, as follows,

$$
\psi_g(\omega) = |G(\omega)|^2 = |F(\omega)H(\omega)|^2 = \psi_f(\omega)|H(\omega)|^2
\tag{2.58}
$$

This is a very useful relationship. It shows us how to find the energy density, and therefore the total energy, of the output of a linear system without first finding the actual time function, $g(t)$.

In practice, given $f(t)$, $\psi_f(\omega)$ is sometimes hard to find using Eq. (2.54). Is it possible to find $\psi_f(\omega)$ from $f(t)$ in a more direct way? The answer is yes. However, another definition is first required. The only reason that we continue at this point is that this new definition will prove very useful in later work.

We define the *autocorrelation* of $f(t)$ as the inverse Fourier Transform of $\psi_f(\omega)$. Thus, if autocorrelation is denoted by $\varphi_f(t)$, we have

$$
\begin{aligned}
\varphi_f(t) &\triangleq \mathcal{F}^{-1}[\psi_f(\omega)] = \mathcal{F}^{-1}[|F(\omega)|^2] \\
\varphi_f(t) &= \mathcal{F}^{-1}[F(\omega)F^*(\omega)]
\end{aligned}
\tag{2.59}
$$

We have used the relationship

$$
|F(\omega)|^2 = F(\omega)F^*(\omega)
$$

where the asterisk (*) indicates complex conjugate.

By the convolution theorem, this last equality becomes

$$\varphi_f(t) = \mathcal{F}^{-1}[F(\omega)]*\mathcal{F}^{-1}[F^*(\omega)] \tag{2.60}$$

We need a simple relationship before simplifying this further. If

$$f(t) \leftrightarrow F(\omega)$$

then

$$f(-t) \leftrightarrow F^*(\omega)$$

This can be easily seen from the Fourier Transform integral.

$$\mathcal{F}[f(-t)] = \int_{-\infty}^{\infty} f(-t)e^{-j\omega t}\, dt$$

Changing variables, we let $T = -t$.

$$\begin{aligned}\mathcal{F}[f(-t)] &= \int_{-\infty}^{\infty} f(T)e^{+j\omega T}\, dT, \\ &= \left[\int_{-\infty}^{\infty} f(t)e^{-j\omega t}\, dt\right]^* = F^*(\omega)\end{aligned} \tag{2.61}$$

We have assumed that $f(t)$ is real.

Finally, using this relationship, Eq. (2.60) can be rewritten as

$$\varphi_f(t) = f(t) * f(-t) \tag{2.62}$$

Since $f(t) * g(t) = g(t) * f(t)$ (i.e., convolution is commutative), we have

$$\varphi_f(-t) = f(-t) * f(t) = \varphi_f(t)$$

and the autocorrelation is an even function of time. This could have been predicted since its Fourier Transform, the energy spectral density, is real and even.

Using the definition of convolution in order to write Eq. (2.63) in its expanded integral form, we have

$$\varphi_f(t) = \int_{-\infty}^{\infty} f(\tau)f(\tau - t)\, d\tau = \int_{-\infty}^{\infty} f(\tau)f(\tau + t)\, d\tau \tag{2.63}$$

The integral of Eq. (2.63) essentially takes $f(\tau)$, shifts it by "t," and compares (correlates) it with the unshifted version. As stated previously, it is often easier to use this integral form to find $\varphi_f(t)$ from the time function, $f(t)$, and then find the energy spectral density by taking the Fourier Transform. Indeed, at this point the *only* apparent use of the autocorrelation is as an intermediate step in finding the energy spectral density. Other applications will be developed later in the study of communications (Chapter 3). In fact, the autocorrelation will form a major tie between this elementary theory and the study of random noise.

Illustrative Example 2.5

A signal, $f(t) = e^{-2t}U(t)$ is passed through an ideal low pass filter with cutoff frequency, $\omega_m = 1$ rad/sec. Find the ratio of output energy to input energy.

Solution

It is first necessary to determine whether $f(t)$ is indeed a finite energy signal. To do this, we check the integral,

$$E_f = \int_{-\infty}^{\infty} |f^2(t)| \, dt = \int_{0}^{\infty} e^{-4t} \, dt = \tfrac{1}{4} < \infty \qquad (2.64)$$

The signal is therefore one with bounded energy. We must now find the energy spectral density of the output signal. Calling this output signal $g(t)$, v'e have

$$\psi_g(\omega) = \psi_f(\omega) \, |H(\omega)|^2$$

where

$$\psi_f(\omega) = |F(\omega)|^2 = \left| \frac{1}{j\omega + 2} \right|^2 = \frac{1}{\omega^2 + 4}$$

$$|H(\omega)|^2 = \begin{cases} 1 & |\omega| < \omega_m = 1 \\ 0 & \text{otherwise} \end{cases} \qquad (2.65)$$

Therefore,

$$\psi_g(\omega) = \begin{cases} \dfrac{1}{\omega^2 + 4} & |\omega| < 1 \\ 0 & \text{otherwise} \end{cases} \qquad (2.66)$$

Finally, the total energy contained in $g(t)$ is given by

$$\frac{1}{\pi} \int_{0}^{\infty} \psi_g(\omega) \, d\omega = \frac{1}{\pi} \int_{0}^{1} \frac{1}{\omega^2 + 4} \, d\omega$$

$$= \frac{1}{2\pi} \tan^{-1} \left(\frac{1}{2} \right) \qquad (2.67)$$

Therefore, the ratio of output to input energy is given by

$$\frac{(1/2\pi) \tan^{-1} \left(\frac{1}{2} \right)}{\frac{1}{4}} = \frac{2}{\pi} \tan^{-1} \left(\frac{1}{2} \right) = 0.29$$

Thus, 71% of the input energy is absorbed by the low pass filter itself. Only 29% reaches the output (i.e., lies in the pass band of the filter).

Signals with Finite Power (Group II)

We have worked so hard in order to derive the previous results for finite energy signals, it would only be just that they be applicable to finite power signals. This is, of course, impossible since the energy of a finite power signal is infinite. We will, however, use the previous results as an intermediate step in the derivation of analogous power formulae.

We start with a finite power signal, $f(t)$, and modify it in such a way that it becomes a finite energy signal. One way of doing this is to truncate $f(t)$ in time. That is, define a new signal, $f_T(t)$ as follows,

$$f_T(t) \triangleq \begin{cases} f(t) & |t| < \dfrac{T}{2} \\ 0 & \text{otherwise} \end{cases} \qquad (2.68)$$

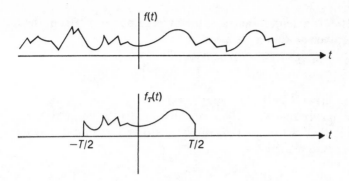

Figure 2.24 A representative $f(t)$ and $f_T(t)$.

Figure 2.24 shows an example of an $f(t)$ and its associated truncated version, $f_T(t)$. If $f(t)$ is well behaved, $f_T(t)$ will be a bounded energy signal for finite T. That is,

$$\int_{-\infty}^{\infty} |f_T(t)|^2 \, dt = \int_{-T/2}^{T/2} |f(t)|^2 \, dt < \infty$$

We can therefore define an energy spectral density of $f_T(t)$.

$$\psi_{f_T}(\omega) = |F_T(\omega)|^2$$

where

$$F_T(\omega) = \int_{-\infty}^{\infty} f_T(t) e^{-j\omega t} \, dt = \int_{-T/2}^{T/2} f(t) e^{-j\omega t} \, dt \qquad (2.69)$$

Now we proceed to some handwaving. This energy spectral density, $\psi_{f_T}(\omega)$, is actually the density of the energy contained in $f(t)$ in the interval $-T/2 < t < T/2$. Therefore, the density of the average power of $f(t)$ in this interval is given by this energy density divided by the length of the interval.

$$\frac{\psi_{f_T}(\omega)}{T} = \text{Density of average power in } -\frac{T}{2} < t < \frac{T}{2}$$

We will call this the average power spectral density of $f(t)$ in the interval $-T/2 < t < T/2$. This is given the symbol, $S_T(\omega)$. Therefore,

$$S_T(\omega) = \frac{\psi_{f_T}(\omega)}{T} = \frac{1}{T} |F_T(\omega)|^2 \qquad (2.70)$$

Since we are really interested in the average power for all time rather than in this finite interval, we let T approach infinity to get the *average power spectral density*, $S(\omega)$, of the finite power signal, $f(t)$.

$$S(\omega) = \lim_{T \to \infty} S_T(\omega) = \lim_{T \to \infty} \frac{1}{T} |F_T(\omega)|^2 \qquad (2.71)$$

where

$$F_T(\omega) = \int_{-T/2}^{T/2} f(t) e^{-j\omega t} \, dt$$

Since $S(\omega)$ is a density function, the average power of $f(t)$ in the frequency band between ω_1 and ω_2 is given by

$$\frac{1}{\pi} \int_{\omega_1}^{\omega_2} S(\omega)\, d\omega \tag{2.72}$$

just as in the case of the energy spectral density. From Eq. (2.71) we can see that the power spectral density of the output of a linear system, $S_g(\omega)$, is related to that of the input, $S_f(\omega)$, in the same way that the energy spectral densities were related.

$$S_g(\omega) = S_f(\omega)|H(\omega)|^2 \tag{2.73}$$

Note that in cases where the corresponding time function is obvious, we will omit the subscript on the density function. For example, when we were dealing with one time function, $f(t)$, we wrote $S(\omega)$ rather than $S_f(\omega)$.

In the power case, we define an autocorrelation, $R_f(t)$, as the inverse Fourier Transform of $S_f(\omega)$.

$$R_f(t) \triangleq \mathcal{F}^{-1}[S_f(\omega)] \tag{2.74}$$

It is unfortunate that both $R(t)$ and $\varphi(t)$ are called the autocorrelation since their definitions are completely different. However, given an $f(t)$ it will become obvious which of the two definitions will apply. If the wrong one is accidentally used, an autocorrelation of either zero or infinity will result.

As before, $R_f(t)$, the autocorrelation, was invented as an intermediate step in finding the power spectral density. It would therefore be nice to be able to find $R_f(t)$ directly from the time function, $f(t)$. The derivation is analogous to that used for finite energy signals. Defining $R_T(t)$ as the inverse transform of $S_T(\omega)$, we have

$$R_T(t) \leftrightarrow S_T(\omega) = \frac{1}{T}|F_T(\omega)|^2$$

$$R_T(t) = \frac{1}{T}\mathcal{F}^{-1}[F_T(\omega)F_T^*(\omega)] \tag{2.75}$$

where the asterisk ($*$) denotes complex conjugate.

From the basic properties of the Fourier Transform, and the convolution theorem, we find

$$R_T(t) = \frac{1}{T}[f_T(t) * f_T(-t)]$$

$$= \frac{1}{T}\int_{-\infty}^{\infty} f_T(\tau)f_T(\tau - t)\, d\tau \tag{2.76}$$

Taking the definition of $f_T(t)$ into account and carefully changing the limits of integration, we get

$$R_T(t) = \frac{1}{T}\int_{-(T/2)+t}^{T/2} f(\tau)f(\tau - t)\, d\tau$$

Finally, letting T approach infinity, we get, for finite t,

$$R(t) = \lim_{T \to \infty} \frac{1}{T} \int_{-T/2}^{T/2} f(\tau) f(\tau - t) \, d\tau \qquad (2.77)$$

Figure 2.25 summarizes the results of this section.

Finite Power Signals	*Finite Energy Signals*
Power spectral density	**Energy spectral density**
$S_f(\omega) = \lim\limits_{T \to \infty} \frac{1}{T} \lvert F_T(\omega) \rvert^2$	$\psi_f(\omega) = \lvert F(\omega) \rvert^2$
where	where·
$F_T(\omega) = \int_{-T/2}^{T/2} f(t) e^{-j\omega t} \, dt$	$F(\omega) = \int_{-\infty}^{\infty} f(t) e^{-j\omega t} \, dt$
Power between ω_1 and ω_2	**Energy between ω_1 and ω_2**
$\frac{1}{\pi} \int_{\omega_1}^{\omega_2} S_f(\omega) \, d\omega$	$\frac{1}{\pi} \int_{\omega_1}^{\omega_2} \psi_f(\omega) \, d\omega$
Total average power	**Total energy**
$\frac{1}{\pi} \int_{0}^{\infty} S_f(\omega) \, d\omega$	$\frac{1}{\pi} \int_{0}^{\infty} \psi_f(\omega) \, d\omega$
Output density-linear system	**Output density-linear system**
$S_g(\omega) = S_f(\omega) \lvert H(\omega) \rvert^2$	$\psi_g(\omega) = \psi_f(\omega) \lvert H(\omega) \rvert^2$
Autocorrelation	**Autocorrelation**
$R_f(t) = \mathscr{F}^{-1}[S_f(\omega)]$	$\varphi_f(t) = \mathscr{F}^{-1}[\psi_f(\omega)]$
$R_f(t) = \lim\limits_{T \to \infty} \frac{1}{T} \int_{-T/2}^{T/2} f(\tau) f(\tau \pm t) \, d\tau$	$\varphi_f(t) = \int_{-\infty}^{\infty} f(\tau) f(\tau \pm t) \, d\tau$

Figure 2.25   Table of results on power and energy.

Illustrative Example 2.6
    Find the total average power of

$$f(t) = A \cos \omega_0 t$$

Solution
    If one attempts to apply the definition,

$$S_T(\omega) = \lim_{T \to \infty} \frac{1}{T} \lvert F_T(\omega) \rvert^2$$

in order to find the power spectral density of $f(t)$, it will be found that the limit does not exist for $\omega = \omega_0$ (try it!). We therefore attempt the autocorrelation route.

$$R(t) = \lim_{T \to \infty} \frac{1}{T} \int_{-T/2}^{T/2} f(\tau) f(\tau + t)\, d\tau$$

$$= \lim_{T \to \infty} \frac{1}{T} \int_{-T/2}^{T/2} A \cos \omega_0 \tau A \cos \omega_0 (\tau + t)\, d\tau \qquad (2.78)$$

Using trigonometric identities, this becomes

$$R(t) = \lim_{T \to \infty} \frac{A^2}{2T} \int_{-T/2}^{T/2} [\cos \omega_0 t + \cos \omega_0 (2\tau + t)]\, d\tau \qquad (2.79)$$

Since $\cos \omega_0 t$ is not a function of the variable of integration, we have

$$R(t) = \lim_{T \to \infty} \frac{A^2}{2T} \left[\cos \omega_0 t \int_{-T/2}^{T/2} d\tau + \frac{\sin \omega_0 (2\tau + t)}{2\omega_0} \Big|_{-T/2}^{T/2} \right]$$

$$R(t) = \lim_{T \to \infty} \frac{A^2}{2} \left[\frac{T \cos \omega_0 t}{T} + \frac{\sin \omega_0 (T + t)}{2\omega_0 T} + \frac{\sin \omega_0 (t - T)}{2\omega_0 T} \right] \qquad (2.80)$$

In the limit as T approaches infinity, $(\sin x)/T$ must go to zero. This is true since, for any value of x, the absolute value of the numerator is bounded by 1, and the denominator is increasing without limit. Therefore,

$$R(t) = \frac{A^2}{2} \cos \omega_0 t \qquad (2.81)$$

Taking the transform of this, we find $S(\omega)$, the power spectral density of $f(t)$.

$$S(\omega) = \mathcal{F}[R(t)] = \frac{A^2 \pi}{2} [\delta(\omega - \omega_0) + \delta(\omega + \omega_0)] \qquad (2.82)$$

The total average power of $f(t)$ is therefore given by

$$\frac{1}{\pi} \int_0^\infty S(\omega)\, d\omega = \frac{1}{\pi} \int_0^\infty \frac{A^2 \pi}{2} [\delta(\omega - \omega_0) + \delta(\omega + \omega_0)]\, d\omega$$

$$\text{Average power} = P_{\text{avg}} = \frac{A^2}{2} \int_0^\infty \delta(\omega - \omega_0)\, d\omega$$

since $\delta(\omega + \omega_0)$ doesn't fall within the range of integration. Finally,

$$P_{\text{avg}} = \frac{A^2}{2} \int_0^\infty \delta(\omega - \omega_0)\, d\omega = \frac{A^2}{2} \qquad (2.83)$$

Illustrative Example 2.7

A signal with power spectral density,

$$S(\omega) = \pi[\delta(\omega - 1) + \delta(\omega + 1)]$$

is the input to the linear system shown in Fig. 2.26. Find the average power of the output voltage, $e_{\text{out}}(t)$.

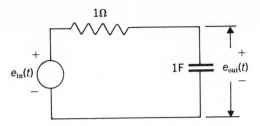

Figure 2.26   Circuit for Illustrative Example 2.7.

Solution

We must first find the power spectral density of $e_{out}(t)$.

$$S_{e_{out}}(\omega) = S_{e_{in}}(\omega) |H(\omega)|^2$$

where

$$H(\omega) = \frac{1/j\omega}{1 + 1/j\omega} = \frac{1}{1 + j\omega}$$

Therefore,

$$|H(\omega)|^2 = \frac{1}{1 + \omega^2}$$

and

$$S_{e_{out}}(\omega) = \frac{\pi}{1 + \omega^2}[\delta(\omega - 1) + \delta(\omega + 1)] \tag{2.84}$$

The average power at the output may now be found.

$$P_{avg} = \frac{1}{\pi} \int_0^\infty S_{e_{out}}(\omega)\, d\omega$$

$$= \int_0^\infty \frac{1}{1 + \omega^2}[\delta(\omega - 1) + \delta(\omega + 1)]\, d\omega \tag{2.85}$$

$$= \frac{1}{1 + 1^2} = \frac{1}{2}$$

The student should verify that the average input power is equal to unity in this case.

2.10   SPECTRUM ANALYSIS

The Fourier Transform does not exist in real life. It is a mathematical tool to aid in the analysis of systems. The Fast Fourier Transform is one technique of approximating the Fourier Transform of a continuous time function in real life. The time signal can be sampled, A/D-converted, and then FFT'ed.

There are severe limitations when one attempts to find the continuous Fourier Transform of a time signal by using an analog system. First, since any system we build must be causal, the best one can hope to do is find the Fourier Transform based upon past input values, that is,

$$\int_{-\infty}^t f(t)e^{-j\omega t}\, dt$$

Suppose now that $f(t)$ forms the input to an ideal (non-causal) narrow band pass filter with $H(\omega)$ as shown in Fig. 2.27. The output transform is given by

$$G(\omega) = \begin{cases} F(\omega) & \text{for } \omega_1 < |\omega| < \omega_2 \\ 0 & \text{otherwise} \end{cases}$$

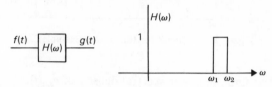

Figure 2.27 Narrow band pass filter.

$g(t)$ is given by the inverse transform of $G(\omega)$,

$$g(t) = \frac{1}{2\pi}\left(\int_{\omega_1}^{\omega_2} F(\omega)e^{j\omega t}\, d\omega + \int_{-\omega_2}^{-\omega_1} F(\omega)e^{j\omega t}\, d\omega \right) \qquad (2.86)$$

If ω_2 is very close to ω_1, we can assume that the integrand is approximately constant over the entire range of integration. Thus, Eq. (2.86) reduces to

$$f(t) = \frac{\omega_2 - \omega_1}{2\pi}(F(\omega_1)e^{j\omega_1 t} + F(-\omega_1)e^{-j\omega_1 t}) \qquad (2.87)$$

Now since $F(-\omega_1) = F^*(\omega_1)$, we have

$$\begin{aligned} g(t) &= \frac{\omega_2 - \omega_1}{\pi}(Re[F(\omega_1)]\cos\omega_1 t + Im[F(\omega_1)]\sin\omega_1 t) \\ &= \frac{\omega_2 - \omega_1}{\pi}|F(\omega_1)|\cos(\omega_1 t + \underline{/F(\omega_1)}) \end{aligned} \qquad (2.88)$$

The magnitude of the output is proportional to the magnitude of the input transform evaluated at ω_1, and the phase is shifted by the phase of $F(\omega_1)$.

In many practical spectrum analyzers, the band pass filter is swept across a range of frequencies (we are oversimplifying), and the magnitude of the output varies approximately with $|F(\omega)|$. There are three primary sources of error. First, while the filter is narrow, the bandwidth is not zero. Second, the filter is causal and non-ideal. Finally, when the center frequency of the filter is varying with time, the output will not necessarily reach its steady state value. Caution must therefore be observed in choosing a bandwidth for the filter and a rate of sweep.

PROBLEMS

2.1. Prove that the circuit shown below, where $v_{in}(t)$ is the input and $v_{out}(t)$ is the output, obeys superposition.

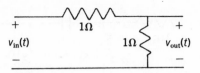

2.2. Does the circuit shown below, with $v_{in}(t)$ as input and $v_{out}(t)$ as output, obey superposition?

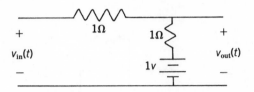

2.3. Which of the following systems are time invariant?

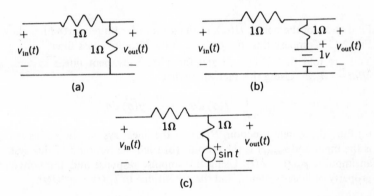

2.4. You are given a system with input $f(t)$ and output $g(t)$. You are further told that when $f(t) = 0$, $g(t)$ is not equal to zero. Show that this system cannot possibly obey superposition.

2.5. Find $H(\omega)$ and $h(t)$ for the following system, where $v_{in}(t)$ is the input and $v_{out}(t)$ is the output.

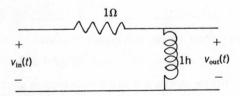

2.6. Find $v_{out}(t)$ when $v_{in}(t)$ is equal to
 (a) $\delta(t)$.
 (b) $U(t)$.
 (c) $e^{-2t}U(t)$.
 (See figure on next page.)

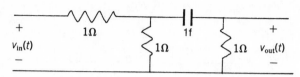

2.7. Consider the ideal low pass filter with system function as shown below. Show that the response of this filter to an input $(\pi/K)\delta(t)$ is the same as its response to $\sin Kt/Kt$.

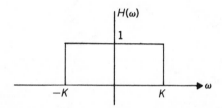

2.8. Find and plot the phase, $H(\omega)$, of the third order Butterworth filter shown in Fig. 2.13. Compare this to the phase of the ideal low pass filter.

2.9. You are told that if $f(t) \longrightarrow g(t)$ for a time invarient linear system, then $df/dt \longrightarrow dg/dt$. Using this fact, show that

$$\int_{-\infty}^{t} f(\sigma)\, d\sigma \longrightarrow \int_{-\infty}^{t} g(\sigma)\, d\sigma$$

2.10. (a) Find the system function, $H(\omega)$, of the linear system shown below. $v_{in}(t)$ is the input and $v_{out}(t)$ is the output. (b) Find the response of this system to an impulse, $v_{in}(t) = \delta(t)$. Using this impulse response and the convolution property of linear systems, find the output due to $v_{in}(t) = e^{-t}U(t)$.

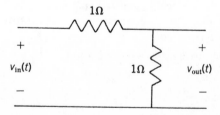

2.11. You are given a linear system with system function

$$H(\omega) = \delta(\omega - \omega_c) + \delta(\omega + \omega_c)$$

You are told that the output due to a certain input is given by

$$g(t) = e^{-3t}U(t)$$

Can you find $f(t)$, the input that caused this output? If not, why?

2.12. In the circuit shown below the input power spectral density is given by $S_{v_i}(\omega)$ as shown. Find the power spectral density of the voltages across the resistors, $S_{v_1}(\omega)$ and $S_{v_2}(\omega)$. Find the total input power and the total power of $v_1(t)$ and $v_2(t)$. Does conservation of power seem to be in trouble? (Hint: Carefully reread the definition of energy.)

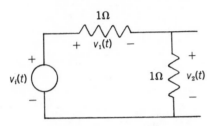 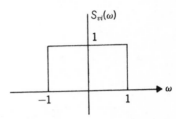

2.13. Find the average power, $\overline{v_{out}^2(t)}$ at the output of the circuit shown if $v_{in}(t)$ is a finite power signal with power density spectrum

(c) $S_{v_i}(\omega)$ as shown:

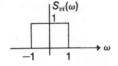 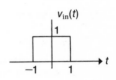

(d) The time signal, $v_{in}(t)$, as shown:

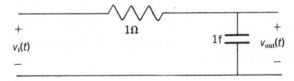

(a) $S_{v_i}(\omega) = K$.

(b) $S_{v_i}(\omega) = \delta(\omega + 1) + \delta(\omega - 1)$.

2.14. If $f(t) = A$, is $f(t)$ a finite power or finite energy signal? Find its autocorrelation and power (or energy) density spectrum.

2.15. If $f(t) = AU(t)$, is $f(t)$ a finite power or finite energy signal? Find its autocorrelation and power (or energy) density spectrum.

2.16. The impulse response of a linear system is

$$h(t) = e^{-2t}U(t)$$

The autocorrelation of its input is $R_f(t) = \delta(t)$.

(a) Find the input power spectral density.

(b) Find the output power spectral density, $S_g(\omega)$.

(c) Find the total output power.

(d) Find the output autocorrelation (use table of transforms or leave in integral form).

2.17. A signal $f(t) = A$ is the input to an ideal band pass filter with system function as shown below. Find the total output power or energy.

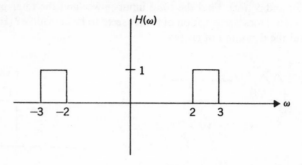

2.18. The signal, $f(t)$, goes through a linear system with transfer function, $H(\omega) = j\omega$. This is a differentiator. If $S_f(\omega)$ is the power spectrum of the input, what is the power spectrum of the output? What is the autocorrelation of the output in terms of $R_f(t)$? (Use the time differentiation property of the Fourier Transform.)

2.19. Find the autocorrelation of $f(t)$, where

$$f(t) = 2 \cos t + 3 \cos 3t + 4 \sin 4t$$

Find the average power of this signal. From this result can you reason what the average power of $f(t)$ would be if $f(t)$ were given by

$$f(t) = \sum_{n=1}^{\infty} a_n \cos n\omega_0 t$$

2.20. Evaluate the transfer function of the active filter shown below. What function does this filter perform?

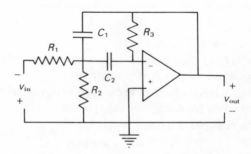

2.21. Convolve the following two sequences together and sketch the resulting sequence:

$$x(m) = \begin{cases} 1 & 0 \le m \le 5 \\ 0 & \text{otherwise} \end{cases}$$

$$h(n) = \begin{cases} 1 & 0 \le n \le 2 \\ 0 & \text{otherwise} \end{cases}$$

2.22. Convolve the following two sequences together and sketch the resulting sequence:

$$x(m) = \begin{cases} 1 & 0 \le m \le 4 \\ 0 & \text{otherwise} \end{cases}$$

$$h(n) = \begin{cases} n & 0 \le n \le 4 \\ 0 & \text{otherwise} \end{cases}$$

2.23. Draw a block diagram of a non-recursive digital filter with $h(n)$ as given in Problem 2.21. Repeat for $h(n)$ as given in Problem 2.22.

2.24. A time function, $x(t) = \cos \pi t$, is sampled once each second to form $x(n)$.
   (a) This $x(n)$ forms the input to a digital filter with $h(n)$ as given in Problem 2.21. Find the output of the filter.
   (b) $x(n)$ forms the input to a filter with $h(0) = h(1) = 2$, and $h(n) = 0$ for all other values of n. Find the output of this filter, and compare with your answer to part (a). Describe the action of this filter.

REFERENCES

GABEL, R. A., and ROBERTS, R. A., *Signals and Linear Systems*, Wiley, New York, 1973.

HAMMING, R. W., *Digital Filters*, Prentice-Hall, Englewood Cliffs, N.J., 1977.

KUO, F., *Network Analysis and Synthesis*, Wiley, New York, 1962.

MILLMAN, J., and HALKIAS, C. C., *Integrated Electronics*, McGraw-Hill, New York, 1972.

SCHWARTZ, M., and SHAW, L., *Signal Processing, Discrete Spectral Analysis, Detection and Estimation*, McGraw-Hill, New York, 1975.

PROBABILITY
AND
RANDOM ANALYSIS
3

3.1 INTRODUCTION

Studying communication systems without taking noise into consideration is analogous to learning how to drive a car by practicing in a huge abandoned parking lot. It is a valuable first step, though it is not very realistic. Were it not for what is known as noise, the communication of signals would be trivial indeed. One could simply build a circuit with a battery and a microphone and hang the wire outside the window. The intended receiver (perhaps on the other side of the globe) would simply hold a wire outside the window, receive a very weak signal, and use lots and lots of amplification.

In real life, the person at the receiver would receive a complete mess made up of an infinitesimal portion of the desired signal mixed with static (automotive ignition, people dialing telephones and turning lights on and off, distant electrical storms, etc.) and also mixed with the signals of all other people wishing to communicate at the same time. The real challenge of communication is therefore that of separating the desired signal from the undesired junk.

Anything other than the desired signal is called *noise*. For example, a radar system attempting to track a particular object will consider the signals returned from other objects to be noise. The point is that the definition of noise goes far beyond the usual assumption of static. If you tune channel 4 on your television and receive a background image of channel 3, that background image is noise.

The most common annoying types of noise do fall into the "static" category and are made up of a combination of numerous sources. In fact, those sources are so numerous that one cannot hope to describe the noise in a deterministic way (i.e., by giving a formula). We must therefore resort to discussing noise in terms of *averages*. The study of averages requires a knowledge of basic probability theory, which justifies this chapter.

It is possible to continue using this text by skipping this entire chapter and then later skipping sections which discuss noise in communication systems. For some people, this represents a better approach to the subject (again, think of the example of learning how to drive a car by first practicing in the absence of other cars).

3.2 BASIC ELEMENTS OF PROBABILITY THEORY

Probability theory can be approached either on a strictly mathematical level or in an empirical setting. The mathematical approach embeds probability theory into the realm of abstract set theory, and is itself somewhat abstract. The empirical approach satisfies one's intuition. Fortunately, this latter approach will be sufficient for our projected applications.

We use the so-called "relative frequency" definition of probability. Before we present this definition, some other related definitions will be required.

An *experiment* is a set of rules governing an operation which is performed.

An *outcome* is a result realized after performing the experiment once.

An *event* is a combination of outcomes.

For example, consider the experiment defined by flipping a single die. There are six possible outcomes, these being any one of the six surfaces of the die facing upward after the performance of the experiment. There are many possible events (64 to be precise). For example, one event would be that of "an even number of dots showing." This event is a combination of the three outcomes; two dots, four dots, and six dots. Another possible event would be "one dot." This latter event is called an *elementary event* since it is actually equal to one of the outcomes.

We can now define what is meant by the probability of an event. Suppose that the experiment is performed N times, where N is very large. Suppose also that in "n" of these N experiments, the outcome belongs to a given event. If N is large enough, the probability of this event is given by the ratio n/N. That is, it is the fraction of times that the event occurred. Formally, we define the probability of an event, A, as

$$Pr\{A\} \triangleq \lim_{N \to \infty} \left[\frac{n_A}{N} \right] \tag{3.1}$$

where n_A is the number of times that the event A occurs in N performances of the experiment. This definition is intuitively satisfying. For example, if one flipped a coin many times, the ratio of the number of heads to the total number of flips would approach $\frac{1}{2}$. We would therefore define the probability of a head to be $\frac{1}{2}$.

Suppose that we now consider two different events, A and B, with probabilities

$$Pr\{A\} = \lim_{N \to \infty} \left[\frac{n_A}{N}\right] \quad \text{and} \quad Pr\{B\} = \lim_{N \to \infty} \left[\frac{n_B}{N}\right]$$

If A and B could not possibly occur at the same time, we call them *disjoint*. The events "an even number of dots" and "two dots" are not disjoint in the die-throwing example, while the events "an even number of dots" and "an odd number of dots" are disjoint.

The probability of event A or event B would be the ratio of the number of times A or B occurs divided by N. If A and B are disjoint, this is seen to be

$$Pr\{A \text{ or } B\} = \lim_{N \to \infty} \frac{n_A + n_B}{N} = Pr\{A\} + Pr\{B\} \tag{3.2}$$

Equation (3.2) expresses the "additivity" concept. That is, if two events are disjoint, the probability of their "sum" is the sum of their probabilities.

Since each of the elementary events (outcomes) is disjoint from every other outcome, and each event is a sum of outcomes, we see that it would be sufficient to assign probabilities to the elementary events only. We could derive the probability of any other event from these. For example, in the die-flipping experiment, the probability of an even outcome is the sum of the probabilities of a "2 dots," "4 dots" and "6 dots" outcome.

Illustrative Example 3.1

Consider the experiment of flipping a coin twice. List the outcomes, events, and their respective probabilities.

Solution

The outcomes of this experiment are (letting H denote heads and T, tails)

$$HH, \quad HT, \quad TH, \quad \text{and} \quad TT$$

We shall assume that somebody has used some intuitive reasoning or has performed this experiment enough times to establish that the probability of each of the four outcomes is $\frac{1}{4}$. There are sixteen events, or combinations of these outcomes. These are

$\{HH\}, \ \{HT\}, \ \{TH\}, \ \{TT\}$
$\{HH, HT\}, \ \{HH, TH\}, \ \{HH, TT\}, \ \{HT, TH\}, \ \{HT, TT\}, \ \{TH, TT\}$
$\{HH, HT, TH\}, \ \{HH, HT, TT\}, \ \{HH, TH, TT\}, \ \{HT, TH, TT\},$
$\{HH, HT, TH, TT\}, \ \text{and} \ \{\varnothing\}$

Note that the comma within the curly brackets is read as "or." Thus, the events {*HH, HT*} and {*HT, HH*} are identical. For completeness we have included the zero event, denoted {∅}. This is the event made up of none of the outcomes.

Using the additivity rule, the probability of each of these events would be the sum of the probabilities of the outcomes comprising each event. Therefore,

$$Pr\{HH\} = Pr\{HT\} = Pr\{TH\} = Pr\{TT\} = \tfrac{1}{4}$$

$$Pr\{HH, HT\} = Pr\{HH, TH\} = Pr\{HH, TT\} = Pr\{HT, TH\}$$
$$= Pr\{HT, TT\} = Pr\{TH, TT\} = \tfrac{1}{2}$$

$$Pr\{HH, HT, TH\} = Pr\{HH, HT, TT\} = Pr\{HH, TH, TT\}$$
$$= Pr\{HT, TH, TT\} = \tfrac{3}{4}$$

$$Pr\{HH, HT, TH, TT\} = 1$$
$$Pr\{\varnothing\} = 0$$

The last two probabilities indicate that the event made up of all four outcomes is the "certain" event. It has probability "1" of occurring since each time the experiment is performed, the outcome must belong to this event. Similarly, the zero event has probability zero of occurring since each time the experiment is performed, the outcome does not belong to the zero event (sometimes called the "null set").

Random Variables

We would like to perform several forms of analysis upon these probabilities. As such, it is not too satisfying to have symbols such as "heads," "tails," and "two dots" floating around. We would much prefer to work with numbers. We therefore will associate a real number with each possible outcome of an experiment. Thus, in the single flip of the coin experiment, we could associate the number "0" with "tails" and "1" with "heads." Similarly, we could just as well associate "π" with heads and "2" with tails.

The mapping (function) which assigns a number to each outcome is called a *random variable*.

With a random variable so defined, many things can now be done that could not have been done before. For example, we could plot the various outcome probabilities as a function of the random variable. In order to make such a meaningful plot, we first define something called the *distribution function*, $F(x)$. If the random variable is denoted by[1] "X," then the distribution function, $F(x)$, is defined by

$$F(x_0) = Pr\{X \le x_0\} \tag{3.3}$$

---

[1]We shall use capital letters for random variables, and lower case letters for the values that they take on. Thus $X = x_0$ means that the random variable, X, is equal to the number, x_0.

We note that the set, $\{X \leq x_0\}$ defines an event, a combination of outcomes.

Illustrative Example 3.2

Assign two different random variables to the "one flip of a die" experiment, and plot the two distribution functions.

Solution

The first assignment which we shall use is the one that is naturally suggested by this particular experiment. That is, we will assign the number "1" to the outcome described by the face with one dot ending up in the top position. We will assign "2" to "two dots," "3" to "three dots," etc. We therefore see that the event $\{X \leq x_0\}$ includes the one dot outcome if x_0 is between 1 and 2. If x_0 is between 2 and 3, the event includes the one dot and the two dot outcomes. Thus the distribution function is easily found, and is shown in Fig. 3.1 (a).

If we now assign the random variable as follows,

Outcome	Random Variable
one dot	1
two dots	π
three dots	2
four dots	$\sqrt{2}$
five dots	11
six dots	5

we find, for example, that the event $\{X \leq 3\}$ is the event made up of three outcomes; one dot, three dots, and four dots. The distribution function is easily found and is shown in Fig. 3.1(b).

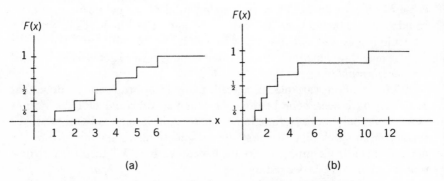

(a)   (b)

Figure 3.1   Distribution functions for Illustrative Example 3.2.

Note that a distribution function can never decrease with increasing argument. This is true since an increase in argument can only add outcomes to the event. We also easily verify that

$$F(-\infty) = 0 \quad \text{and} \quad F(+\infty) = 1 \tag{3.4}$$

HOW TO FILL IN YOUR PLAY SLIP

- Mark all entry boxes with a heavy horizontal mark like this ▬
- Do not use red ink. If you make a mistake on a board, simply mark the VOID box and start again on another board. Marking the VOID box will void the Lotto™ game selection on that board and any associated Lotto Extra™ selection.
- This Play Slip enables you to play Lotto and Lotto Extra. The Play Slip has 7 boards on it, so you can enter your selected draw(s) up to 7 times on this slip.

GUIDANCE
HOW TO PLAY THE LOTTO™ GAME

1. **Select your Lotto numbers** by placing a heavy horizontal mark in 6 boxes on a board of 1 to 49. Each board represents 1 selection of numbers. Alternatively, mark the Lucky Dip® box to have your numbers randomly generated.

2. **Select which draws you wish to enter.** Mark [wed] to enter the Wednesday draw only, mark [sat] to enter the Saturday draw only or mark [both] to enter both draws. These boxes are located at the bottom of the Play Slip.

3. **Select how many weeks you wish to enter** your selections for. You can play up to 8 weeks in advance for the Wednesday, Saturday or both draws, by marking the appropriate box at the bottom of the Play Slip (£1.00 per board per draw). If you do not mark any of the number of weeks boxes, you will only play for the next draw(s) you have selected (see 2 above).

HOW TO WIN – LOTTO

On each Lotto draw date there will be a random drawing of 6 Main Numbers and 1 Bonus Number from the numbers 1 to 49.

IF ANY OF THE 6-NUMBER LOTTO SELECTIONS ON YOUR TICKET:	PRIZE
Matches the first 6 Main Numbers drawn, in any order:	win or share the jackpot prize.
Matches any 5 of the first 6 Main Numbers drawn, in any order, plus the Bonus Number:	win a Match 5 + Bonus Number* prize.
Matches any 5, 4, or 3 numbers from the 6 Main Numbers drawn, in any order:	win the Match 5, Match 4 or Match 3 prize respectively.

*The Bonus Number is valid only for the "Match 5 + Bonus Number" prize.

HOW TO PLAY THE LOTTO EXTRA™ GAME

Lotto Extra is a separate game from Lotto and the draw is held immediately after the Lotto draw on Wednesday and Saturday nights. Lotto Extra is a jackpot only game, so play for another chance to win a big jackpot.

1. **To play Lotto Extra you must first make your Lotto selection** on a board (by either selecting your own numbers or marking the Lucky Dip® box).

2. **To enter the same 6 numbers** that you selected for 1 of your Lotto selections, simply mark the "Same numbers" box in the yellow Lotto Extra section on the associated Lotto game board where you have made your Lotto game selection.

3. **To enter six new numbers,** mark the Lucky Dip® box in the yellow Lotto Extra section on the Lotto game board where you have made your Lotto game selection. Your number selection will then be randomly generated.

4. Your Lotto Extra selection(s) will be entered for the same draw day(s) and number of weeks as your Lotto game selection(s). Lotto Extra costs £1.00 per board per draw.

HOW TO WIN – LOTTO EXTRA

On each Lotto Extra draw date there will be a random drawing of 6 numbers from 1 to 49.

IF ANY 6-NUMBER LOTTO EXTRA SELECTION ON YOUR TICKET:	PRIZE
Matches the 6 Lotto Extra numbers drawn, in any order:	win or share the jackpot prize.

OBTAINING AND CHECKING YOUR TICKET

You will receive your ticket from your Lotto Retailer. **You are responsible for checking the numbers selected, the number of selections, draw days and draw date(s) and games on your ticket before you leave the Retailer to ensure they are the ones you have selected.**

ALLOCATION OF PRIZE MONEY

The promoter of the games determines a prize structure for each game, the text of which may be obtained from The National Lottery® and at points of sale, and allocates for that purpose a sum not less than 45% of all the sales for each draw. The amount of each prize as a result of the draw is calculated in accordance with the applicable prize structure.

HOW TO CLAIM YOUR PRIZE

Prizes must be claimed in accordance with the Rules and Procedures applicable to the relevant game.

The information contained in the above section is for guidance purposes only

THE LOTTO AND LOTTO EXTRA GAME RULES AND PROCEDURES

- The National Lottery Rules for draw-based games and the respective Procedures for the Lotto and Lotto Extra games, as amended from time to time, form a contract between you and Camelot Group plc and govern these games and the sale of the ticket to you. Prize winnings and all aspects of the games are subject to these Rules and Procedures. You may inspect copies at any Lotto Retailer or obtain copies by telephoning the National Lottery Line on 0845 910 0000.
- The ticket issued is the only valid proof of your number selection(s) and is the only valid receipt for claiming a prize. THE PLAY SLIP IS NOT A VALID RECEIPT.
- The promoter accepts no responsibility for the entries on the face of the ticket differing from the entries on Camelot Group plc's central computer.
- Tickets shall not be sold to, nor prizes claimed by, persons under the age of 16.
- The Lotto and Lotto Extra games are run and promoted by Camelot Group plc under licences granted by the National Lottery Commission.

Strålfors   7M09043B

Don't live a little, live a Lotto™

Mark your choice like this: ▬ ⑮ Lotto Extra costs £1 per play

Board A £1
(1)	(2)	(3)	(4)	(5)	(6)	(7)	(8)	(9)	(10)	(11)	(12)
(13)	(14)	(15)	(16)	(17)	(18)	(19)	(20)	(21)	(22)	(23)	(24)
(25)	(26)	(27)	(28)	(29)	(30)	(31)	(32)	(33)	(34)	(35)	(36)
(37)	(38)	(39)	(40)	(41)	(42)	(43)	(44)	(45)	(46)	(47)	(48)

(49) Lucky Dip ▢ VOID | Lotto EXTRA Same numbers ▢ or Lucky Dip ▢

Board B £1
(1)	(2)	(3)	(4)	(5)	(6)	(7)	(8)	(9)	(10)	(11)	(12)
(13)	(14)	(15)	(16)	(17)	(18)	(19)	(20)	(21)	(22)	(23)	(24)
(25)	(26)	(27)	(28)	(29)	(30)	(31)	(32)	(33)	(34)	(35)	(36)
(37)	(38)	(39)	(40)	(41)	(42)	(43)	(44)	(45)	(46)	(47)	(48)

(49) Lucky Dip ▢ VOID | Lotto EXTRA Same numbers ▢ or Lucky Dip ▢

Board C £1
(1)	(2)	(3)	(4)	(5)	(6)	(7)	(8)	(9)	(10)	(11)	(12)
(13)	(14)	(15)	(16)	(17)	(18)	(19)	(20)	(21)	(22)	(23)	(24)
(25)	(26)	(27)	(28)	(29)	(30)	(31)	(32)	(33)	(34)	(35)	(36)
(37)	(38)	(39)	(40)	(41)	(42)	(43)	(44)	(45)	(46)	(47)	(48)

(49) Lucky Dip ▢ VOID | Lotto EXTRA Same numbers ▢ or Lucky Dip ▢

Board D £1
(1)	(2)	(3)	(4)	(5)	(6)	(7)	(8)	(9)	(10)	(11)	(12)
(13)	(14)	(15)	(16)	(17)	(18)	(19)	(20)	(21)	(22)	(23)	(24)
(25)	(26)	(27)	(28)	(29)	(30)	(31)	(32)	(33)	(34)	(35)	(36)
(37)	(38)	(39)	(40)	(41)	(42)	(43)	(44)	(45)	(46)	(47)	(48)

(49) Lucky Dip ▢ VOID | Lotto EXTRA Same numbers ▢ or Lucky Dip ▢

Board E £1
(1)	(2)	(3)	(4)	(5)	(6)	(7)	(8)	(9)	(10)	(11)	(12)
(13)	(14)	(15)	(16)	(17)	(18)	(19)	(20)	(21)	(22)	(23)	(24)
(25)	(26)	(27)	(28)	(29)	(30)	(31)	(32)	(33)	(34)	(35)	(36)
(37)	(38)	(39)	(40)	(41)	(42)	(43)	(44)	(45)	(46)	(47)	(48)

(49) Lucky Dip ▢ VOID | Lotto EXTRA Same numbers ▢ or Lucky Dip ▢

Board F £1
(1)	(2)	(3)	(4)	(5)	(6)	(7)	(8)	(9)	(10)	(11)	(12)
(13)	(14)	(15)	(16)	(17)	(18)	(19)	(20)	(21)	(22)	(23)	(24)
(25)	(26)	(27)	(28)	(29)	(30)	(31)	(32)	(33)	(34)	(35)	(36)
(37)	(38)	(39)	(40)	(41)	(42)	(43)	(44)	(45)	(46)	(47)	(48)

(49) Lucky Dip ▢ VOID | Lotto EXTRA Same numbers ▢ or Lucky Dip ▢

Board G £1
(1)	(2)	(3)	(4)	(5)	(6)	(7)	(8)	(9)	(10)	(11)	(12)
(13)	(14)	(15)	(16)	(17)	(18)	(19)	(20)	(21)	(22)	(23)	(24)
(25)	(26)	(27)	(28)	(29)	(30)	(31)	(32)	(33)	(34)	(35)	(36)
(37)	(38)	(39)	(40)	(41)	(42)	(43)	(44)	(45)	(46)	(47)	(48)

(49) Lucky Dip ▢ VOID | Lotto EXTRA Same numbers ▢ or Lucky Dip ▢

Draws
Choose which draw: both wed sat | Choose how many weeks you wish to play for: 2 3 4 5 6 7 8

FULL INSTRUCTIONS ON REVERSE

LEL1-0/7%

Density Function

Whenever a function exists, somebody is going to come along and differentiate it. In the particular case of $F(x)$, the derivative has great significance. We therefore define

$$p_X(x) \triangleq \frac{dF(x)}{dx} \tag{3.5}$$

and call $p_X(x)$ the *density function* of the random variable, X. By definition of the derivative we see that

$$F(x_0) = \int_{-\infty}^{x_0} p_X(x)\, dx \tag{3.6}$$

The random variable can be used to define any event. For example, $\{x_1 < X \leq x_2\}$ defines an event. Since the events $\{X \leq x_1\}$ and $\{x_1 < X \leq x_2\}$ are disjoint, one should be able to use the additivity principle to prove that

$$Pr\{x_1 < X \leq x_2\} = Pr\{X \leq x_2\} - Pr\{X \leq x_1\} \tag{3.7}$$

Therefore in terms of the density, we have

$$\begin{aligned} Pr\{x_1 < X \leq x_2\} &= \int_{-\infty}^{x_2} p_X(x)\, dx - \int_{-\infty}^{x_1} p_X(x)\, dx \\ &= \int_{x_1}^{x_2} p_X(x)\, dx \end{aligned} \tag{3.8}$$

We can now see why $p_X(x)$ was called a density function. The probability that X is between any two limits is given by the integral of the density function between these two limits.

The properties of the distribution function indicate that the density function can never be negative and that the integral of the density function over infinite limits must be unity.

The examples given previously (die and coin) would result in densities which contain impulse functions. A more common class of experiments gives rise to random variables whose density functions are continuous. This class is logically called the class of *continuous random variables*. Two very common density functions are presented below.

The first, and most common, density encountered in the real world is called the *Gaussian density function*. It arises whenever a large number of factors (with some broad restrictions) contribute to an end result, as in the case of static previously discussed. It is defined by

$$p_X(x) = \frac{1}{\sqrt{2\pi}\,\sigma} \exp\left[\frac{-(x-m)^2}{2\sigma^2}\right] \tag{3.9}$$

where "m" and "σ" are given constants. This density function is sketched in Fig. 3.2. From the figure one can see that the parameter "m" dictates the center position and "σ", the spread of the density function. In order to find probabilities of various events, the density must be integrated. Since this

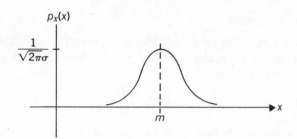

Figure 3.2 The Gaussian density function.

particular density function cannot be integrated in closed form, the integral has been calculated on a digital computer and tabulated under the name *error function* (*see* Appendix III).

Illustrative Example 3.3

A *binary* communication system is one which sends only two possible messages. The simplest form of binary system is one in which either *zero* or *one* volt is sent. Consider such a system in which the transmitted voltage is corrupted by additive atmospheric noise. If the receiver receives anything above $\frac{1}{2}$ volt, it assumes that a *one* was sent. If it receives anything below $\frac{1}{2}$ volt, it assumes that a *zero* was sent. Measurements have shown that if one volt is transmitted, the received signal level is random and has a Gaussian density with unit mean and $\sigma = \frac{1}{2}$. Find the probability that a transmitted *one* will be interpreted as a *zero* at the receiver.

Solution

The received signal level has density (assigning the random variable V as equal to the voltage level)

$$p_V(v) = \frac{1}{\sqrt{2\pi[\frac{1}{2}]}} \exp\left[\frac{-(v-1)^2}{2(\frac{1}{2})^2}\right]$$

Assuming that anything received above a level of $\frac{1}{2}$ is called "1" (and anything below $\frac{1}{2}$ is called "0"), the probability that a transmitted "1" will be interpreted as a "0" at the receiver is simply the probability that the random variable V is less than $\frac{1}{2}$. This is given by

$$\int_{-\infty}^{1/2} p_V(v)\,dv = \frac{1}{\sqrt{2\pi[\frac{1}{2}]}} \int_{-\infty}^{1/2} \exp\left[-2(v-1)^2\right] dv$$

Reference to a table of error functions (*see* Appendix III) indicates that this integral is approximately equal to 0.16. Thus, on the average, one would expect 16 out of every 100 transmitted 1's to be misinterpreted as 0's at the receiver.

The second commonly encountered density function is the so-called *uniform density function*. This is sketched in Fig. 3.3, where x_0 is a given

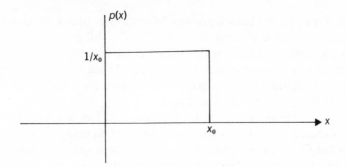

Figure 3.3   The uniform density function.

parameter. As an example of a situation in which this density arises, suppose you were asked to turn on a sinusoidal generator. The output of the generator would be of the form

$$v(t) = A \cos(\omega_0 t + \theta) \qquad (3.10)$$

Since the absolute time at which you turn on the generator is random, it would be reasonable to expect that θ is uniformly distributed between 0 and 2π. It would therefore have the density shown in Fig. 3.3 with $x_0 = 2\pi$.

A third commonly encountered density function is the *Rayleigh* density function, described by

$$p_x(x) = \begin{cases} \dfrac{x}{\sigma^2} \exp\left\{\dfrac{-x^2}{2\sigma^2}\right\} & x > 0 \\ 0 & x < 0 \end{cases} \qquad (3.11)$$

where σ is a given constant. Figure 3.4 shows several representative sketches of this density function.

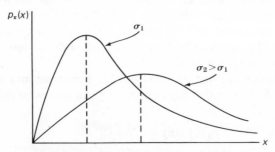

Figure 3.4   The Rayleigh density function.

The Rayleigh density function is related to the Gaussian density function. In fact, the square root of the sum of the squares of two Gaussian distributed random variables is, itself, Rayleigh. Thus, if we transform from rectangular to polar coordinates, the radius is given by

$$r = \sqrt{x^2 + y^2}$$

Therefore, if x and y are both Gaussian, in most cases (the restriction is one of *independence*, a term we have not yet defined) r will be Rayleigh. As an ideal example, suppose you were throwing darts at a target on a dart board and that the horizontal and vertical components of your error were Gaussian distributed (neglecting gravity effects). The distance from the center of the target would then be Rayleigh distributed.

If we were to consider r^2 instead of r, we would find that variable to be *chi-square* distributed. That is, a chi-square distribution results from summing the squares of Gaussian variables. If we sum two such variables, the result is chi-square with 2 *degrees of freedom*. In general, if

$$z = x_1^2 + x_2^2 + x_3^2 + \cdots + x_n^2$$

and $x_1, x_2, \ldots, x_n$ are Gaussian with $\sigma = 1$, z will be chi-square with n degrees of freedom. The form of the density is given by

$$p(z) = \begin{cases} \dfrac{(z)^{n/2-1}}{2^{n/2}(n/2 - 1)!} \exp\left(\dfrac{-z}{2}\right) & z > 0 \\ 0 & z < 0 \end{cases} \tag{3.12}$$

In Equation (3.12), "!" is the *factorial*.

What if more than one parameter were required to describe the outcome of an experiment? For example in the die-tossing experiment, consider the output as the face showing *and* the time at which the die hits the floor. We would therefore have to define two random variables. The density function would have to be two dimensional, and would be defined by

$$Pr\{x_1 < X \leq x_2 \quad \text{and} \quad y_1 < Y \leq y_2\} = \int_{y_1}^{y_2} \int_{x_1}^{x_2} p(x, y) \, dx \, dy \tag{3.13}$$

The question now arises as to whether or not we can tell if one random variable has any effect upon the other. In the die experiment, if we knew the time at which the die hit the floor would it tell us anything about which face was showing? This question leads naturally into a discussion of *conditional* probabilities.

Let us abandon the density functions for a moment and speak of two events, A and B. The probability of event A given that event B has occurred is defined by

$$Pr[A/B] \triangleq \frac{Pr\{A \text{ and } B\}}{Pr\{B\}} \tag{3.14}$$

Thus, for example, if A represented two dots appearing in the die experiment and B represented an even number of dots, the probability of A given B would (intuitively) be $\frac{1}{3}$. That is, given that either 2, 4, or 6 dots appeared, the probability of 2 dots is $\frac{1}{3}$. Now using Eq. (3.14) the probability of A and B is the probability of getting two *and* an even number of dots simultaneously. This is simply the probability of 2 dots, that is, $\frac{1}{6}$. The probability of B is the probability of 2, 4, or 6 dots, which is $\frac{1}{2}$. The ratio is $\frac{1}{3}$ as expected.

Similarly we could have defined event A as "an even number of dots" and event B as "an odd number of dots." The event "*A and B*" would therefore be the zero event, and $Pr[A/B]$ would be zero. This is reasonable since the probability of an even outcome given that an odd outcome occurred is clearly zero.

It is now logical to define independence as follows: Two events, A and B, are said to be *independent* if

$$Pr[A/B] = Pr\{A\} \tag{3.15}$$

The probability of A given that B occurred is simply the probability of A. Knowing that B has occurred tells nothing about A. Plugging this in Eq. (3.14) shows that independence implies,

$$Pr\{A \text{ and } B\} = Pr\{A\}Pr\{B\} \tag{3.16}$$

Now going back to the density functions, this indicates that two variables, X and Y, are independent if

$$p(x, y) = p_X(x)p_Y(y) \tag{3.17}$$

In words, the joint density of two independent random variables is the product of their individual densities.

Illustrative Example 3.4

A coin is flipped twice. Four different events are defined.

A is the event of getting a head on the first flip.
B is the event of getting a tail on the second flip.
C is the event of a match between the two flips.
D is the elementary event of a head on both flips.

Find $Pr\{A\}, Pr\{B\}, Pr\{C\}, Pr\{D\}, Pr[A/B]$, and $Pr[C/D]$. Are A and B independent? Are C and D independent?

Solution

The events are defined by the following combination of outcomes,

$$\{A\} = \{HH, HT\}$$
$$\{B\} = \{HT, TT\}$$
$$\{C\} = \{HH, TT\}$$
$$\{D\} = \{HH\}$$

Therefore, as was done in Illustrative Example 3.1,

$$Pr\{A\} = Pr\{B\} = Pr\{C\} = \tfrac{1}{2}$$
$$Pr\{D\} = \tfrac{1}{4}$$

In order to find $Pr[A/B]$ and $Pr[C/D]$ we use Eq. (3.14),

$$Pr[A/B] = \frac{Pr\{A \text{ and } B\}}{Pr\{B\}}$$

and

$$Pr[C/D] = \frac{Pr\{C \text{ and } D\}}{Pr\{D\}}$$

The event $\{A \text{ and } B\}$ is $\{HT\}$.
The event $\{C \text{ and } D\}$ is $\{HH\}$.
Therefore,

$$Pr[A/B] = \frac{\frac{1}{4}}{\frac{1}{2}} = \frac{1}{2}$$

$$Pr[C/D] = \frac{\frac{1}{4}}{\frac{1}{4}} = 1$$

Since $Pr[A/B] = Pr\{A\}$, the event of a head on the first flip is independent of that of a tail on the second flip. Since $Pr[C/D] \neq Pr\{C\}$, the event of a match and that of two heads are not independent.

Illustrative Example 3.5

X and Y are each Gaussian random variables and they are independent of each other. What is their joint density?

Solution

$$p_X(x) = \frac{1}{\sqrt{2\pi}\sigma_1} \exp\left[-\frac{(x - m_1)^2}{2\sigma_1^2}\right] \tag{3.18}$$

$$p_Y(y) = \frac{1}{\sqrt{2\pi}\sigma_2} \exp\left[-\frac{(y - m_2)^2}{2\sigma_2^2}\right] \tag{3.19}$$

$$p(x, y) = \frac{1}{2\pi\sigma_1\sigma_2} \exp\left[-\frac{(x - m_1)^2}{2\sigma_1^2}\right] \exp\left[-\frac{(y - m_2)^2}{2\sigma_2^2}\right] \tag{3.20}$$

Functions of a Random Variable

"Everybody talks about the weather, but nobody does anything." We, as communication experts, would be open to the same type of accusation if all we ever did was make statements such as "there is a 42% probability of the noise being annoying." A significant part of communication engineering involves itself with changing noise from one form to another in the hopes that the new form will be less annoying than the old. We must therefore study the effects of processing upon random phenomena.

Consider a function of a random variable, $y = g(x)$, where X is a random variable with known density function. Since X is random, Y is also random. We therefore ask what the density function of Y will look like. The event $\{x_1 < X \leq x_2\}$ corresponds to the event $\{y_1 < Y \leq y_2\}$,[2] where

$$y_1 \triangleq g(x_1) \quad \text{and} \quad y_2 \triangleq g(x_2)$$

That is, the two events are identical since they include the same outcomes. (We are assuming for the time being that $g(x)$ is a single valued function.)

---

[2] We are assuming that $y_1 < y_2$ if $x_1 < x_2$ ($g(x)$ is monotonic). That is, $dy/dx > 0$ in this range. If this is not the case, the inequalities must be reversed.

Since the events are identical, their respective probabilities must also be equal,

$$Pr\{x_1 < X \le x_2\} = Pr\{y_1 < Y \le y_2\} \tag{3.21}$$

and in terms of the densities,

$$\int_{x_1}^{x_2} p_X(x)\, dx = \int_{y_1}^{y_2} p_Y(y)\, dy \tag{3.22}$$

If we now let x_2 get very close to x_1 Eq. (3.22) becomes

$$p_X(x_1)\, dx = p_Y(y_1)\, dy \tag{3.23}$$

and finally,

$$p_Y(y_1) = \frac{p_X(x_1)}{dy/dx} \tag{3.24}$$

If, on the other hand $y_1 > y_2$, we would find (do it)

$$p_Y(y_1) = -\frac{p_X(x_1)}{dy/dx} \tag{3.25}$$

We can account for both of these cases by writing

$$p_Y(y_1) = \frac{p_X(x_1)}{|dy/dx|} \tag{3.26}$$

Finally, writing $x_1 = g^{-1}(y_1)$, and realizing that y_1 can be set equal to any value, we have

$$p_Y(y) = \frac{p_X[g^{-1}(y)]}{|dy/dx|} \tag{3.27}$$

If the function $g(x)$ is not monotone, then the event $\{y_1 < Y \le y_2\}$ corresponds to several intervals of the variable X. For example, if $g(x) = x^2$, then the event $\{1 < Y \le 4\}$ is the same as the event $\{1 < X \le 2\}$ or $\{-1 > X \ge -2\}$. Therefore,

$$\int_1^2 p_X(x)\, dx + \int_{-2}^{-1} p_X(x)\, dx = \int_1^4 p_Y(y)\, dy$$

In terms of the density functions this would mean that $g^{-1}(y)$ has several values. Denoting these values as x_a and x_b, then

$$p_Y(y) = \left.\frac{p_X(x)}{|dy/dx|}\right|_{x=x_a} + \left.\frac{p_X(x)}{|dy/dx|}\right|_{x=x_b} \tag{3.28}$$

Illustrative Example 3.6

A random voltage, v, is put through a full wave rectifier. v is uniformly distributed between -2 volts and $+2$ volts. Find the density of the output of the full wave rectifier.

Solution

Calling the output y, we have $y = g(v)$, where $g(v)$ and the density of V (letting the random variable be the same as the value of voltage) are sketched in Fig. 3.5.

Figure 3.5 $g(v)$ and $p_V(v)$ for Illustrative Example 3.6.

At every value of V, $|dg/dv| = 1$. For $y > 0$, $g^{-1}(y) = \pm y$. For $y < 0$, $g^{-1}(y)$ is undefined. That is, there are no values of v for which $g(v)$ is negative. Using Eq. (3.28), we find

$$p_Y(y) = p_X(y) + p_X(-y) \qquad (y > 0) \qquad (3.29)$$
$$p_Y(y) = 0 \qquad\qquad\qquad (y < 0) \qquad (3.30)$$

Finally, after plugging in values, we find $p_Y(y)$ as sketched in Fig. 3.6.

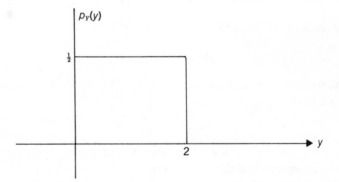

Figure 3.6   Resulting density for Illustrative Example 3.6.

Density Function of the Sum of Two Random Variables

Suppose that $z = x + y$, where x and y are random variables which are statistically independent of each other. We wish to find the density function of z in terms of the densities of x and y. The result will be useful in later work.

We start by finding the distribution function of the sum variable, z. The probability that z is less than some quantity z_0 is the probability that the sum $x + y < z_0$. This is the probability that x and y fall within the shaded region of Fig. 3.7. This probability is given by the double integral of the joint density for x and y:

$$P_z(z_0) = \int_{-\infty}^{+\infty} \int_{-\infty}^{z_0 - y} p(x, y) \, dx \, dy \qquad (3.31)$$

Since x and y are assumed to be independent, $p(x, y)$ is given by the product

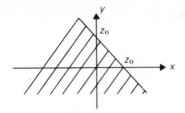

Figure 3.7   Region for which $x + y < z_0$.

of the individual densities:

$$P_z(z_0) = \int_{-\infty}^{+\infty} \int_{-\infty}^{z_0-y} p_x(x)p_y(y)\, dx\, dy$$
$$= \int_{-\infty}^{+\infty} p_y(y) \int_{-\infty}^{z_0-y} p_x(x)\, dx\, dy \tag{3.32}$$

We now differentiate to find the density function:

$$p_z(z_0) = \frac{dP_z(z_0)}{dz_0}$$
$$= \int_{-\infty}^{+\infty} p_y(y)p_x(z_0 - y)\, dy \tag{3.33}$$

Equation (3.33) has a very simple interpretation. That is, the density of the sum of two independent random variables is equal to the convolution of the two individual densities.

From Chapter 1, we know that if it were desired to avoid computation of the convolution, we could take Fourier Transforms (which, in this case, would be known as *characteristic functions*) and multiply these transforms together to get the transform of the resulting density function.

Expected Values

Picture yourself as a teacher who has just given an examination. How would you average the resulting grades?

You would add them all together and divide by the number of grades. Thus if an experiment is performed many times, the average of the random variable resulting would be found in the same way.

Let x_i, $i = 1, 2, \ldots, M$ represent the possible values of the random variable, and let n_i represent the number of times x_i occurs as the (random variable assigned to the) outcome. The average of the random variable after N performances of the experiment would therefore be

$$X_{\text{avg}} = \frac{1}{N} \sum_{i=1}^{M} n_i x_i$$
$$= \sum_{i=1}^{M} \frac{n_i}{N} x_i \tag{3.34}$$

Since x_i has ranged over all possible values of the random variable, we observe

$$\sum_{i=1}^{M} n_i = N$$

As N approaches infinity, n_i/N becomes the definition of $Pr\{x_i\}$. Therefore,

$$X_{\text{avg}} = \sum_{i=1}^{M} x_i Pr\{x_i\} \tag{3.35}$$

This average value is sometimes called the *mean* or *expected value* of X, and is often given the symbol m_x.

Now suppose that we wish to find the average value of a continuous random variable. We can use the previous result (Eq. (3.35)) if we first round off the continuous variable to the nearest multiple of Δx. Thus, if X is between $k\Delta x - \frac{1}{2}\Delta x$ and $k\Delta x + \frac{1}{2}\Delta x$, we shall round it off to $k\Delta x$. The probability of X being in this range is given by

$$\int_{k\Delta x-(1/2)\Delta x}^{k\Delta x+(1/2)\Delta x} p_X(x)\, dx \tag{3.36}$$

which, if Δx is small, is approximately equal to $p_X(k\Delta x)\Delta x$. Therefore, using Eq. (3.35), the expected value of X is

$$\sum_{k=-\infty}^{\infty} k\Delta x p_X(k\Delta x)\, \Delta x \approx X_{\text{avg}} \tag{3.37}$$

As Δx approaches zero, Eq. (3.37) becomes (by definition of the integral)

$$X_{\text{avg}} = m_x = \int_{-\infty}^{\infty} x p_X(x)\, dx \tag{3.38}$$

Now consider a function of the random variable, $y = g(x)$. Suppose we wish to find the expected value of Y. This is given by Eq. (3.38) as

$$Y_{\text{avg}} = \int_{-\infty}^{\infty} y p_Y(y)\, dy \tag{3.39}$$

which, using Eq. (3.27), becomes

$$Y_{\text{avg}} = \int_{-\infty}^{\infty} y p_X[g^{-1}(y)]\, \frac{dy}{|dy/dx|} \tag{3.40}$$

which, after substituting for y, becomes

$$[g(x)]_{\text{avg}} = \int_{-\infty}^{\infty} g(x) p_X(x)\, dx \tag{3.41}$$

This is an extremely significant equation. It tells us that in order to find the expected value of a function of X, we simply integrate the function weighted by the *density of X*.

We will switch between several different notational forms for the average value operator. These are

$$[g(x)]_{\text{avg}} = \overline{g(x)} = E\{g(x)\}$$

Using Eq. (3.41), the expected value of x^2 is given by

$$E\{x^2\} = \int_{-\infty}^{\infty} x^2 p_X(x) \, dx \tag{3.42}$$

Illustrative Example 3.7

X is uniformly distributed as shown in Fig. 3.8. Find $E\{x\}$, $E\{x^2\}$, $E\{\cos x\}$, and $E\{(x - m_x)^2\}$.

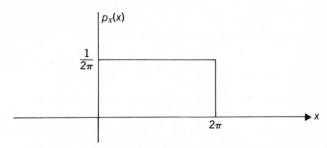

Figure 3.8   Density function of X for Illustrative Example 3.7.

Solution

$$E\{x\} = \int_{-\infty}^{\infty} x p_X(x) \, dx = \frac{1}{2\pi} \int_0^{2\pi} x \, dx = \pi \tag{3.43}$$

$$E\{x^2\} = \int_{-\infty}^{\infty} x^2 p_X(x) \, dx = \frac{1}{2\pi} \int_0^{2\pi} x^2 \, dx = \frac{4}{3}\pi^2 \tag{3.44}$$

$$E\{\cos x\} = \int_{-\infty}^{\infty} \cos x p_X(x) \, dx = \frac{1}{2\pi} \int_0^{2\pi} \cos x \, dx = 0 \tag{3.45}$$

$$E\{(x - \pi)^2\} = \int_{-\infty}^{\infty} (x - \pi)^2 p_X(x) \, dx = \frac{1}{2\pi} \int_0^{2\pi} (x - \pi)^2 \, dx = \frac{\pi^2}{3} \tag{3.46}$$

The last expected value in Example 3.7 is called the *variance* of X, and it gives a measure of how far we can expect the variable to deviate from its mean value. That is, if $E\{(x - m_x)^2\}$ is large, it means that on the average the square of the difference between X and its mean value will be large. This is the same as saying that the density function is spread out.

Thus if you were told that a variable X has mean 1 and variance 0.1, and that a variable Y has mean 1 and variance 10, you would intuitively say that on the average X will be closer to 1 than Y is. This is true even though the average of both variables is exactly equal to 1.

The concept of mean and variance of a random variable will prove to be a major building block in the work to follow.

Illustrative Example 3.8

X is a Gaussian distributed random variable with density function

$$p_X(x) = \frac{1}{\sqrt{2\pi}\,\sigma} e^{-(x-m)^2/2\sigma^2}$$

(a)   Show that $p_x(x)$ integrates to unity.

(b)   Find $E\{x\}$.

(c)   Find $E\{(x - m_x)^2\}$.

Solution

(a) We first write the integral directly:

$$\frac{1}{\sqrt{2\pi}\,\sigma} \int_{-\infty}^{+\infty} e^{-(x-m)^2/2\sigma^2} \, dx$$

Making a change of variables, $y = (x - m)/\sigma$, we have

$$\frac{1}{\sqrt{2\pi}\,\sigma} \int_{-\infty}^{+\infty} e^{-(x-m)^2/2\sigma^2} \, dx = \frac{1}{\sqrt{2\pi}} \int_{-\infty}^{+\infty} e^{-y^2/2} \, dy$$

This integral cannot be evaluated in closed form for general limits of integration. In the case of infinite limits, the integral can be evaluated. Although the infinite limit integral can be found in most any table of integrals, it is instructive to show a "trick" for evaluating it. Define

$$I = \int_{-\infty}^{+\infty} e^{-y^2/2} \, dy \qquad (3.47)$$

Then

$$\begin{aligned}
I^2 &= \int_{-\infty}^{+\infty} \int_{-\infty}^{+\infty} e^{-x^2/2} e^{-y^2/2} \, dx \, dy \\
&= \int_{-\infty}^{+\infty} \int_{-\infty}^{+\infty} e^{-1/2(x^2+y^2)} \, dx \, dy
\end{aligned} \qquad (3.48)$$

Transforming to polar coordinates, we have

$$x^2 + y^2 = r^2$$

and

$$dx \, dy = r \, dr \, d\theta$$

$$I^2 = \int_0^{2\pi} \int_0^{\infty} e^{-r^2/2} r \, dr \, d\theta \qquad (3.49)$$

Since the integrand is not a function of θ, Eq. (3.49) becomes

$$I^2 = 2\pi \int_0^{\infty} r e^{-r^2/2} \, dr \qquad (3.50)$$

The integrand of Eq. (3.50) is the derivative of $-e^{-r^2/2}$. Therefore,

$$I^2 = 2\pi(-e^{-r^2/2} \big|_0^{\infty}) = 2\pi$$

and

$$I = \sqrt{2\pi}$$

Finally,

$$\frac{1}{\sqrt{2\pi}} I = 1$$

which is the desired result.

(b)

$$E\{x\} = \frac{1}{\sqrt{2\pi}\,\sigma} \int_{-\infty}^{+\infty} x e^{-(x-m)^2/2\sigma^2}\,dx$$

$$= \frac{1}{\sqrt{2\pi}\,\sigma} \int_{-\infty}^{+\infty} (x-m) e^{-(x-m)^2/2\sigma^2}\,dx$$

$$+ \frac{m}{\sqrt{2\pi}\,\sigma} \int_{-\infty}^{+\infty} e^{-(x-m)^2/2\sigma^2}\,dx$$

$$= \frac{\sigma}{\sqrt{2\pi}} e^{-(x-m)^2/2\sigma^2}\Big|_{-\infty}^{+\infty} + m$$

$$= 0 + m = m$$

Therefore the "*m*" in the formula for the Gaussian density is actually the mean value of the random variable. This should have been obvious from the symmetry displayed in Fig. 3.2.

(c)

$$E\{(x-m)^2\} = \frac{1}{\sqrt{2\pi}\,\sigma} \int_{-\infty}^{+\infty} (x-m)^2 e^{-(x-m)^2/2\sigma^2}\,dx$$

This can be integrated by parts. Let

$$u = x - m$$
$$dv = (x-m) e^{-(x-m)^2/2\sigma^2}\,dx$$

Then

$$du = dx$$
$$v = -e^{-(x-m)^2/2\sigma^2}$$

$$E\{(x-m)^2\} = \frac{1}{\sqrt{2\pi}\,\sigma} \int_{-\infty}^{+\infty} e^{-(x-m)^2/2\sigma^2}\,dx$$

$$- (x-m)\sigma^2 e^{-(x-m)^2/2\sigma^2}\Big|_{-\infty}^{+\infty}$$

$$= \sigma^2$$

Therefore, "σ^2" in the formula for the Gaussian density is the variance of the random variable.

A simple relationship may be derived involving the mean, variance, and second moment of any random variable. It is worth mentioning since one often has already calculated the mean and second moment and must find the variance. With the use of this relationship, no additional integration is required.

We start with the definition of variance:

$$\sigma_x^2 = E\{(x - E\{x\})^2\}$$

Expanding this, and using the fact that the expected value of a sum is the sum of the expected values, we have

$$\sigma_x^2 = E\{x^2 - 2E\{x\}E\{x\} + (E\{x\})^2\}$$
$$= E\{x^2\} - (E\{x\})^2 \tag{3.51}$$

Note that $E\{x\}$ is non-random and is treated as a constant in the above deriva-tion. Equation (3.51) is the desired result. That is, the variance can be found by subtracting the square of the mean from the second moment.

3.3 STOCHASTIC PROCESSES

Until this point, we have considered only single random variables. All aver-ages (and related parameters) were simply numbers. In this section we shall, in effect, add another dimension to our study; the dimension of time. Instead of talking of numbers only, we will now be able to characterize random func-tions. The advantages of such a capability should be obvious. Our approach to random functions' analysis will begin with the consideration of discrete-time functions (i.e., sampled), since these will prove to be a simple extension of random variables.

Imagine a single die being flipped 1,000 times. Let X_i be the random variable assigned to the outcome of the ith flip. We now list the 1,000 values of the random variables,

$$x_1 x_2 x_3, \ldots, x_{999} x_{1,000} \tag{3.52}$$

For example, if we assign the random variable as equal to the number of dots on the top face after flipping the die, a typical list might resemble that shown in Eq. (3.53):

$$463514253145, \ldots \tag{3.53}$$

Suppose that we now listed all possible sequences of random variables. We would have a collection of $6^{1,000}$ entries, each one resembling that shown in Eq. (3.53). This collection will be known as the *ensemble* of possible outcomes. This ensemble, together with associated statistical properties, forms a *stochastic process*. In this particular example, the process is discrete valued and discrete time.

If we were to view one digit, say the third entry, we would have a random variable. In this example the random variable would represent that assigned to the outcome of the third flip of the die.

We can completely describe the above stochastic process by specifying the 1,000-dimensional probability density function,

$$p(x_1, x_2, x_3, \ldots, x_{999}, x_{1,000})$$

In many instances it will prove sufficient to specify only the first and second order averages (moments). That is, we specify

$$E\{X_i\} \quad \text{and} \quad E\{X_i X_j\}$$

for all i and j.

We would now like to extend this example to the case of an infinite num-ber of random variables in each list and an infinite number of lists in the ensemble. It may be helpful to refer back to the above simple example from time to time.

The most general form of the stochastic process would result if our simple experiment could yield an infinite range of values as the outcome, and if the sampling period approached zero (i.e., the die is flipped faster and faster).

Another way to arrive at a stochastic process is to perform a simple experiment, but to each outcome assign a *time function* instead of a number. As a simple example, consider the experiment defined by picking a 6-volt DC generator from a virtually infinite inventory in a warehouse. The generator voltage is then measured as a function of time on an oscilloscope. This waveform, $v(t)$, will be called a *sample function* of the process. The waveform will not be a perfect constant of 6 volts value, but will wiggle around due to imperfections in the generator construction and due to radio pickup when the wires act as an antenna. There are therefore an infinite number of possible sample functions of the process. This infinite number of samples forms the ensemble. Each time we choose a generator and measure its voltage, we get a sample function from this infinite ensemble.

If we were to sample the voltage at a specific time, say $t = t_0$, the sample, $v(t_0)$, can be considered as a random variable. Since $v(t)$ is assumed to be continuous with time, there are an infinity of random variables associated with the process.

Summing this up and changing notation slightly, let $x(t)$ represent a stochastic process. Then $x(t)$ can be thought of as being an infinite ensemble of all possible sample functions. For every specific value of time, $t = t_0$, $x(t_0)$ is a random variable.

As in the simple example which opened this section, we can first hope to characterize the process by a joint density function of the random variables. Unfortunately, since there are an infinite number of random variables this density function would be infinite dimensional. We therefore resort to first and second moment characterization of the process.

The first moment is given by the mean value.

$$m(t) \triangleq E\{x(t)\} \tag{3.54}$$

The second moments are given by

$$R_{xx}(t_1, t_2) \triangleq E\{x(t_1)x(t_2)\} \tag{3.55}$$

for all t_1 and t_2.

At the moment we must think of these averages as being taken over the ensemble of possible time function samples. That is, in order to find $m(t_0)$, we would average all members of the ensemble at time t_0. In practice, we could measure the voltage of a great number of the generators at time t_0 and average the resulting numbers. In the case of the DC generators, one would intuitively expect this average not to depend upon t_0. Indeed, most processes which we will consider have mean values which are independent of time.

We define a process whose overall statistics are independent of time as a *stationary* process. Given a stationary process, $x(t)$, then the process described by $x(t - T)$ will have the same statistics independent of the value of T. Clearly, for a stationary process, $m(t_0)$ will not depend upon t_0.

If the process is stationary, we see that

$$R_{xx}(t_1, t_2) = E\{x(t_1)x(t_2)\} = E\{x(t_1 - T)x(t_2 - T)\} \qquad (3.56)$$

Since Eq. (3.56) applies for all values of T, let $T = t_1$. Then,

$$R_{xx}(t_1, t_2) = E\{x(0)x(t_2 - t_1)\}$$

This indicates that the second moment, $R_{xx}(t_1, t_2)$ does not depend upon the actual values t_1 and t_2, but only upon the difference, $t_2 - t_1$. That is, the left hand time point can be placed anywhere, and as long as the right hand point is separated from this by $t_2 - t_1$, the second moment will be $R_{xx}(t_1, t_2)$. We state this mathematically as

$$R_{xx}(t_1, t_2) = R_{xx}(t_2 - t_1) \qquad (3.57)$$

for a stationary process.

The function $R_{xx}(t_2 - t_1)$ is called the *autocorrelation* of the process, $x(t)$. The word autocorrelation was used earlier in this text. We shall soon see the reason for using the same word again.

In the work to follow, we will replace the argument, $t_2 - t_1$ by "τ".

$R_{xx}(t_2 - t_1)$ tells us most of what we need to know about the process. In particular, it gives us some idea as to how fast a particular sample function can change with time. If t_2 and t_1 are sufficiently far apart such that $x(t_1)$ and $x(t_2)$ are independent random variables, the autocorrelation reduces to

$$\begin{aligned} R_{xx}(t_2 - t_1) &= E\{x(t_1)x(t_2)\}, \\ &= E\{x(t_1)\}E\{x(t_2)\} = m^2 \end{aligned} \qquad (3.58)$$

For most processes encountered in communications, the mean value is zero ($m = 0$). In this case, the value of "t" at which $R_{xx}(t)$ goes to zero represents the time over which the process is "correlated." If two samples are separated by more than this length of time, one sample has no effect upon the other.

Illustrative Example 3.9

You are given a stochastic process, $x(t)$, with mean value m_x and autocorrelation, $R_{xx}(\tau)$. Obviously, the process is stationary. If not, the mean and autocorrelation could not have been given in this form. Find the mean and autocorrelation of a process $y(t)$, where

$$y(t) = x(t) - x(t - T)$$

Solution

To solve problems of this type we need only recall the definition of the mean and autocorrelation and the fact that taking expected values is a linear operation. Proceeding,

$$
\begin{aligned}
m_y &= E\{y(t)\} = E\{x(t) - x(t - T)\} \\
&= E\{x(t)\} - E\{x(t - T)\} \\
&= m_x - m_x = 0
\end{aligned} \tag{3.59}
$$

$$
\begin{aligned}
R_{yy}(\tau) &= E\{y(t)y(t + \tau)\} \\
&= E\{[x(t) - x(t - T)][x(t + \tau) - x(t + \tau - T)]\} \\
&= E\{x(t)x(t + \tau)\} - E\{x(t)x(t + \tau - T)\} \\
&\quad - E\{x(t - T)x(t + \tau)\} + E\{x(t - T)x(t + \tau - T)\} \\
&= R_{xx}(\tau) - R_{xx}(\tau - T) - R_{xx}(\tau + T) + R_{xx}(\tau) \\
&= 2R_{xx}(\tau) - R_{xx}(\tau - T) - R_{xx}(\tau + T)
\end{aligned} \tag{3.60}
$$

We note that the process, $y(t)$ is stationary. If this were not the case, both m_y and R_{yy} would be functions of t.

Suppose you were asked to find the average value of the voltage in the DC generator example. You would have to measure the voltage (at any given time) of many generators and then compute the average. Once you were told that the process is stationary, you would probably be tempted to take one generator and average its voltage over a large time interval. You would expect to get 6 volts as the result of either technique. That is, you would reason that

$$
m_v = E\{v(t)\} = \lim_{T \to \infty} \frac{1}{T} \int_{-T/2}^{T/2} v_i(t)\, dt \tag{3.61}
$$

Here, $v_i(t)$ is one sample function of the ensemble. Actually this reasoning will not always be correct. Suppose, for example, that one of the generators was burned out. If you happened to choose this particular generator, you would find that $m_v = 0$, which is certainly not correct.

If any single sample of a stochastic process contains all of the information (statistics) about the process, we call the process *ergodic*. Most processes which we deal with will be ergodic. The generator example is ergodic as long as none of the generators is exceptional (e.g., burned out).

Clearly a process which is ergodic must also be stationary. This is true since, once we allow that the averages can be found from one time sample, these averages can no longer be a function of the time at which they are computed. Conversely, a stationary process need not be ergodic. (e.g., the burned-out generator example ... convince yourself this process would still be stationary.)

If a process is ergodic, the autocorrelation can be found from any time sample,

$$
R_{xx}(\tau) = E\{x(t)x(t + \tau)\} = \lim_{T \to \infty} \frac{1}{T} \int_{-T/2}^{T/2} x(t)x(t + \tau)\, dt \tag{3.62}
$$

We can now see why the functional notation, $R(\tau)$, and the name "autocorrelation" were used. Equation (3.62) is exactly the same as the equation for $R(\tau)$ found in Section 2.9. The earlier equation held for deterministic finite power signals.

Since the equations are identical, we can borrow all of our previous results. Thus, the power spectral density of a stochastic process is given by

$$S(\omega) = \mathfrak{F}[R(t)] = \int_{-\infty}^{\infty} R(t)e^{-j\omega t}\, dt \tag{3.63}$$

and the total average power by

$$P_{\text{avg}} = \frac{1}{\pi}\int_{0}^{\infty} S(\omega)\, d\omega \tag{3.64}$$

Here, the word "average" takes on more meaning than it did for determinsitic signals. Note that since

$$R(t) = \frac{1}{2\pi}\int_{-\infty}^{\infty} S(\omega)e^{j\omega t}\, d\omega \tag{3.65}$$

then

$$P_{\text{avg}} = \frac{1}{\pi}\int_{0}^{\infty} S(\omega)\, d\omega = R(0) \tag{3.66}$$

The observation that $R(0)$ is the average power makes a great deal of sense since

$$R(0) = E\{x^2(t)\} \tag{3.67}$$

If a stochastic process now forms the input to a linear system the output will also be a stochastic process. That is, each sample function of the input process yields a sample function of the output process. The autocorrelation of the output process is found as in Chapter 2. (*See* Fig. 3.9.)

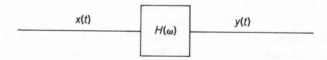

Figure 3.9   A stochastic process as the input to a linear system.

$$S_{yy}(\omega) = S_{xx}(\omega)\,|H(\omega)|^2 \tag{3.68}$$

Taking inverse transforms,

$$R_{yy}(t) = R_{xx}(t) * h(t) * h(-t) \tag{3.69}$$

We therefore have a way of finding the average power of the output of a system when the input is a stochastic process. Since most noise waveforms can be thought of as being samples of a stochastic process, the above analysis is critical to any noise reduction scheme analysis.

Illustrative Example 3.10

A received signal is made up of two components, signal and noise.

$$r(t) = s(t) + n(t)$$

The signal can be considered as a sample of a random process since random amplitude fluctuations are introduced by turbulence in the air. You are told that the autocorrelation of the signal process is

$$R_s(\tau) = 2e^{-|\tau|}$$

The noise is a sample function of a random process with autocorrelation,

$$R_n(\tau) = e^{-2|\tau|}$$

You are told that both processes have zero mean value, and that they are independent of each other.

Find the autocorrelation and total power of $r(t)$.

Solution

From the definition of autocorrelation,

$$
\begin{aligned}
R_r(\tau) &\triangleq E\{r(t)r(t+\tau)\} \\
&= E\{[s(t) + n(t)][s(t+\tau) + n(t+\tau)]\} \\
&= E\{s(t)s(t+\tau)\} + E\{s(t)n(t+\tau)\} \\
&\quad + E\{s(t+\tau)n(t)\} + E\{s(t+\tau)n(t+\tau)\}
\end{aligned}
\tag{3.70}
$$

Since the signal and noise are independent,

$$E\{s(t+\tau)n(t)\} = E\{s(t+\tau)\}E\{n(t)\} = 0$$

and

$$E\{s(t)n(t+\tau)\} = E\{s(t)\}E\{n(t+\tau)\} = 0$$

Finally, Eq. (3.70) becomes

$$
\begin{aligned}
R_r(\tau) &= R_s(\tau) + R_n(\tau) \\
&= 2e^{-|\tau|} + e^{-2|\tau|}
\end{aligned}
\tag{3.71}
$$

The total power of $r(t)$ is simply $R_r(0) = 3$.

3.4 WHITE NOISE

Let $x(t)$ be a stochastic process with a constant power spectral density (*see* Fig. 3.10).

This process contains all frequencies to an equal degree. Since white light is composed of all frequencies (colors), the process described above is known as *white noise*.

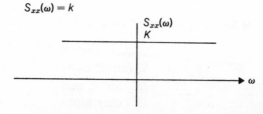

$S_{xx}(\omega) = k$

Figure 3.10 Power spectrum of white noise.

The autocorrelation of white noise is the inverse Fourier Transform of K, which is simply $R_{xx}(t) = K\delta(t)$.

The average power of white noise, $R(0)$, is infinity. It therefore cannot exist in real life. However, many types of noise encountered can be assumed to be approximately white.

Illustrative Example 3.11

White noise forms the input to an RC circuit (approximation to a low pass filter). Find the autocorrelation and power spectral density of the output of this filter. (*See* Fig. 3.11.)

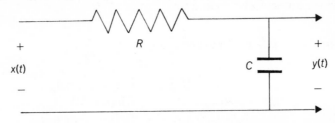

Figure 3.11   Circuit for Illustrative Example 3.11.

Solution

$$S_{yy}(\omega) = S_{xx}(\omega)|H(\omega)|^2 = \frac{K}{1 + \omega^2 C^2 R^2} \tag{3.72}$$

$$R_{yy}(t) = \frac{K}{2RC} \exp\left[\frac{-|t|}{RC}\right] \tag{3.73}$$

Illustrative Example 3.12

Repeat Illustrative Example 3.11 for an ideal low pass filter with cutoff frequency, ω_m. (*See* Fig. 3.12.)

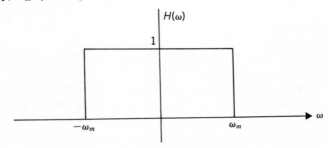

Figure 3.12   Low pass filter for Illustrative Example 3.12.

Solution

$$\begin{aligned} S_{yy}(\omega) &= S_{xx}(\omega)|H(\omega)|^2 \\ &= \begin{cases} K & |\omega| < \omega_m \\ 0 & \text{otherwise} \end{cases} \end{aligned} \tag{3.74}$$

This is shown in Fig. 3.13.

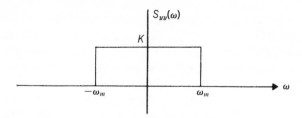

Figure 3.13   Output power spectrum for Illustrative Example 3.12.

The autocorrelation is the inverse transform of the power spectral density and is given by

$$R_{yy}(t) = K\frac{\sin \omega_m t}{\pi t} \tag{3.75}$$

In the previous example, suppose that the input process had a power spectrum as shown in Fig. 3.14. with $\omega_1 \geq \omega_m$. The output process would be identical to that found in Illustrative Example 3.12. Therefore, if a processing system exhibits an upper cutoff frequency (all real systems approximately do) and the input noise has a flat spectrum up to this cutoff frequency, we make no errors if we consider the input noise to be white. This will simplify the required mathematics considerably.

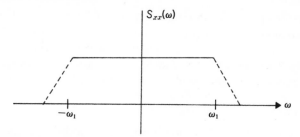

Figure 3.14   Spectrum of an input to the filter of Illustrative Example 3.12 which would yield the same output spectrum as white noise input did.

Since $R(t)$ is zero for $t \neq 0$, two samples of white noise are independent of each other even if they are taken very closely together.[3] That is, knowing the sample value of white noise at one instant of time tells us absolutely nothing about its value an instant later. From a practical standpoint, this is a

---

[3]Actually the zero correlation ($x(t)$ and $x(t + \tau)$ are "uncorrelated") is a necessary, but not sufficient, condition for independence of the two random variables. The only complete definition of independence relates to conditional probabilities or, equivalently, to the joint density function. However, one can show that two random variables which are jointly Gaussian are independent as long as they are uncorrelated. Therefore as long as we deal with Gaussian processes, we can use the words "uncorrelated" and "independent" interchangeably.

sad situation. It would seem to make the elimination of noise more difficult. Although this is true, the "independence of samples" assumption serves to greatly simplify the analysis of many complex processing systems.

Up to this point we have said nothing about the actual probability distributions of the process. We have only talked about the first and second moments. Each random variable, $x(t_0)$, has a certain probability distribution. For example, it can be either Gaussian or uniformly distributed. By only considering the means and second moments we are not telling the whole story. Two completely different random variables may possess the same mean and variance.

Perhaps this last idea will become clearer if we draw an analogy with simple dynamics. The equations for the first and second moment of a random variable are

$$E\{x\} = \int_{-\infty}^{\infty} xp(x)\,dx \qquad (3.76)$$

$$E\{x^2\} = \int_{-\infty}^{\infty} x^2p(x)\,dx \qquad (3.77)$$

These are of the same form as the equations used to find the center of gravity (centroid) and the moment of inertia of a body. Certainly two bodies can be completely different in shape yet have the same center of gravity and moment of inertia.

We are rescued from this dilemma by the observation that most processes encountered in real life are Gaussian. This means that all relevant densities (joint and single) are Gaussian. Once we know that a random variable has a Gaussian distribution, the density function is completely specified by giving its mean and second moment.

Thermal and Shot Noise

The assumption of *white noise* is not too far off base for several common forms of noise. *Thermal noise* and *shot noise* are two physically encountered types of noise.

Thermal noise is produced by the random motion of electrons in a medium. The intensity of this motion increases with increasing temperature and is zero only at a temperature of absolute zero. If a resistor is hooked directly to the input of a sensitive oscilloscope, a random pattern will be displayed on the screen. The power spectral density of this random process can be shown to be of the form

$$S(\omega) = \frac{A\omega}{e^{B\omega} - 1} \qquad (3.78)$$

where A and B are constants that depend upon temperature and other physical constants. Figure 3.15 is a sketch of $S(\omega)$. For low frequencies, $S(\omega)$ is almost constant. We can therefore usually approximate thermal noise as white noise.

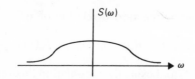

Figure 3.15   Spectral density of thermal noise.

Shot noise occurs since, although we think of current as being continuous, it is actually a discrete phenomenon. In fact, current occurs in discrete pulses each time an electron moves between two points. A plot of current as a function of time would show a waveform which wiggles around what we conventionally think of as the current plot. An example is shown in Fig. 3.16. This

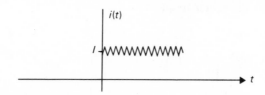

Figure 3.16   Shot noise current waveform.

variation of current around the average value is known as shot noise. An analysis (see the references) would show that, for frequencies of interest to us, shot noise can usually be thought of as being white noise.

3.5   NARROWBAND NOISE—QUADRATURE COMPONENTS

In studies of noise in communication systems, we encounter noise that is band limited to some range of frequencies. Such a noise waveform would result if any noise signal (white, for example), were passed through a band pass filter. It will prove useful to us to have a trigonometric expansion for such so called narrowband noise signals. The actual form will be

$$n(t) = x(t) \cos \omega_0 t - y(t) \sin \omega_0 t \tag{3.79}$$

where $n(t)$ is the noise waveform and ω_0 is a frequency (often the center) within the band occupied by the noise. Since sine and cosine vary by 90°, $x(t)$ and $y(t)$ are known as the "quadrature components" of the noise waveform.

Explicit solution of Eq. (3.79) for $x(t)$ and $y(t)$ is most easily done by resorting to the Hilbert Transform. For the present, we shall simply sketch the derivation.

Equation (3.79) can be solved to get

$$x(t) = n(t) \cos \omega_0 t + n(t) * \frac{1}{\pi t} \sin \omega_0 t$$

$$y(t) = n(t) * \frac{1}{\pi t} \cos \omega_0 t - n(t) \sin \omega_0 t \tag{3.80}$$

from which we can derive the autocorrelation of $x(t)$ and $y(t)$,

$$R_x(\tau) = R_y(\tau) = R_n(\tau) \cos \omega_0 \tau + R_n(\tau) * \frac{1}{\pi t} \sin \omega_0 \tau \qquad (3.81)$$

Finally, from Eq. (3.81) and using the modulation theorem (see Problem 1.18), we get the power spectral density relationship,

$$S_x(\omega) = S_y(\omega) = S_n(\omega - \omega_0) + S_n(\omega + \omega_0) \qquad (3.82)$$

Equation (3.82) is the key result which will enable us to calculate the effects of noise upon AM and FM communication systems.

Illustrative Example 3.13

Express the following three narrowband noise processes in quadrature form using ω_0 as the center frequency.

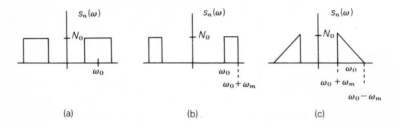

(a)   (b)   (c)

Solution

Using Eq. (3.82), we can immediately sketch the power spectral densities of $x(t)$ and $y(t)$, where the noise is expressed as

$$n(t) = x(t) \cos \omega_0 t - y(t) \sin \omega_0 t$$

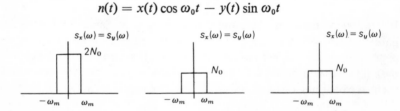

3.6 SIGNAL TO NOISE RATIO

In most communication studies, the probabilistic parameter of interest to us is the average power. In fact, the quantity of major interest is the ratio of signal power to the average noise power. This ratio is, quite logically, called the *signal to noise ratio* and is abbreviated as S/N.

In a good number of communication situations, the purpose of processing a received signal is to increase the signal to noise ratio. That is, we receive a signal including unwanted noise and we wish to somehow process this function so as to increase the S/N.

Figure 3.17 shows the block diagram of a system with the input S/N and output S/N indicated. The ratio of these two S/Ns gives some measure of the effectiveness of the system. This ratio, designated ΔSNR, is defined as the *signal to noise improvement* of the system:

$$\Delta\text{SNR} = \frac{(S/N)_{\text{in}}}{(S/N)_{\text{out}}} \tag{3.83}$$

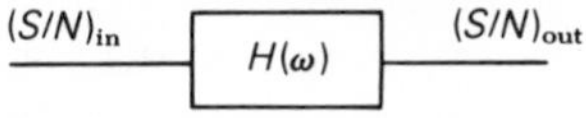

Figure 3.17 System for definition of signal to noise improvement.

The ratio is often expressed in *decibels*, or dB, defined by

$$\Delta\text{SNR}_{\text{dB}} = 10 \log_{10}(\Delta\text{SNR}) \tag{3.84}$$

3.7 NOISE GENERATORS

There exist situations in which it is desirable to generate noise. This appears odd since, in most communication applications, the name of the game is noise reduction. Why, then, would one ever be interested in intentionally generating noise?

One obvious situation is in the laboratory. In a controlled environment, one often wishes to add known noise to a signal in order to investigate the performance of various systems.

Since white noise has a constant power spectral density, it serves as a good test signal (just as the impulse function did) in the analysis of certain filter parameters.

White noise is useful for accurate time synchronization. Suppose, for example, that white noise experiences a time delay going through a system. If we cross-correlate the system output with its input. the result is an impulse shifted by the exact amount of delay. This is true since the autocorrelation of (ideal) white noise is an impulse.

A generated noise signal is often a good starting point in developing random signals with certain characteristics. For example, filtered noise can be used for synthesizing electronic music.

The simplest way to construct a noise generator is to start with a natural source of noise. If the signal appearing across the terminals of a resistor were amplified, the result could serve as a source of white (thermal) noise. In practice, there are far better starting points than resistors. Diodes and, in particular, hot cathodes in vacuum tubes are good sources.

A second type of noise generator is known as a *pseudonoise* generator. Such generators simulate noise with a known, periodic signal. The signal has characteristics approaching those of white noise if the period is made large.

That is, the autocorrelation of pseudonoise is a sharp pulse approaching an impulse. The pseudonoise generators possess an important advantage over the true noise generator. Every time the generator is turned on, the exact same signal is produced. Indeed, two separate generators produce identical signals (provided, of course, that the settings are identical). One application for which this last observation is important is remote time measurements. Suppose that it was desired to accurately measure the time delay in transmission from a ship to shore. If a pseudonoise signal is transmitted from ship to shore and, when received, is correlated with an identical signal which started at the same time, a sharp peak occurs at the time of transmission delay. Although accurate clocks do exist, it is difficult to start two remote generators at precisely the same time. A more likely distance measuring system would beam a signal from ship to shore and reflect it back to the ship. The total delay would then be twice the one-way transmission delay.

Pseudonoise generators start by forming a pseudonoise binary pulse train. This pulse train may then be filtered. The pseudonoise binary sequence is discussed in Chapter 8.

PROBLEMS

3.1. Three coins are tossed at the same time. List all possible outcomes of this experiment. List five representative events. Find the probabilities of the following events:

$$\{\text{all tails}\} \qquad \{\text{one head only}\} \qquad \{\text{three matches}\}$$

3.2. If a perfectly balanced die is rolled, find the probability that the number of spots on the face turned up is greater than or equal to 2.

3.3. An urn contains 4 white balls and 7 black balls. An experiment is performed in which three balls are drawn out in succession without replacing any. List all possible outcomes and assign probabilities to each one.

3.4. In Problem 3.3 you are told that the first two draws are white balls. Find the probability that the third is also a white ball.

3.5. An urn contains 4 red balls, 7 green balls and 5 white balls. Another urn contains 5 red balls, 9 green balls and 2 white balls. One ball is drawn from each urn. What is the probability that both balls will be of the same color?

3.6. Three people, A, B, and C live in the same neighborhood and use the same bus line to go to work. Each of the three has a probability of $\frac{1}{4}$ of making the 6:10 bus, a probability of $\frac{1}{2}$ of making the 6:15 and a probability of $\frac{1}{4}$ of making the 6:20 bus. Assuming independence, what is the probability that they all take the same bus?

3.7. The probability density function of a certain voltage is given by

$$p_v(v) = ve^{-v}U(v)$$

where $U(v)$ is the unit step function.

(a) Sketch this probability density function.
(b) Sketch the distribution function of v.
(c) What is the probability that v is between 1 and 2 volts?

3.8. Find the density of $y = |x|$ given that $p(x)$ is as shown below. Also find the mean and variance of both y and x.

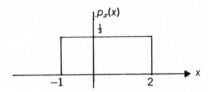

3.9. Find the mean and variance of x where the density of x is shown below.

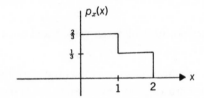

3.10. The density function of x is shown below. A random variable, y, is related to x as shown. Determine $p_y(y)$.

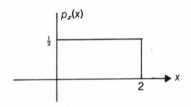 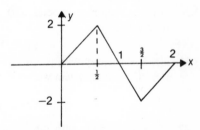

3.11. Find the expected value of $y = \sin x$ if x is uniformly distributed between 0 and 2π.

3.12. You are told that the mean rainfall per year is 4″ in California and that the variance in the amount of rain per year is 1. Can you tell, from this information, what the density of rainfall is? If so, roughly sketch this density function.

3.13. The random variable x is Rayleigh distributed as in Eq. (3.11). Find the mean, second moment, and variance.

3.14. The random variable x is uniformly distributed between $x = -1$ and $x = +1$. The random variable y is uniformly distributed between $y = 0$ and $y = +5$. x and y are independent of each other. A new variable, z, is formed as the sum of x and y.
(a) Find the probability density function of z.
(b) Find the mean and variance of x, y, and z.

(c) Find a general relationship among the means of x, y, and z.

(d) Repeat part (c) for the variance.

3.15. A random variable x has the following density function:

$$p_x(x) = Ae^{-2|x|}$$

(a) What is the value of A?

(b) Find the probability that x is between -3 and $+3$.

(c) Find the variance of x.

3.16. A voltage is known to be Gaussian distributed with mean value of 4. When this voltage is impressed across a 4-ohm resistor, the average power dissipated is 8 watts. Find the probability that the voltage exceeds 2 volts at any instant of time.

3.17. k is a random variable. It is uniformly distributed in the interval between -1 and $+1$. Sketch several possible samples of the process

$$x(t) = k \sin 2t$$

3.18. $x(t)$ is a stationary process with mean value 1 and autocorrelation

$$R(\tau) = e^{-|\tau|}$$

Find the mean and autocorrelation of the process $y(t) = x(t - 1)$.

3.19. $x(t)$ is a stationary process with zero mean value. Is the process

$$y(t) = tx(t)$$

stationary? Find the mean and autocorrelation of $y(t)$.

3.20. Given a stationary process, $x(t)$, find the autocorrelation of $y(t)$ in terms of $R_x(\tau)$, where

$$y(t) = x(t - 1) + \sin 2t$$

3.21. $x(t)$ is stationary with mean value "m." Find the mean value of the output, $y(t)$, of a linear system with $h(t) = e^{-t}U(t)$ and $x(t)$ as input. That is,

$$y(t) = \int_0^x h(\tau)x(t - \tau) \, d\tau$$

3.22. $x(t)$ is white noise with autocorrelation $R_x(\tau) = \delta(\tau)$. It forms the input to an ideal low pass filter with cutoff frequency ω_m. Find the average power of the output, $E\{y^2(t)\}$.

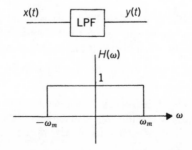

3.23. Use the result of Illustrative Example 3.9 in order to find the autocorrelation of the process $y(t) = dx/dt$ in terms of the autocorrelation of $x(t)$. (Hint: You will need the definition of the derivative,

$$\frac{dx}{dt} \triangleq \lim_{T \to 0} \frac{x(t) - x(t-T)}{T}$$

and l'Hospital's rule.)

3.24. Use an approach similar to that of Problem 3.23 in order to find

$$E\{x'(t)x(t+\tau)\} \quad \text{and} \quad E\{x'(t+\tau)x(t)\}$$

where $x'(t) = dx/dt$.

3.25. Find $R_e(\tau)$ in terms of $R_i(\tau)$ where $i(t)$ and $e(t)$ are related by the following circuit:
That is, $e(t) = Ri(t) + L \, di/dt$.

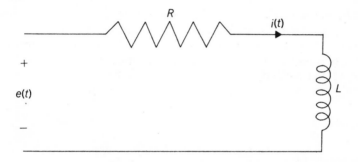

3.26. Given a constant, "a," and a random variable, "ω," with density $p_\omega(\omega)$, form

$$x(t) = ae^{j\omega t}$$

Find $R_x(\tau)$ and show that $S_x(\omega) = 2\pi |a|^2 p_\omega(\omega)$. (Hint: For complex $x(t)$, use $R_x(\tau) = E\{x(t+\tau)x^*(t)\}$.)

3.27. Find the autocorrelation and power spectral density of the wave

$$v(t) = A \cos(\omega_c t + \theta)$$

where A and ω_c are not random, and θ is uniformly distributed between 0 and 2π. (Hint: Use the definition of autocorrelation and the definition of expected value. You may also need the identity $\cos x \cos y = \frac{1}{2} \cos(x+y) + \frac{1}{2} \cos(x-y)$.)

3.28. Express the following narrowband noise processes in quadrature form. For each process, choose two different center frequencies and sketch the power spectral density of $x(t)$ and $y(t)$.

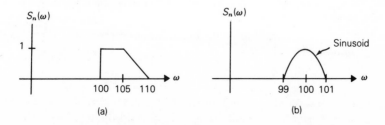

3.29. In a given communication system, the signal is $f(t) = 20 \cos t$. Noise of power spectral density $S_n(\omega) = e^{-3|\omega|}$ is added to the signal, and the resulting sum forms the input to a filter, $H(\omega)$.

(a) Find the S/N at the input to the filter.

(b) If the filter is low pass with $H(\omega) = 1$ for $|\omega| < 2$, find the signal to noise improvement and express this in decibels.

(c) If the filter is band pass with $H(\omega) = 1$ for $0.9 < |\omega| < 1.1$, find the noise figure and express this in decibels.

3.30. You are given the following system composed of two cascaded filters:

$f(t)$ and $n(t)$ are as given in Problem 3.29. $H_1(\omega)$ is a low pass filter with $H_1(\omega) = 1$ for $|\omega| < 2$. $H_2(\omega)$ is a bandpass filter with $H_2(\omega) = 1$ for $0.9 < |\omega| < 1.1$. Find the noise figure of $H_1(\omega)$ and the signal to noise improvement of $H_2(\omega)$, and express these in decibels. Now find the overall signal to noise improvement of the entire system and compare this to the individual signal to noise improvement of the two filters.

REFERENCES

BENDAT, J. S., *Principles and Applications of Random Noise Theory*, Wiley, New York, 1958.

COOPER, G. R., and McGILLEM, C. D., *Probabilistic Methods of Signal and System Analysis*, Holt, Rinehart and Winston, New York, 1971.

DAVENPORT, W., JR., and ROOT, W., *Random Signals and Noise*, McGraw-Hill, New York, 1958.

DOOB, J. L., *Stochastic Processes*, Wiley, New York, 1953.

FELLER, W., *An Introduction to Probability Theory and its Applications*, Vol. II, Wiley, New York, 1966.

LATHI, B. P., *Random Signals and Communication Theory*, International Textbook Co., Scranton, Pennsylvania, 1968.

PAPOULIS, A., *Probability, Random Variables, and Stochastic Processes*, McGraw-Hill, New York, 1965.

SCHWARTZ, M., and SHAW, L., *Signal Processing*, McGraw-Hill, New York, 1958.

TAUB, H., and SCHILLING, D. L., *Principles of Communication Systems*, McGraw-Hill, New York, 1971.

VAN DER ZIEL, A., *Noise: Sources, Characterization, Measurement*, Prentice-Hall, Englewood Cliffs, N.J., 1970.

AMPLITUDE MODULATION SYSTEMS

4

In the majority of communication situations, the information to be transmitted is in the form of signals suitable for human ears (no reference to taste or censorship is intended). This will undoubtedly change as data transmission becomes prevalent.

Figure 4.1 shows a typical speech waveform. One can observe that, in general, it is a mess. Since the exact waveform which we wish to transmit is not known, how can anything be said about the system required to transmit it? Asked in a different way, does any similarity exist among various information signals which will help us to design a transmission system?

Figure 4.1 A representative speech waveform.

In the case of the speech (or music) waveform, the answer to this question comes in the form of recalling something learned in high school physics. The human ear can only respond to (hear) signals with frequencies below about 15,000 hertz. Thus, if our eventual goal is the reception of audio type signals, it is fair to assume that the Fourier Transform of the time signal is zero for $|\omega| > 2\pi \times 15,000$ rad/sec. That is,

$$F(\omega) = 0 \qquad (|\omega| > \omega_m)$$

where

$$\omega_m = 2\pi \times 15,000$$

One might argue that the vocal chords, or other sound generators, are capable of generating frequency components above 15,000 Hz, even though the ear cannot hear them. This is indeed correct. However, if one of these signals were passed through a low pass filter with cutoff frequency of 15,000 Hz, the output of the filter, if fed into a loudspeaker, would sound exactly the same as the input. We are therefore justified in assuming that our "information" signals are band limited to an upper frequency of 15,000 Hz. While this may seem to be an insignificant observation, this one assumption will prove sufficient for the design of reasonably sophisticated communications systems.

"What about data transmission?" you ask. In that case, the upper frequency limitation is supplied by amplifiers and other electronic devices (recall the "flashlight test" from the discussion of the sampling theorem). Although this upper frequency limitation is usually considerably higher than that imposed by the human ear, it is nevertheless present and definable. The signal frequency cutoff restriction therefore applies to all information signals of interest, though the actual upper cutoff frequency depends upon the source and the eventual application of the signal.

For all of the future discussion, we will therefore assume that the signal to the transmitted (the "information" signal) has a Fourier Transform which is zero above a certain frequency. We will call this upper frequency limit, ω_m, for "ω-maximum." Even if ω_m is high (the word "high" has no meaning by itself, so let us interpret it as "high with respect to 15,000 Hz"), we will call signals of this type "low-frequency band limited signals."

In the case of short range transmission, as in the path between the pre-amp and amplifier or between the amplifier and speakers of a hi-fi, these low frequency signals are usually sent through wires. For longer distances, this is often not desirable since wires require "rights of way" meaning that towers must be built or trenches dug. Also, in the case of transmission through wires, one must specify the location of every terminal. For example, in the case of television, the wire would have to terminate in the home of every prospective viewer.

For these reasons, transmission through the air is often used instead of wire transmission. If a low frequency signal were simply applied to the input of an antenna, it would not radiate very far. In fact, in order to efficiently transfer the signal to the air (i.e., impedance match), the antenna would have to be many miles long for audio signals!

Even if one could efficiently transmit this signal through the air, a serious problem would arise. Suppose it were desirable to transmit more than one signal at a time, as it certainly is. That is, suppose that more than one of the dozens of local radio stations wished to transmit broadcasts simultaneously. They would each have antennas several miles long on top of their studios, and they would pollute the air with many audio signals. The poor listener would erect a receiving antenna several miles high, and get some sort of weighted sum (depending upon relative distances from the different studios to the antenna) of all of the signals. Since the only information the listener (receiver) has about the signals is that they are all low frequency band limited to the same upper cutoff frequency, he would have absolutely no way of separating one station from all of the others.

For the above two reasons, that is, efficient radiation and station separation, it is desirable to modify the low frequency signal before sending it from one point to another. An added bonus arises if the modified signal is less susceptible to noise than the original signal, but we're getting ahead of the game.

The most common method of accomplishing this modification is to use the low frequency signal to modulate (modify the parameters of) another signal. Most commonly, this other signal is a pure sine wave.

The following sections will analyze several different possibilities. During each analysis, the accomplishment of the primary objectives of efficient radiation and station separation will become obvious.

4.1 MODULATION

We start with a pure sinusoid, $f_c(t)$, called the *carrier* wave. It is given this name since it is used to carry the information signal from the transmitter to the receiver.

$$f_c(t) = A \cos(\omega_c + \theta) \tag{4.1}$$

We stated that the information signal will be used to modulate, or vary one of the parameters of this signal. Examination of Eq. (4.1) illustrates that there are three parameters which may be varied; the amplitude, A; the frequency, ω_c; and the phase, θ. Using the information signal to vary A, ω_c, or θ leads to amplitude modulation, frequency modulation, or phase modulation respectively.

For each of these three cases we will show that the two objectives of modulating are achieved. In addition we will have to illustrate a third property. That is, the information signal which we shall denote $f(t)$ must be uniquely recoverable from the modulated carrier wave. It wouldn't be of much use to modify $f(t)$ for efficient transmission if we could not reproduce $f(t)$ accurately at the receiver.

4.2 AM USING HIGH SCHOOL MATH ONLY

Prior to a thorough analysis of AM using the Fourier Transform, it may be instructive to give the subject a once-over using only trigonometry.

Suppose first that a talented performer is whistling a single tone at a frequency of 1 kHz. It is desired that this (uninteresting) signal be transmitted through the air. From Section 4.1, we know that, without modification, the signal will transmit very poorly. We decide that shifting of the frequency to something near 1 MHz would improve the transmission capabilities. Since it is unreasonable to ask the performer to whistle at a frequency of 1 MHz, we investigate electronic techniques for performing this frequency shift.

You may recall a trigonometric identity:

$$\cos (A \pm B) = \cos A \cos B \mp \sin A \sin B \tag{4.2}$$

From this, we see that (perform the derivation)

$$\cos \omega_1 t \cos \omega_2 t = \tfrac{1}{2} \cos (\omega_1 + \omega_2)t + \tfrac{1}{2} \cos (\omega_1 - \omega_2)t \tag{4.3}$$

Therefore, if we wish to shift our 1-kHz sinusoid to a frequency near 1 MHz, we need simply multiply the sinusoid by another sine wave of about 1 MHz and then filter out the undesired term (*see* Fig. 4.2). This process is known as *heterodyning*.

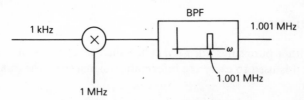

Figure 4.2 Heterodyning.

Let's simplify things and try eliminating the filter in Fig. 4.2. That is, we consider transmitting the product of the 1-kHz sinusoid with the 1-MHz sinusoid. This new signal has frequencies of 1.001 and 0.999 MHz. Both of these new frequencies will transmit efficiently.

One important job remains if we are to be convinced that this is a useful way of transmitting our 1-kHz whistle. We must be able to recover the original whistle at the receiver.

Since the AM wave is formed by shifting the frequency of the sound signal, the original signal is recovered by performing yet another shift. We must shift the received frequency *down* by 1 MHz, and we do this by multiplying the received signal by a sinusoid of frequency 1 MHz. (*See* Fig. 4.3.)

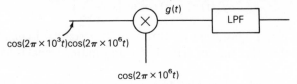

Figure 4.3   Recovering original signal from AM.

The result of this multiplication is the signal

$$g(t) = \cos(2\pi \times 10^3 t) \cos^2(2\pi \times 10^6 t) \tag{4.4}$$

But from Eq. (4.3) the square of a cosine is equal to $\frac{1}{2}$ plus a cosine of twice the frequency. We therefore have

$$g(t) = \tfrac{1}{2}\cos(2\pi \times 10^3 t) + \tfrac{1}{2}\cos(2\pi \times 10^3 t)\cos(2\pi \times 2 \times 10^6 t) \tag{4.5}$$

The second term in Eq. (4.5) is composed of two frequencies—2.001 and 1.999 MHz. If a low pass filter is added as shown in Fig. 4.3, these two relatively high frequencies are easily rejected, leaving only the original 1-kHz whistle.

We have therefore shown that a sinusoid of low frequency can be transmitted through the air by first multiplying it by a sinusoid of high frequency. The product is an AM wave, and it can be transmitted. At the receiver, the reverse operation is performed to recover the original signal.

A more complex audio signal than a whistle can be thought of as being composed of a sum of sinusoids. In this manner, the above trigonometric analysis can be extended to the general case.

The next section considers the general case but does so using the Fourier Transform rather than trigonometric identities. The results are more general, and, indeed, with a little bit of comfort with Fourier Transforms, one can extend the analysis more readily.

4.3   DOUBLE SIDEBAND SUPPRESSED CARRIER AMPLITUDE MODULATION

The case of amplitude modulation is characterized by a modulated signal waveform of the form $f_m(t)$.

$$f_m(t) = A(t)\cos(\omega_c t + \theta) \tag{4.6}$$

ω_c and θ are constants and $A(t)$, the amplitude, varies somehow in accordance with $f(t)$.

In the following analysis, we shall assume that $\theta = 0$ for simplicity. This will not affect any of the basic results since the angle actually corresponds to a time shift of θ/ω_c. That is, we can rewrite $f_m(t)$ as

$$f_m(t) = A(t) \cos \omega_c\left(t + \frac{\theta}{\omega_c}\right) \tag{4.7}$$

A time shift is not considered as distortion in a standard communications system since the exact time of arrival of a transmitted signal is usually of no importance.

If somebody asked you how to vary $A(t)$ in accordance with $f(t)$, the simplest answer you could suggest would be to make $A(t)$ equal to $f(t)$. Thus, the modulated signal would be of the form

$$f_m(t) = f(t) \cos \omega_c t \tag{4.8}$$

This type of modulated signal is given the name *double sideband suppressed carrier amplitude modulation* for reasons that will become clear in Section 4.7.

This simple guess for $A(t)$ does indeed satisfy the criteria demanded of a communication system. The easiest way to illustrate this fact is to express $f_m(t)$ in the frequency domain, that is, to find its Fourier Transform.

Suppose that we call the Fourier Transform of $f(t)$, $F(\omega)$. Note that we are requiring nothing more of $F(\omega)$ other than its being equal to zero for frequencies above some cutoff frequency, ω_m. We sketch $F(\omega)$ as in Fig. 4.4 since this represents an easy sketch to reproduce. This does not mean that $F(\omega)$ must be of the shape shown. The sketch is meant only to indicate the transform of a general low frequency band limited signal.

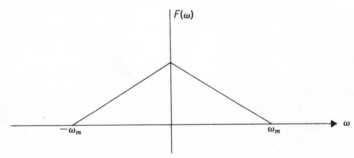

Figure 4.4   General form of $F(\omega)$.

From Chapter 1, it should be clear that the transform of $f_m(t)$, $F_m(\omega)$, is given by

$$F_m(\omega) = \mathfrak{F}[f(t) \cos \omega_c t] = \mathfrak{F}\left[\frac{f(t)e^{-j\omega_c t}}{2} + \frac{f(t)e^{+j\omega_c t}}{2}\right]$$

$$F_m(\omega) = \tfrac{1}{2}[F(\omega + \omega_c) + F(\omega - \omega_c)] \tag{4.9}$$

This transform is sketched as Fig. 4.5.

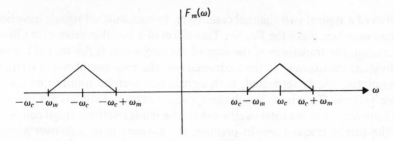

Figure 4.5 $F_m(\omega)$, the transform of $f_m(t)$.

Observation of Fig. 4.5 and the acceptance of a result carried over from electromagnetic field theory indicates that the first objective of a communication system has been met. The observation is that the frequencies present in $f_m(t)$ fall in the range between $\omega_c - \omega_m$ and $\omega_c + \omega_m$. The result from electromagnetics is that an efficient radiator (antenna) must have a length that is of the order of the wavelength of the signal to be transmitted. This is what we used to get the figure of many miles for an audio signal antenna. Since wavelength is inversely proportional to frequency, we can make the wavelength as short as desired by raising the frequency of a wave. Since the lowest frequency present in $f_m(t)$ is $\omega_c - \omega_m$, and this can be made as large as desired by choosing ω_c large, the signal can be tailored to any length antenna. In addition, as ω_c increases, the range of frequencies occupied by $f_m(t)$ (i.e., the bandwidth) becomes smaller relative to ω_c. This also leads to simpler antenna construction.

As an example, in the audio case even though $f(t)$ has frequencies between 0 and 15,000 Hz, $f_m(t)$ can be made to have only those frequencies between 985,000 and 1,015,000 Hz by choosing a carrier frequency of 1 MHz. This would lend itself to efficient transmission.

The second objective is that of channel separability. We see that if one information signal modulates a sinusoid of frequency ω_{c1} and another information signal modulates a sinusoid of frequency ω_{c2}, the Fourier Transforms of the two modulated carriers will be separated in frequency provided that ω_{c1} and ω_{c2} are not too close together. Figure 4.6 shows a

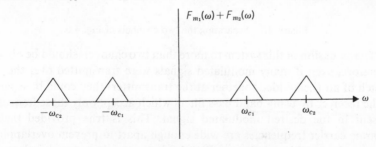

Figure 4.6   An example of two-channel AM.

sketch of a typical two channel case. Here, two modulated signals have been added together. Since the Fourier Transform of a function represents a linear operation, the transform of the sum of the two waves is the sum of the two individual transforms. If the frequencies of the two modulated waveforms are not too widely separated, both signals can even be transmitted via the same antenna. That is, although the optimum antenna length is not the same for both channels, the total bandwidth can be made relatively small compared to the carrier frequencies. In practice, the antenna is useable over a range of frequencies rather than just being effective at a single frequency. If the latter were the case, radio broadcasting could not exist.

As an example, you don't have to readjust your car antenna when you tune across the AM dial. The effectiveness of the antenna does not vary greatly from one frequency limit to the other. However, if you switch to the FM band which encompasses a much higher range of carrier frequencies than AM, the radio manufacturer recommends shortening the car antenna to about 30 inches.

Our study of band pass filters should have convinced us that whenever the transforms of two signals occupy non-overlapping ranges of frequency, the signals can be separated from each other by means of band pass filters. Thus, a system such as that shown in Fig. 4.7 could be used to separate the two modulated carriers from each other in the case presented in Fig. 4.6.

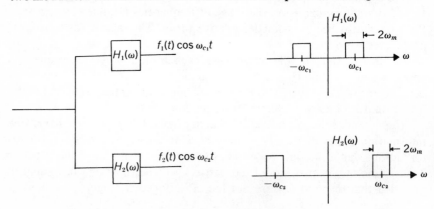

Figure 4.7 Separating the two channels of Fig. 4.6.

The extension of this system to more than two channels should be obvious. Therefore, even if many modulated signals were transmitted over the same stretch of air (i.e., added together at the transmitter) they could be separated at the receiver by using band pass filters which accept only those frequencies present in the desired modulated signal. This is true provided that the separate carrier frequencies are wide enough apart to prevent overlapping of the transforms. Observation of Fig. 4.6 indicates that if each information signal is low frequency band limited to ω_m, the adjacent carrier frequencies must be separated by at least $2\omega_m$ to avoid overlapping.

The process of stacking many stations in separate "frequency slots" is called *frequency division multiplexing* (*FDM*), and is the system used in all standard broadcast transmission.

Before going on, we should again note that all of our sketches of Fourier Transforms seem to indicate that they are real functions of ω. This would require that the corresponding time functions be all even functions, a very severe restriction indeed. However, examination of the formulae and derivations will indicate that this assumption was never made. It is simply used in the sketches to avoid clouding the picture with phase diagrams. Indeed, the sketches are included only to give an intuitive feel for what is happening. If the reader feels that he must be precise throughout, he can assume that all of our drawings are actually plots of the magnitudes of the Fourier Transforms and that the phase plots were inadvertently omitted.

Illustrative Example 4.1

An information signal, $f(t) = (\sin t)/t$ is used to amplitude modulate a carrier of frequency 20 rad/sec. Sketch the AM wave and its Fourier Transform.

Solution

The AM wave is given by the time function

$$f_m(t) = f(t)\cos 20t$$
$$= \frac{\sin t}{t}\cos 20t \tag{4.10}$$

This is sketched in Fig. 4.8.

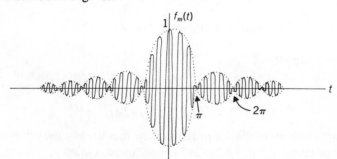

Figure 4.8 $f_m(t)$ for Illustrative Example 4.1.

We note that when the carrier, $\cos 20t$, is equal to 1, $f_m(t) = f(t)$, and when $\cos 20t = -1$, $f_m(t) = -f(t)$. Therefore, the information signal, $f(t)$, and its mirror image, $-f(t)$, can be used as a type of outline to guide in the sketching of the waveform.

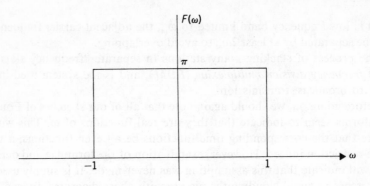

Figure 4.9 $F(\omega)$, the Transform of $(\sin t)/t$.

From Chapter 1, the Transform of $f(t) = (\sin t)/t$ is as sketched in Fig. 4.9.

From Eq. (4.9), the transform of the modulated wave is

$$F_m(\omega) = \tfrac{1}{2}[F(\omega - 20) + F(\omega + 20)] \qquad (4.11)$$

This is sketched as Fig. 4.10.

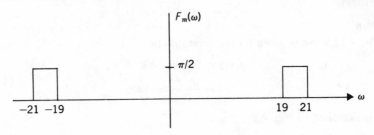

Figure 4.10 $F_m(\omega)$ for $F(\omega)$ of Fig. 4.9.

Illustrative Example 4.2

Define the dot product of $f_1(t)$ with $f_2(t)$ as the convolution of the two functions.

$$f_1(t) \cdot f_2(t) \triangleq f_1(t) * f_2(t) \qquad (4.12)$$

Using this definition of the dot product, show that the two modulated carrier signals illustrated in Fig. 4.6 are orthogonal to each other.

Solution

We shall define

$$f_{m1}(t) = f_1(t) \cos \omega_{c1}(t)$$
$$f_{m2}(t) = f_2(t) \cos \omega_{c2}(t)$$
$$f_{m1}(t) * f_{m2}(t) = \int_{-\infty}^{\infty} f_{m1}(\tau) f_{m2}(t - \tau)\, d\tau$$

Since we do not know the exact time functions, we cannot evaluate this convolution integral. However, for this special case, the evaluation is trivial via the time convolution theorem.

$$f_{m1}(t) * f_{m2}(t) \longleftrightarrow F_{m1}(\omega)F_{m2}(\omega)$$

In Fig. 4.6 we see that the product of the two transforms is zero since, at values of ω for which $F_{m1}(\omega)$ is not equal to zero, $F_{m2}(\omega)$ is zero and vice versa. We are assuming that the carrier frequencies are wide enough apart so that the individual transforms do not overlap.

Therefore,

$$f_{m1}(t) * f_{m2}(t) = \mathcal{F}^{-1}[0] = 0 \qquad (4.13)$$

and $f_{m1}(t)$ is orthogonal to $f_{m2}(t)$. Actually, this analogy with the vector case can be carried much further. The set of time functions with transforms shown in Fig. 4.11 can be thought of as analogous to unit vectors. The actual $f_{m1}(t)$

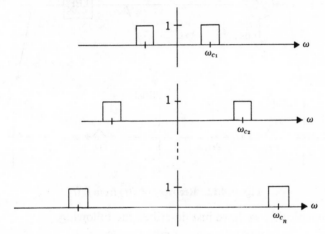

Figure 4.11 Unit vectors for Illustrative Example 4.2.

and $f_{m2}(t)$ can be regarded as the coefficients. The results of Section 1.1 can therefore indicate how to send many signals at once, and how to recover one signal from the total sum (i.e., evaluate one of the coefficients).

The details will not be covered here as we will take a different approach. This example was meant to stimulate the interested student to thinking about the meaning and various applications of the principle of orthogonality. This principle will exhibit itself many times throughout the study of communications, but particularly in the area of noise reduction systems.

We have indicated that an AM wave of the type discussed could be transmitted efficiently, and that more than one information signal could be sent at the same time and still allow for separation at the receiver. An important property that must still be verified is that the information signal, $f(t)$, can be recovered from the modulated waveform, $f_m(t)$. We proceed to do that now.

Since $F_m(\omega)$ was derived from $F(\omega)$ by shifting all of the frequency components of $f(t)$ by ω_c, we can recover $f(t)$ from $f_m(t)$ by shifting the frequencies again by the same amount, but this time in the opposite direction.

The frequency shifting theorem tells us that multiplication of $f(t)$ by a sinusoid shifts the Fourier Transform both up and down in frequency. Thus multiplying $f_m(t)$ by cos $\omega_c t$ will shift the Fourier Transform back down to its low-frequency position (sometimes called its DC position since direct current represents a zero frequency sinusoid). This multiplication will also shift the transform of $f(t)$ up to a position centered about a frequency of $2\omega_c$, but this part can easily be rejected by using a low pass filter. This process is illustrated in Fig. 4.12.

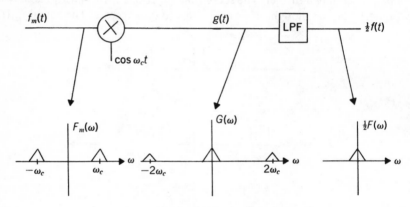

Figure 4.12   Recovery of $f(t)$ from $f_m(t)$.

Mathematically, we have just described the following,

$$[f_m(t)] \cos \omega_c t = [f(t) \cos \omega_c t] \cos \omega_c t$$
$$= f(t) \cos^2 \omega_c t \qquad (4.14)$$
$$= \tfrac{1}{2}[f(t) + f(t) \cos 2\omega_c t]$$

The output of the low pass filter is therefore $\tfrac{1}{2}f(t)$ which represents an undistorted version of $f(t)$. Quite logically, this process of recovering $f(t)$ from the modulated waveform is known as *demodulation*. We have taken this moment now to discuss demodulation rather than waiting until Section 4.5 where we will have more to say about it. Indeed, if $f(t)$ could not be recovered from $f_m(t)$, there would be no reason to go on to Section 4.4.

4.4 MODULATORS

Although this text is primarily intended to present system theory rather than practical realizations, it is instructive to consider the actual techniques of building a modulator or demodulator. We say this is instructive since the modulator represents the first time we have come across a system that is not linear time invariant.

Why is modulation not linear time invariant? We can use the basic properties of systems to illustrate that this must be the case. Recall that any linear time invariant system has an output whose Fourier Transform is the product of the Fourier Transform of the input with $H(\omega)$. Thus, if the Fourier Transform of the input, $F(\omega)$, is zero at some value of ω, the Fourier Transform of the output must also be zero at this frequency. Indeed, a general property of linear time invariant systems is that they cannot generate any output frequencies that do not appear in the input. Going one step further, the systems can always be described by linear differential equations. No matter how many times we integrate or differentiate a sine wave, it never changes frequency. Thus a linear time invariant system is incapable of creating any new frequencies that do not exist as part of its input.

We now ask whether any linear time invariant system can have $f(t)$ as an input and $f_m(t)$ as an output. In other words, is there any $H(\omega)$ for which $F_m(\omega) = F(\omega)H(\omega)$ as in Fig. 4.13?

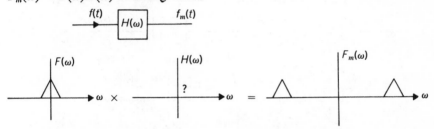

Figure 4.13   Can a linear time invariant system modulate?

Since $F_m(\omega)$ is non-zero at frequencies for which $F(\omega) = 0$, there exists no $H(\omega)$ that can be used as a modulator.

The problem arises because the form of AM we are discussing is linear, but not time invariant. To show linearity, we note

$$f_1(t) \rightarrow f_1(t) \cos \omega_c t$$
$$f_2(t) \rightarrow f_2(t) \cos \omega_c t$$
$$af_1(t) + bf_2(t) \rightarrow [af_1(t) + bf_2(t)] \cos \omega_c t$$
$$= af_1(t) \cos \omega_c t + bf_2(t) \cos \omega_c t$$

We shall see that this linearity makes the analysis much simpler. (This will become clear when we study FM, a type of modulation which is not even linear.)

Synthesis of a system which is not linear time invariant is, in general, virtually impossible. We shall attempt to simplify the synthesis, at least in the case of modulation, to a point at which the physical system realization seems intuitively realizable.

We state the problem of amplitude modulation as follows: Given $f(t)$ and $\cos \omega_c t$, we wish to form the product, $f(t) \cos \omega_c t$.

The methods of accomplishing this can be divided into three general areas. We shall assign the names gated, square law, and waveshape modulators to these three classes.

The *gated modulator* takes advantage of an observation that can be made from the sampling theorem. That is, if $f(t)$ is sampled (multiplied) with a periodic wave of fundamental frequency ω_c, the Fourier Transform of the sampled wave will consist of the original Fourier Transform shifted and repeated at multiples of ω_c in frequency. In the case of the sampling theorem, we were interested in the Fourier Transform "hump" centered about $\omega = 0$, and we rejected the others with a low pass filter in order to recover $f(t)$. In amplitude modulation, we are interested in the component of the Fourier Transform centered about $\omega = \omega_c$, and we shall reject all of the other components using a band pass filter.

As a special case, consider the sampling function, $s(t)$, as shown in Fig. 4.14. If $f(t)$ is multiplied by $s(t)$, we have

$$f(t)s(t) = f(t)\left[\frac{1}{2} + \frac{2}{\pi}\cos\omega_c t - \frac{2}{3\pi}\cos 3\omega_c t + \cdots\right] \qquad (4.15)$$

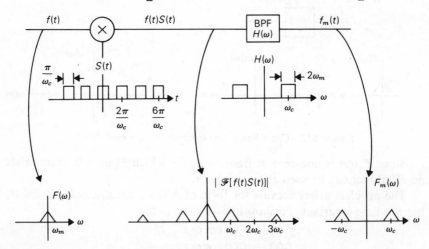

Figure 4.14 The gated modulator.

where we have replaced $s(t)$ by its Fourier Series representation. When the above expression is multiplied out, each term will be of the form

$$a_n f(t) \cos n\omega_c t \qquad (4.16)$$

which is actually an amplitude modulated wave whose carrier frequency is $n\omega_c$. A band pass filter of bandwidth $2\omega_m$ and centered about ω_c will sift out the desired term from all of the undesired terms (harmonics). For this particular $s(t)$, the term will be $(2/\pi)f(t)\cos\omega_c t$ which is, indeed, the desired AM wave. The only requirement is that of the sampling theorem to insure that the separate sections of the Fourier Transform do not overlap. That is, $\omega_c > 2\omega_m$.

We will see later that for all cases of interest, ω_c is intentionally chosen to be much much greater than ω_m, so this restriction is easily met. For example, in standard AM broadcast, ω_m is $2\pi \times 5000$ rad/sec and ω_c is something in the vicinity of $2\pi \times 1,000,000$ rad/sec, the exact value depending upon the desired station.

The entire gated modulating process is illustrated in Fig. 4.14.

A reasonable question would now be "So what? What have we accomplished?" Observation of Fig. 4.14 indicates that we must somehow generate $s(t)$ from $\cos \omega_c t$ (what we were given), perform a multiplication, and also build a band pass filter. Why did we not just take $f(t)$ and $\cos \omega_c t$ and put them into a multiplier as shown in Fig. 4.15?

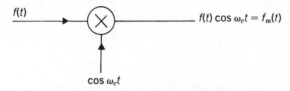

Figure 4.15   An AM modulator.

The answer is that there are very few devices which can take two continuous analog signals and multiply them together accurately.

How then does the gated modulator avoid the multiplication problem? Since one of the functions, $s(t)$, is always either equal to zero or one, the multiplication process can be thought of as a gating process. That is, whenever $s(t) = 1$, the output is equal to the input. When $s(t) = 0$, the output is zero. Multiplication by an $s(t)$ of this type corresponds to periodically opening and closing a gate, or switch in a circuit. Consider the circuit shown in Fig. 4.16.

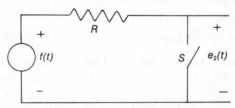

Figure 4.16   A gating circuit.

If the operator sits at the circuit and periodically opens and closes switch S, one can see that $e_2(t)$ will be of the desired form, $f(t)S(t)$.

Unfortunately, the switch has to be opened and closed at a rate which is often greater than a million times every second ($\omega_c/2\pi$ times per second). Mechanical switches are therefore usually ruled out. The obvious alternative is to use an electrical switch.

In practice, one often uses a solid state device operating between cutoff and saturation as an effective electronic switch. For our purpose here, a simple diode is sufficient. The circuit of Fig. 4.17 is known as a diode bridge modulator. It accomplishes the gating in the same way as the circuit of Fig. 4.16.

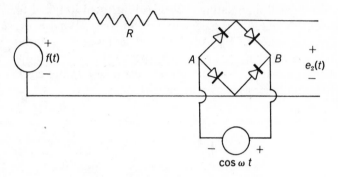

Figure 4.17 The diode bridge modulator.

When $\cos \omega t > 0$, the point "B" is at a higher potential than point "A." In this condition, all four diodes are open circuited, and the circuit is equivalent to that of Fig. 4.16 with switch S open. When $\cos \omega t < 0$, point "A" is at a higher potential than point "B," and all four diodes are short circuits. In that case, the circuit is equivalent to that of Fig. 4.16 with switch S closed. Therefore the $\cos \omega t$ source essentially opens and closes the switch at a rate of $2\pi\omega$ times per second. The only limit to this rate is imposed by the non-idealness of practical diodes (i.e., capacitance).

There is another type of amplitude modulator which is conceptually simpler than the gated modulator. This is known as the *square law modulator*. The square law modulator takes advantage of the fact that the square of the sum of two functions contains their product as the cross term in the expansion.

$$[f_1(t) + f_2(t)]^2 = f_1^2(t) + f_2^2(t) + 2f_1(t)f_2(t) \qquad (4.17)$$

Therefore, a summer and squarer could be used to multiply two functions together provided that the $f_1^2(t)$ and the $f_2^2(t)$ terms could be rejected from the result.

Considering the special case where $f_1(t) = f(t)$, the information signal, and $f_2(t) = f_c(t)$, the carrier sinusoid, we have

$$[f(t) + f_c(t)]^2 = f^2(t) + f_c^2(t) + 2f(t)f_c(t) \qquad (4.18)$$

The third term in Eq. (4.18) is the desired modulated signal, and if the other two terms can be rejected, the modulation is accomplished.

We have already seen that functions whose Fourier Transforms occupy different intervals along the frequency axis can always be separated from each other. The mechanism of this separation is the band pass filter. We would therefore be overjoyed if we could show that the three terms on the right side of Eq. (4.18) have non-overlapping Fourier Transforms.

We note that $f_c^2(t)$ can be explicitly written,

$$f_c^2(t) = \cos^2 \omega_c t = \tfrac{1}{2}[1 + \cos 2\omega_c t] \tag{4.19}$$

Figure 4.18 shows the sum of the transforms of $f_c^2(t)$ and $2f(t)f_c(t)$, both of which have been found in previous work.

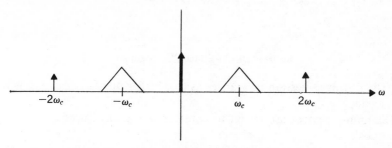

Figure 4.18   Sketch the transform of $f_c^2(t) + 2f(t)f_c(t)$.

We do not know the transform of $f(t)$ exactly since we do not wish to restrict this to any particular information signal. How can we then hope to know anything about the transform of $f^2(t)$? The answer lies in the convolution theorem. Since $f^2(t)$ is the product of $f(t)$ with itself, the transform of this will be the convolution of $F(\omega)$ with $F(\omega)$.

$$f^2(t) \longleftrightarrow \frac{1}{2\pi}[F(\omega) * F(\omega)] \tag{4.20}$$

Illustrative Example 1.6 showed that, if $F(\omega)$ is low frequency band limited to ω_m, then $F(\omega)$ convolved with itself is low frequency band limited to $2\omega_m$.

$$F(\omega) * F(\omega) = 0 \qquad (|\omega| > 2\omega_m) \tag{4.21}$$

We can now sketch the transform of all terms in Eq. (4.18). This is done in Fig. 4.19. There has been no attempt to show the exact shape of $F(\omega) * F(\omega)$, just as the sketch of $F(\omega)$ is not meant to represent its exact shape. The range of frequency occupied is the only important parameter for this study.

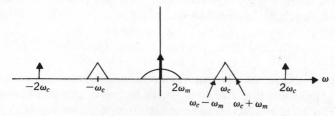

Figure 4.19   Transform of $[f(t) + f_c(t)]^2$.

As long as $\omega_c > 3\omega_m$, the components of Fig. 4.19 do not overlap, and the desired AM wave can be recovered using a band pass filter centered at ω_c with bandwidth equal to $2\omega_m$. This yields a square law modulator block diagram as shown in Fig. 4.20.

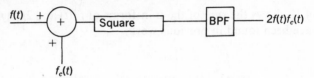

Figure 4.20 The square law modulator.

The obvious question now arises as to whether or not one can construct this device.

Summing devices are trivial to construct since any linear circuit obeys superposition.

Square law devices are not quite as simple. Any practical non-linear device has an output over input relationship which can be expanded in a Taylor series. This assumes that no energy storage is taking place. That is, the output at any time depends only upon the value of the input at that same time, and not upon any earlier input values.

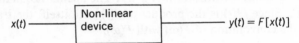

Figure 4.21 A non-energy storing non-linear device.

Thus,

$$y(t) = F[x(t)]$$
$$= a_0 + a_1 x(t) + a_2 x^2(t) + a_3 x^3(t) + \cdots$$

The term we are interested in is the $a_2 x^2(t)$ term in the above expansion. If we could somehow find a way to separate this term from all of the others, it would then be possible to use any non-linear device in place of the squarer in Fig. 4.20. We now optimistically investigate this possibility of separation.

It would be beautiful if the transforms of each of the terms in the Taylor series were non-overlapping in frequency. It would even be sufficient if the transform of each term other than that of the square term fell outside of the interval $\omega_c - \omega_m$ to $\omega_c + \omega_m$ on the frequency axis.

Beautiful as this might be, it is not the case. The transform of the linear term, assuming $x(t) = f(t) + f_c(t)$, has an impulse sitting right in the middle of the interval of interest. The cube term contains $f^2(t) \cos \omega_c t$, which consists of $F(\omega) * F(\omega)$ shifted up around ω_c in the frequency domain. Indeed, every higher order term will have a component sitting right on top of the desired transform, and therefore inseparable from it by means of filters.

All we have indicated is that not just any non-linear device can be used in our modulator. The device must be an essentially pure squarer. Fortunately, such devices do exist approximately in real life. A simple semiconductor diode can be used as a squarer when confined to operate in certain ranges. Indeed, the modulator shown in Fig. 4.22 is often used (with minor modifications).

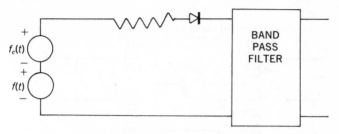

Figure 4.22   A diode square law modulator.

The previous discussion has shown that the modulator of Fig. 4.20 requires an essentially pure square law device. This was true since the means of separating desired and undesired terms was the band pass filter. The terms therefore had to be non-overlapping in frequency. More exotic techniques do exist for separating two signals. One such technique allows the non-linear device of Fig. 4.21 to contain odd powers in addition to the square term.

Suppose, for example, we build the square law modulator and a cube term is present. We already found that an undesired term of the form $f^2(t) \cos \omega_c t$ appears in the output. Suppose that we now build another modulator, but use $-f(t)$ as the input instead of $+f(t)$. The term $f^2(t) \cos \omega_c t$ will remain unchanged, while the desired term, $f(t) \cos \omega_c t$ will change sign. If we subtract the two outputs, the $f^2(t) \cos \omega_c t$ will be eliminated to leave the desired modulated waveform. Some thought will convince one that this technique will eliminate all product terms resulting from odd powers in the non-linear device. Such a modulator is called a *balanced modulator*, and one form is sketched in Fig. 4.23.

The third method of constructing a modulator is more direct than the previous two techniques. We shall call this the *waveshape modulator*. The modulation is performed as an actual waveshape multiplication of two signals in an electronic circuit.

One possible realization is drawn in Fig. 4.24. In this case, the carrier sinusoid is introduced into the emitter of a transistor. The transistor is in the common base configuration. The information signal, $f(t)$, is introduced as a variation in the B^+ supply voltage, E_{cc}. The relatively slow changes in E_{cc} have the effect of moving the load line and changing the quiescent point of the circuit.

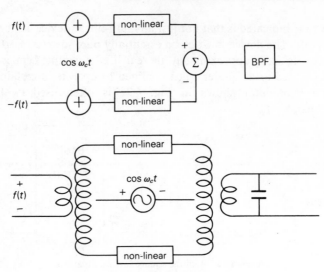

Figure 4.23 A balanced AM modulator.

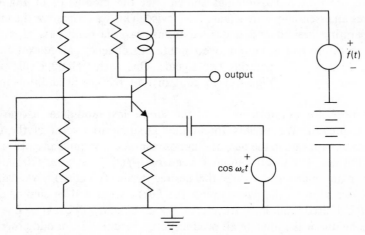

Figure 4.24 A transistor waveshape modulator.

Without any further analysis, it should not be too surprising to find that if the transistor is not permitted to saturate, the output will contain the desired AM waveform.

4.5 DEMODULATORS

We have previouly stated that $f(t)$ is recovered from $f_m(t)$ by "remodulating" $f_m(t)$ and then passing the result through a low pass filter. This yields the demodulator system block diagram of Fig. 4.25.

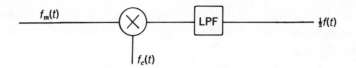

Figure 4.25   An AM demodulator.

Since the multiplier in the above figure looks no different from the multiplier used in the modulator, one would expect that the gated and square law modulators can again be used.

Illustrative Example 4.3

Show that the gated modulator can be used as a demodulator.

Solution

We wish to consider $f_m(t)S(t)$, where $S(t)$ is periodic and given by a Fourier series expansion,

$$S(t) = \sum_{n=0}^{\infty} a_n \cos n\omega_c t \tag{4.22}$$

Expanding the product, we have

$$\begin{aligned}
f_m(t)S(t) &= f(t) \cos \omega_c t \sum_{n=0}^{\infty} a_n \cos n\omega_c t \\
&= f(t) \sum_{n=0}^{\infty} a_n \cos \omega_c t \cos n\omega_c t \\
&= f(t) \sum_{n=0}^{\infty} a_n \tfrac{1}{2} [\cos (n-1)\omega_c t + \cos (n+1)\omega_c t]
\end{aligned} \tag{4.23}$$

Writing out the first few terms, and sketching their transforms, we have

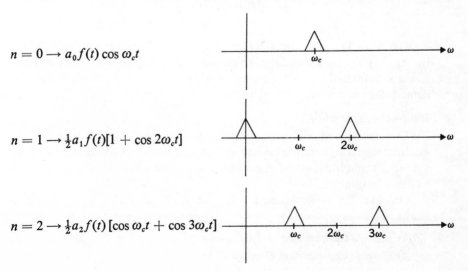

$n = 0 \longrightarrow a_0 f(t) \cos \omega_c t$

$n = 1 \longrightarrow \tfrac{1}{2}a_1 f(t)[1 + \cos 2\omega_c t]$

$n = 2 \longrightarrow \tfrac{1}{2}a_2 f(t)[\cos \omega_c t + \cos 3\omega_c t]$

Viewing the transform of each of these parts, we see that each one represents $F(\omega)$ shifted to some multiple of ω_c, except for the term $\frac{1}{2}a_1 f(t)$. This has as its transform

$$\frac{1}{2}a_1 F(\omega)$$

and it can be separated by use of a low pass filter. We have therefore accomplished demodulation. A block diagram of the gated demodulator is shown in Fig. 4.26.

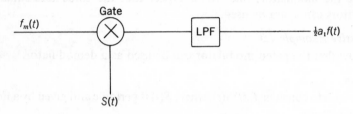

Figure 4.26 The gated demodulator.

Illustrative Example 4.4

Show that the square law modulator *cannot* be used to demodulate AM.

Solution

We wish to show that $[f_m(t) + f_c(t)]^2$ does not contain a separable term of the form, $f(t)$. Expanding the squared sum, we have

$$
\begin{aligned}
[f_m(t) + f_c(t)]^2 &= [f(t)\cos\omega_c t + \cos\omega_c t]^2 \\
&= \cos^2\omega_c t[1 + f(t)]^2 \\
&= \tfrac{1}{2}[1 + \cos 2\omega_c t][1 + 2f(t) + f^2(t)] \\
&= \tfrac{1}{2}[1 + 2f(t) + f^2(t) + \cos 2\omega_c t \\
&\quad + 2f(t)\cos 2\omega_c t + f^2(t)\cos 2\omega_c t]
\end{aligned}
\qquad (4.24)
$$

The desired term in Eq. (4.24) is $f(t)$. It should be clear from Fig. 4.18 that the term $\frac{1}{2}f^2(t)$ has a transform which lies right on top of $F(\omega)$, and can therefore not be filtered out. A square law modulator can therefore not be used for demodulation.

Illustrative Example 4.5

This example should give some practice in applying the principles of modulation and demodulation to a real system. We have already demonstrated that modulators and demodulators can be built, so we will again revert to block diagrams.

Consider the system sketched in Fig. 4.27, with $F_1(\omega)$ and $F_2(\omega)$ as shown.

(a) Sketch the transform of $f_3(t)$.
(b) Sketch the transform of $g(t)$.

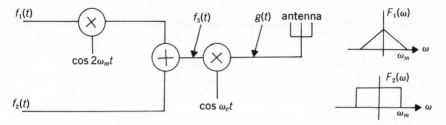

Figure 4.27 System for Illustrative Example 4.5.

(c) If $f_1(t)$ and $f_2(t)$ were to represent the left and right channels of a stereo broadcast, sketch one possible realization of a complete receiver.

Solution

$f_3(t)$ is given by

$$f_3(t) = f_1(t) \cos 2\omega_m t + f_2(t) \qquad (4.25)$$

The transform of this is sketched in Fig. 4.28. $g(t)$ is given by

$$g(t) = f_3(t) \cos \omega_c t \qquad (4.26)$$

and its transform is sketched in Fig. 4.29. We note that, while $f_3(t)$ could not be efficiently transmitted through the air, $g(t)$ can be, with proper choice of ω_c.

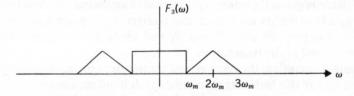

Figure 4.28 $F_3(\omega)$ for Illustrative Example 4.5.

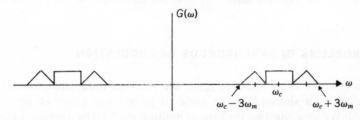

Figure 4.29 $G(\omega)$ for Illustrative Example 4.5.

The demodulation system is sketched in Fig. 4.30. It essentially reverses the operations of the original system, with the addition of filtering to separate the two "channels." The student should be able to sketch the Fourier Transform of the signal at each point in the system of Fig. 4.30.

In many ways, this system is similar to the system used for FM stereo multiplex broadcasting. We will say more about this in Chapter 5.

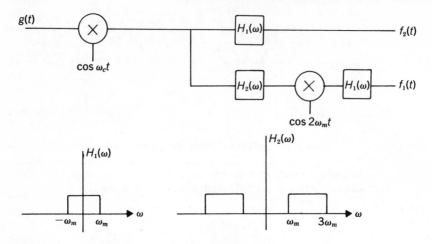

Figure 4.30   Receiver for Illustrative Example 4.5.

Even though the waveshape and gated demodulation schemes appear to be acceptable, significant problems are encountered in practice. In general, we have essentially said "just multiply $f_m(t)$ by cos $\omega_c t$ and low pass filter in order to recover $f(t)$ from the modulated waveform." We now state that this type of demodulator is given the name, *synchronous demodulator* (or detector). The word "synchronous" is used since the receiver must possess an oscillator which is at exactly the same frequency and phase as the carrier oscillator which is located in the transmitter.

In actual operation, the receiver and transmitter are usually separated by many miles, and this matching becomes difficult if not impossible.

Let us first see what happens if either the phase or the frequency of the oscillator in the receiver differs by a small amount from that of the transmitter.

4.6 PROBLEMS IN SYNCHRONOUS DEMODULATION

In synchronous demodulation, $f_m(t)$, the received modulated waveform is multiplied by a sinusoid of the same frequency and phase as the carrier sinusoid. We now consider the case of multiplying $f_m(t)$ by a cosine waveform which deviates in frequency by $\Delta\omega$, and in phase by $\Delta\theta$ from the desired values of frequency and phase. Performing the necessary trigonometric operations, we have

$$f_m(t) \cos\left[(\omega_c + \Delta\omega)t + \Delta\theta\right]$$
$$= f(t) \cos \omega_c t \cos\left[(\omega_c + \Delta\omega)t + \Delta\theta\right]$$
$$= f(t)\left[\frac{\cos(\Delta\omega t + \Delta\theta)}{2} + \frac{\cos\left[(2\omega_c + \Delta\omega)t + \Delta\theta\right]}{2}\right] \quad (4.27)$$

Since the expression in Eq. (4.27) forms the input to the low pass filter of the synchronous demodulator, the output of this filter will be that given in Eq. (4.28). This is true since the second term of Eq. (4.27) has a Fourier Transform which is centered about a frequency of $2\omega_c$, and is therefore rejected by the low pass filter.

$$\text{Filter output} = f(t) \frac{\cos (\Delta\omega t + \Delta\theta)}{2} \tag{4.28}$$

If $\Delta\omega$ and $\Delta\theta$ are both equal to zero, Eq. (4.28) simply becomes $f(t)$, and the demodulator gives the desired output. Assuming that these deviations cannot be made equal to zero, we are stuck with the output given by Eq. (4.28).

To see the physical manifestantions of this, let us first assume that $\Delta\theta = 0$. The output then becomes

$$f(t) \frac{\cos (\Delta\omega t)}{2} \tag{4.29}$$

$\Delta\omega$ is usually (hopefully) small, and the result will be a slowly varying amplitude (beating) of $f(t)$. If, for example, $f(t)$ were an audio signal, this effect would be highly annoying since the volume would periodically vary from zero to a maximum and back to zero again. One could simulate this in a standard radio by continually turning the volume control clockwise and then counterclockwise once every $2\pi/\Delta\omega$ seconds.

As a more specific example, if $\omega_c = 10^6$ rad/sec, one would certainly not consider an error in adjusting the frequency of the demodulator oscillator as being unreasonably large if it were only 1 rad/sec. If anything, the opposite is true. That is, 1 rad/sec is an optimistically small deviation. This would occur if the receiver oscillator were set to either 999,999 or 1,000,001 rad/sec instead of the desired 1,000,000 rad/sec of the transmitter oscillator. The result would be heard as a continuous variation in volume from maximum to zero *once every 2π seconds*. This is clearly an intolerable situation. Even if one could adjust the demodulation oscillator until no beating is heard, it is not unusual for an oscillator to drift in frequency over an interval of time. The example cited above would represent a drift of only 0.0001 % in frequency!

The above analysis is enough to doom synchronous demodulation to highly limited use. However, for completeness, we shall examine the effects of $\Delta\theta$. This can best be done separately from the effects of $\Delta\omega$. We therefore assume that the frequency is miraculously adjusted perfectly, and $\Delta\omega = 0$. The output of the demodulator is then given by

$$f(t) \frac{\cos (\Delta\theta)}{2} \tag{4.30}$$

This is not as annoying as the previous problem since $\cos (\Delta\theta)$ is a constant. We can usually compensate for this factor by varing the volume control (amplification) of the receiver. However, a deviation in phase can end up being just as deadly as a deviation in frequency. If the phase factor gets too small (i.e., $\Delta\theta$ approaches 90°), we may not be able to add enough amplification, and might therefore lose the signal. A more serious problem will turn out to be that of the signal getting so small that it gets lost in the background noise, but this is getting ahead of the game.

Is there any way to get around these problems? Wouldn't it be nice if the so-called local carrier (i.e., the cosine wave required at the receiver to demodulate) could somehow be derived from, or generated by, the incoming signal? If this were possible, we would be assured of having the frequency and phase properly adjusted.

Let's be intuitive for a minute. The Fourier Transform of the arriving modulated waveform is redrawn in Fig. 4.31.

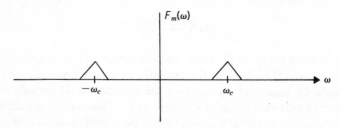

Figure 4.31 Transform of AM wave.

Stare at this transform for a while, and ask yourself whether it possesses enough information to uniquely specify ω_c.

Looking at the portion lying around $\omega = \omega_c$, we see that it is symmetrical about $\omega = \omega_c$. This is true since the original $F(\omega)$ was an even function (i.e., symmetrical about $\omega = 0$). Therefore, in terms of $F_m(\omega)$, one can find ω_c as the point at which $F_m(\omega)$ can be folded upon itself. That is, the symmetry mentioned above serves to point out ω_c as its midpoint. We therefore claim that, intuitively, the incoming waveform possesses sufficient information to allow unique determination of the exact frequency, ω_c. The construction of a system to "detect" this symmetry point is not at all simple.

Pay close attention now as we hand-wave our way through one possible system (Fig. 4.32).[1]

Given $f_m(t) = f(t) \cos \omega_c t$, we wish to somehow generate the term $\cos \omega_c t$. Let us first square $f_m(t)$,

$$f_m^2(t) = f^2(t) \cos^2 \omega_c t, \tag{4.31}$$

$$= \frac{f^2(t)}{2} + \frac{f^2(t) \cos 2\omega_c t}{2} \tag{4.32}$$

---

[1] The "hand-waving" description of this system is not critical to the continuity of this chapter, and the timid student may therefore skip the next few pages.

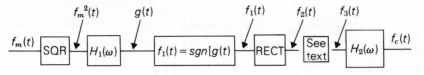

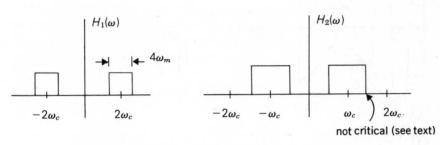

Figure 4.32 A system for generating $f_c(t)$ at the receiver, given $f_m(t)$.

If this waveform is passed through a band pass filter with center frequency $2\omega_c$ and bandwidth $4\omega_m$, the $\frac{1}{2}f^2(t)$ term will be rejected to yield $g(t)$.

$$g(t) = \frac{f^2(t)\cos 2\omega_c t}{2} \tag{4.33}$$

Figure 4.33 sketches this and other significant waveforms to follow for a representative example.

We note that, since $f^2(t)$ is always positive, $g(t)$ will be positive when $\cos 2\omega_c t$ is positive, and negative when $\cos 2\omega_c t$ is negative. We now define the "sign function" as follows,

$$sgn(t) = \begin{cases} 1 & t > 0 \\ -1 & t < 0 \end{cases} \tag{4.34}$$

From Fig. 4.33, it should now be clear that $sgn[g(t)]$ will be a square wave of frequency $2\omega_c$. The student can accept the fact that the operation of taking $sgn[g(t)]$ is relatively easy to perform. It is sometimes called "infinite clipping" since it is like clipping the waveform at a very low level and then amplifying the result.

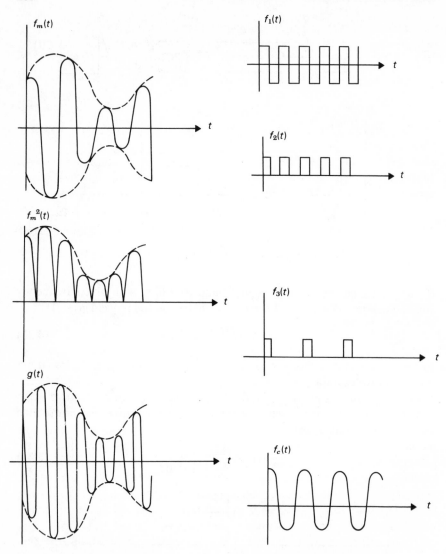

Figure 4.33 Typical waveforms for the system of Fig. 4.32.

We must now halve the fundamental frequency of this square wave. This can be done by using a device known as a bistable multivibrator (flip-flop). In case the reader is unfamiliar with this, we will describe a slightly modified approach. $f_1(t)$ is half-wave-rectified to yield the pulse train shown as $f_2(t)$. We then put $f_2(t)$ through a system which ignores every second pulse. The result is the pulse train $f_3(t)$. The Fourier Transform of this pulse train is a train of impulse functions at frequencies which are multiples of ω_c. A band pass filter which passes ω_c but rejects $2\omega_c$ will yield a sinusoid of exactly ω_c

frequency. This resulting signal can now be used in the synchronous detector with no worries about $\Delta\omega$ being anything other than identically zero.

4.7   AM TRANSMITTED CARRIER

The previous system, while quite simple in theory, is reasonably complex to build. If standard broadcast radio had to resort to it, the garden variety pocket or table radio would probably be much more expensive than it is today. The question arises as to whether or not there exists a simpler method of demodulating an amplitude modulated signal. The answer lies in an examination of the waveshapes present.

Figure 4.34 shows what might be a typical amplitude modulated waveform. Since the carrier frequency, ω_c, is usually much greater than ω_m, the

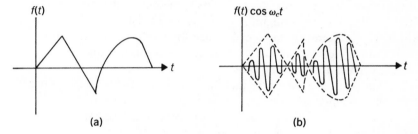

(a)   (b)

Figure 4.34   An AM waveform.

maximum frequency component of the information, one can conclude that the "outline" of the waveform in Fig. 4.34(b) varies much more slowly than the cosine carrier wave does (in practice, the carrier would be of a much higher frequency than that indicated in Fig. 4.34(b)). We would therefore be justified in sketching the modulated waveform as an outline with the interior regions just shaded in. That is, the cosine carrier varies up and down so much faster than does its amplitude, the individual cycles of the carrier are indistinguishable in an accurate sketch.

If, instead of representing a voltage waveform as a function of time, this curve represented the actual shape of a wire or stiff piece of cord, one can envisage a cam (any object) moving along the top surface. If the cam is attached by means of a shock absorber, or viscous damper device, it will approximately follow the upper outline of the curve (*see* Fig. 4.35). This is true since the shock absorber will not allow the cam to respond to the rapid carrier oscillations. This is much the same as the behavior of an automobile suspension system while the car is travelling on a bumpy road.

The higher the carrier frequency, the more smoothly the cam will describe the upper outline provided that it can respond fast enough to follow the shape of the outline itself. This upper outline, or boundary of the curve, is

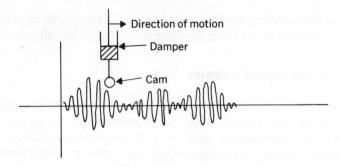

Figure 4.35 A mechanical outline follower.

called the *envelope* of the waveform. As long as ω_c is much greater than ω_m, our intuition allows us to define this new parameter. Actually, the definition of the envelope of any waveform can be made quite exact and rigorous (*see* Section 4.9) using very straightforward Fourier Transform analysis techniques which we have already developed. Since for modulation applications ω_c is always much greater than ω_m, we need not resort to this rigorous approach, but can depend upon our intuition.

At this point, we accept the fact that the electrical analog to the mechanical cam system does exist and is very simple to construct. The critical question is whether the envelope of the modulated signal represents the information signal that we wish to recover.

If should be quite clear from Fig. 4.34 that the envelope of $f(t) \cos \omega_c t$ is given by $|f(t)|$, the absolute value of $f(t)$. Taking the absolute value of a waveform represents a very severe form of distortion. If you don't believe this, consider the simple example of a pure sine wave. Examination of Fig. 4.36 shows that the absolute value of a pure sine wave is a periodic function

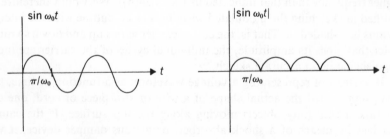

Figure 4.36 The absolute value of a sine wave.

whose period is one half of the period of the original sine wave. Fourier Series analysis tells us that it also contains components at multiples of $2\omega_c$ in frequency. If the original sinusoid represented a pure musical tone, the absolute value would sound like a "raspy" tone one octave higher than the original tone.

Suppose now that $f(t)$ happened to always be non-negative? If this were the case, the absolute value of $f(t)$ would be the same as $f(t)$, and the envelope of $f_m(t)$ would equal $f(t)$. Unfortunately, most information signals of interest have an average value of zero. They must therefore possess both positive and negative excursions.

All is not lost. One can always add a constant to $f(t)$, where the constant is chosen large enough so that the sum is always non-negative. Therefore, instead of using $f(t)$ as the information signal, we use

$$\hat{f}(t) = f(t) + A \qquad (4.35)$$

where A is chosen such that

$$[f(t) + A] \geq 0 \qquad \text{for all } t \qquad (4.36)$$

The modulated waveform is now of the form

$$f_m(t) = [A + f(t)] \cos \omega_c t \qquad (4.37)$$

This is sketched in Fig. 4.37 for the same example as shown in Fig. 4.34.

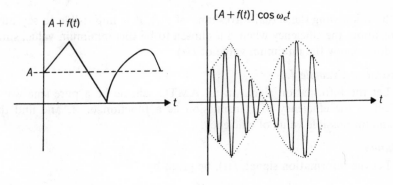

Figure 4.37 Example of Fig. 4.34 with a constant added to $f(t)$.

If we use a device which follows the envelope in order to demodulate $f_m(t)$, the result will be a signal of the form $A + f(t)$. Since the actual desired information signal was simply $f(t)$, we may wish to reject the "A" term. It is always quite simple to get rid of a constant term since it represents zero frequency. For example, a capacitor looks like an open circuit to DC (zero frequency), so a series capacitor will reject any DC term. Equation (4.37) can be rewritten as

$$f_m(t) = A \cos \omega_c t + f(t) \cos \omega_c t \qquad (4.38)$$

The addition of the constant to $f(t)$ can therefore be viewed as the addition of an unmodulated carrier term to $f_m(t)$. That is, since the generation of a sinusoid exactly like the carrier was difficult to perform at the receiver, we have essentially resorted to sending it along with the modulated signal. Since part of the transmitted signal is the carrier itself, this form of amplitude modulation is called *double sideband AM transmitted carrier*. Until we encounter an example of modulation which is not "double sideband," we will omit the words. Thus we will call this type of signal AM transmitted carrier, or AMTC. We can now digress and mention that the form of modulation that we were talking about until now is called *AM suppressed carrier*, or AMSC, since no carrier term was explicitly transmitted.

Since synchronous detection is so difficult to perform and we have prematurely accepted the fact that envelope demodulation is easy to perform, why not always transmit the carrier, that is, use AMTC instead of AMSC?

The answer lies in efficiency. The power of the signal, $f_m(t)$ increases, and is a minimum when $A = 0$ (i.e., for suppressed carrier transmission). The usable information power remains the same regardless of the value of A. One would therefore like to use the minimum value of A consistent with the envelope restriction. That is, choose A such that

$$\min_t [A + f(t)] = 0 \qquad (4.39)$$

Without knowing the actual waveshape of $f(t)$, it is impossible to say anything about the efficiency when A is chosen to be this minimum value, since we don't know the minimum value of $f(t)$.

Illustrative Example 4.6

Let the information signal in an AMTC scheme be a pure sine wave. Calculate the efficiency of transmission as a function of A, and find the maximum possible value of efficiency.

Solution

Let the information signal, $f(t)$, be given by

$$f(t) = K \cos \omega_m t$$

and the carrier frequency be ω_c rad/sec. The AMTC waveform is given by Eq. (4.38),

$$\begin{aligned} f_m(t) &= [A + K \cos \omega_m t] \cos \omega_c t \\ &= A \cos \omega_c t + \tfrac{1}{2}K[\cos (\omega_c + \omega_m)t + \cos (\omega_c - \omega_m)t] \end{aligned} \qquad (4.40)$$

The Fourier Transform of $f_m(t)$ is sketched in Fig. 4.38.

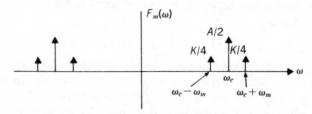

Figure 4.38 $F_m(\omega)$ for Illustrative Example 4.6.

Using the techniques of Chapter 2, the power of this waveform is given by

$$P_{\text{avg}} = \frac{A^2}{2} + \frac{1}{2}\left[\frac{K}{2}\right]^2 + \frac{1}{2}\left[\frac{K}{2}\right]^2$$
$$= \frac{A^2}{2} + \frac{K^2}{4} \qquad (4.41)$$

The usable signal power is contained in the term

$$f(t)\cos\omega_c t$$

and in this case, is equal to $K^2/4$. The system efficiency is therefore given by the ratio of usable power to total power. It is often denoted by the symbol the Greek letter "eta," (η).

$$\eta = \frac{\frac{1}{4}K^2}{\frac{1}{2}A^2 + \frac{1}{4}K^2} \qquad (4.42)$$

η increases as A decreases. We must now ask what the minimum value of A can be. Since the maximum negative excursion of $f(t)$ is $f(t) = -K$, the minimum value of A that will guarantee that $f(t) + A$ is always non-negative is K. For this choice of A, the efficiency is

$$\eta = \frac{\frac{1}{4}K^2}{\frac{1}{2}K^2 + \frac{1}{4}K^2} = \frac{1}{3} = 33\% \qquad (4.43)$$

Even though the maximum possible efficiency (for a pure sine wave information signal) is only 33 % as compared with 100 % for the AMSC case, AMTC is used in almost all standard broadcasting. Envelope demodulation (detection) must certainly be an excellent salesperson!

4.8 THE ENVELOPE DETECTOR

We shall now consider a general amplitude modulated transmitted carrier (AMTC) signal and see how one could actually construct an envelope detector (the electrical analog to the cam system).

Figure 4.39 shows the Fourier Transform of a general AMTC signal where the signal $f(t)$ was chosen to be the low frequency band limited signal we were treating earlier.

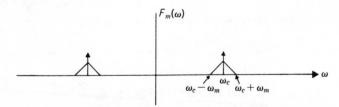

Figure 4.39 Transform of AMTC wave.

Figure 4.40 shows a block diagram of one possible detection system. The rectifier in Fig. 4.40 can either be full wave or half wave. We shall analyze

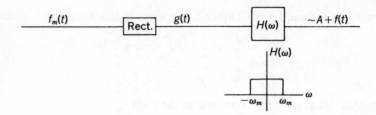

Figure 4.40 Detector for AMTC.

both cases simultaneously. For the full wave rectifier,

$$g(t) = \begin{cases} f_m(t) & f_m(t) > 0 \\ -f_m(t) & f_m(t) < 0 \end{cases} \tag{4.44}$$

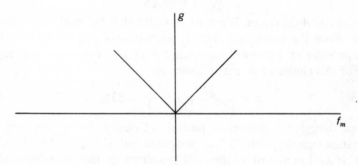

For the half wave rectifier,

$$g(t) = \begin{cases} f_m(t) & f_m(t) > 0 \\ 0 & f_m(t) < 0 \end{cases} \tag{4.45}$$

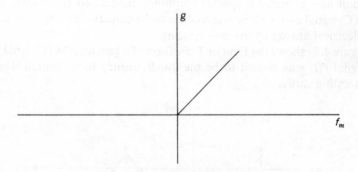

The output of the full wave rectifier is given by

$$\begin{aligned} |f_m(t)| &= |[A + f(t)] \cos \omega_c t| \\ &= |A + f(t)| |\cos \omega_c t| \\ &= [A + f(t)] |\cos \omega_c t| \end{aligned} \tag{4.46}$$

We have eliminated the absolute value sign around $A + f(t)$ since we assume that A is chosen so that this term is never negative. Thus its absolute value is equal to the function itself.

The only difference if a half wave rectifier were used would be that the second term in Eq. (4.46) would be the half wave rectified version of $\cos \omega_c t$ instead of the full wave version. Both the half wave and full wave rectified forms of $\cos \omega_c t$ are sketched in Fig. 4.41. Both functions shown in Fig. 4.41

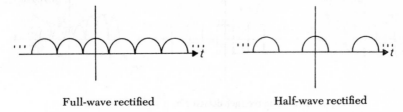

Full-wave rectified   Half-wave rectified

Figure 4.41   Rectified forms of $\cos \omega_c t$.

are periodic functions, and can therefore be expanded in Fourier Series expansions. The half wave version has a fundamental frequency of ω_c, and the full wave, $2\omega_c$ since the period has been cut in half. The output of the rectifier is therefore given by $g_1(t)$ for the full wave case, and by $g_2(t)$ for the half wave case.

$$g_1(t) = [A + f(t)] \sum_{n=0}^{\infty} a_n \cos 2n\omega_c t$$

$$g_2(t) = [A + f(t)] \sum_{n=0}^{\infty} a'_n \cos n\omega_c t$$

(4.47)

The prime notation is used for a_n in $g_2(t)$ to indicate that the Fourier Series coefficients are not the same as those of $g_1(t)$, If we write out the first few terms of each of these series, it should be clear that a low pass filter can be used to reject all but the desired term.

$$g_1(t) = a_0[A + f(t)] + a_1[A + f(t)] \cos 2\omega_c t$$
$$+ a_2[A + f(t)] \cos 4\omega_c t + \cdots$$

(4.48a)

$$g_2(t) = a'_0[A + f(t)] + a'_1[A + f(t)] \cos \omega_c t$$
$$+ a'_2[A + f(t)] \cos 2\omega_c t + \cdots$$

(4.48b)

If $g_1(t)$ is put through a low pass filter with cutoff frequency ω_m, the output is $a_0[A + f(t)]$. If $g_2(t)$ were the input to the same filter, the output would be $a'_0[A + f(t)]$. In both cases, the output of the low pass filter is in the form of a constant multiplying the desired signal, $A + f(t)$. We have therefore built an envelope detector using only a rectifier and low pass filter.

An alternate explanation of the operation of this detector is possible if one realizes that full wave rectification of a signal of the form $g(t) \cos \omega_c t$, where $g(t)$ is never negative, is equivalent to multiplication by a square wave of frequency ω_c. That is, multiply $g(t) \cos \omega_c t$ by $+1$ when the function is positive, and by -1 when it is negative. Figure 4.42 shows an example of this equivalency. This makes the detector analogous to the gated synchronous detector studied earlier. The same analysis applies to the case of the half wave rectifier. The only change is that the multiplying (gating) function is now in the form shown in Fig. 4.43.

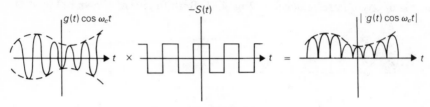

Figure 4.42   The rectifier detector as a gated demodulator.

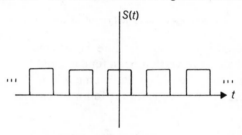

Figure 4.43   Gating function for half-wave rectifier detector.

We make an interesting observation at this point. While the above system has simulated what appears to be the envelope of $f_m(t)$, this is not exactly the case. It has actually yielded a_0 times the exact envelope (a_0' in the half wave rectifier case). While this subtle difference does not represent distortion of the signal, it does indicate that the exact mechanism of operation is basically different from that of the mechanical cam.

If you were asked to construct this system, you might make a lucky mistake, as the first person to attempt this problem might have done. You may, in effect, build an actual analog to the cam instead of the system of Fig. 4.40.

In Fig. 4.44 we illustrate a simple half-wave rectifier and a simple (non-ideal) low pass filter. One might just blindly put these two together and think

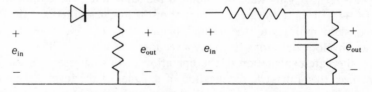

Figure 4.44   A half-wave rectifier and low pass filter.

that he had constructed the system which we just analyzed. Unfortunately, the idea of just putting two systems together to develop the composite properties is totally wrong. It doesn't even apply for simple linear circuits (can you prove that the rectifier is non-linear from the definition of linearity?). As a trivial example, consider the two circuits of Figs. 4.45(a) and 4.45(b). The

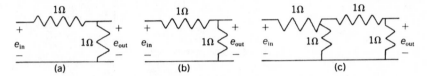

Figure 4.45   Incorrect cascading of systems.

composite is shown as Fig. 4.45(c). The first circuit essentially multiplies the input by $\frac{1}{2}$, as does the second circuit. The composite, instead of multiplying the input by $\frac{1}{4}$, possesses the relationship $e_{out}/e_{in} = \frac{1}{5}$. The concept of cascading systems without any regard to isolation is equally inapplicable to more complicated circuits.

In any case, a simple circuit constructed using the fallacious reasoning turned out to be the cheapest, simplest, and easiest to build envelope detector that could be dreamed possible. It can be built for about 15¢ retail, and is used in all standard household receivers. Unfortunately, it is virtually impossible to analyze in detail. Fortunately, a detailed analysis is totally unnecessary.

The device of which we are speaking is shown in Fig. 4.46.

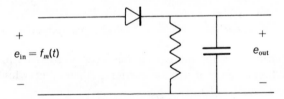

Figure 4.46   The envelope detector.

We claim that the circuit in Fig. 4.46 is an exact electrical analog of the cam system of Fig. 4.35. The capacitor represents the mass of the cam. The resistor is analogous to the viscous damping, or shock absorber. The diode takes into account the fact that the wire in Fig. 4.35 (bumpy road?) can only push up on the cam. It cannot pull the cam down.

Now that our intuition should be satisfied, we shall attempt a partial analysis of the circuit. As a first step in the approximate analysis of the envelope detector, consider the well known peak detector circuit shown in Fig. 4.47. If the input to this peak detector is $f_m(t)$, the output will be as shown. That is, $f_m(t)$ can never become greater than $e_{out}(t)$ since this would imply a

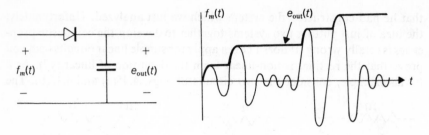

Figure 4.47   The peak detector.

forward voltage across the diode. Similarly, the capacitor voltage, $e_{out}(t)$, can never decrease since there is no path through which the capacitor can discharge. The output, $e_{out}(t)$, is therefore always equal to the maximum past value of the input, $f_m(t)$. Note that, in order to exaggerate the effect, the carrier frequency has been shown much lower than it would actually be in practice.

If a "discharging" resistor is now added (*see* Fig. 4.48) the output voltage will exponentially decay toward zero, until such time as the input tries to

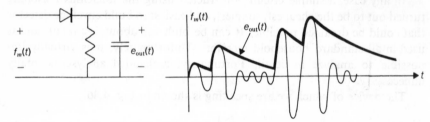

Figure 4.48   Addition of a discharging resistor.

exceed the output. Again, the carrier has been sketched with a very low frequency. We can see that, as the carrier frequency gets high, $e_{out}(t)$ will approach the envelope of $f_m(t)$.

Ideally, we would like the output, $e_{out}(t)$, to be able to decay as fast as the envelope of $f_m(t)$ changes. Any faster decay will cause $e_{out}(t)$ to start resembling the fast carrier oscillations.

Since the shortest time in which the envelope can go from its maximum to its minimum value is π/ω_m, the time constant of the RC circuit should be made to be the same order of magnitude as this value (usually about $\pi/5\omega_m$). This analysis is admittedly sloppy, but the large difference between ω_c and ω_m permits considerable leeway. In practice, it is better to choose the RC time constant to be a little too short rather than too long. A time constant which is too large would result in $e_{out}(t)$ completely missing some of the peaks of $f_m(t)$, as it does in Fig. 4.47. On the other hand, if the time constant is too small, the result is that $e_{out}(t)$ has a ripple around the actual envelope waveshape. This ripple represents a high-frequency wave since it occurs at the peaks of the carrier. It is therefore composed of frequencies around ω_c, the carrier frequency. An exact analysis would show that it is quite simple to remove this ripple with a low pass filter. In practice one doesn't even bother since the radio speaker or, indeed, the human ear could never respond to frequencies of the order of ω_c.

4.9 ENVELOPES AND PRE-ENVELOPES OF WAVEFORMS

The previous section has led us to an intuitive definition of the envelope of a restricted class of waveforms. Given $f(t) \cos \omega_c t$, we said that if the maximum frequency component of $f(t)$ is much less than ω_c then the envelope of $f(t) \cos \omega_c t$ is given by the absolute value of $f(t)$. We made this intuitive definition on the basis of a sketch of the waveform and an intuitive feel for what we wanted to mean by envelope.

This definition is not very satisfying to anyone who desires a firm mathematical development of the analysis. We therefore rigorously define what is meant by the envelope of any general waveform. Since this is being labelled as a definition, it requires no further justification. However, we shall show that the mathematical and intuitive definition are identical for that class of function for which the intuitive approach applies.

This technique of defining something in a rigorous way such that it agrees with intuition for special cases is very common. We will see it again in our approach to frequency modulation, where a rigorous definition of frequency will be required.

The *pre-envelope* (sometimes called "complex envelope" or "analytic function") of a waveform is defined as the complex time function whose Fourier Transform is given by

$$2U(\omega)F(\omega) \tag{4.49}$$

where $U(\omega)$ is the unit step function and $F(\omega)$ is the transform of the original waveform. That is, the transform of the pre-envelope is zero for negative ω and is equal to twice the Fourier Transform of the original function for positive ω. The symbol, $z(t)$, is commonly used to represent the pre-envelope function. It should be clear from the properties of the transform that $z(t)$ could not possibly be a real time function.

Illustrative Example 4.7

Find the pre-envelope of $f(t) = \cos \omega_0 t$.

Solution

The Fourier Transform of $f(t) = \cos \omega_0 t$ is given by

$$F(\omega) = \pi[\delta(\omega - \omega_0) + \delta(\omega + \omega_0)]$$

The transform of the pre-envelope is therefore given by

$$Z(\omega) = 2\pi\delta(\omega - \omega_0)$$

and the pre-envelope, denoted as $z(t)$, is given by $e^{j\omega_0 t}$, the inverse transform of $Z(\omega)$.

In elementary sinusoidal steady state analysis, a system input of $\cos \omega_0 t$ is rarely carried through equations. Instead, circuits are solved for an input of $e^{j\omega_0 t}$, and the real part of the resulting output is taken. The reasons for this approach are not relevant to our present discussion. We simply note that this technique amounts to the substitution of the pre-envelope of the input for the actual input.

We now wish to express $z(t)$ in the time domain in terms of a general $f(t)$. Starting with $Z(\omega) = 2F(\omega)U(\omega)$, we use the time convolution theorem to get

$$z(t) = 2f(t) * \mathcal{F}^{-1}[U(\omega)] \tag{4.50}$$

The inverse Fourier Transform of $U(\omega)$ is found much the same way as the transform of $U(t)$ was found (Eq. 1.93). It is perfectly acceptable to look this up in the table of transform pairs in Appendix II of this text.

$$U(\omega) \longleftrightarrow \frac{1}{2}\delta(t) + \frac{-1}{2\pi jt}$$

Finally,

$$z(t) = 2f(t) * \frac{1}{2}\delta(t) + 2f(t) * \frac{-1}{2\pi jt}$$

$$z(t) = f(t) + \frac{j}{\pi} \int_{-\infty}^{\infty} \frac{f(\tau)}{t - \tau} d\tau \tag{4.51}$$

We note that the real part of $z(t)$ is the original time function, $f(t)$, just as the real part of $e^{j\omega_0 t}$ is $\cos \omega_0 t$ in Illustrative Example 4.7.

The imaginary part of $z(t)$ is given by the convolution of $f(t)$ with $1/\pi t$. This convolution with $1/\pi t$ comes up quite often, and it is given the name "the *Hilbert Transform* of $f(t)$."

We now *define* the envelope of $f(t)$ as the magnitude of the pre-envelope, $z(t)$.

Illustrative Example 4.8

Find the envelope of $f(t) = \cos \omega_0 t$.

Solution

The pre-envelope of $f(t)$ was found in Illustrative Example 4.7 to be

$$z(t) = e^{j\omega_0 t} \tag{4.52}$$

The magnitude of this is equal to 1. Therefore, the envelope of $f(t) = \cos \omega_0 t$ is a constant, 1. We already knew this from our intuitive definition of envelope since this time function represents an unmodulated carrier wave.

Illustrative Example 4.9

Find the envelope of $f_m(t) = f(t) \cos \omega_c t$.

Solution

We first must find the pre-envelope of $f(t) \cos \omega_c t$.

$$z_m(t) = f_m(t) + \frac{j}{\pi}\left[f_m(t) * \frac{1}{t}\right] \qquad (4.53)$$

We shall perform the indicated convolution in the frequency domain by multiplying the two transforms together. The transform of $1/\pi t$ is $-j\,sgn(\omega)$. Therefore, the transform of the second term in Eq. (4.53) (i.e., the imaginary part of $z_m(t)$) is given by the product of $F_m(\omega)$ with $-j\,sgn(\omega)$. Recall that

$$F_m(\omega) = \frac{F(\omega - \omega_c) + F(\omega + \omega_c)}{2}$$

The product of this transform with $-j\,sgn(\omega)$ is given by

$$\frac{F(\omega - \omega_c) - F(\omega + \omega_c)}{2j} \qquad (\omega_m \leq \omega_c) \qquad (4.54)$$

as shown in Fig. 4.49. This is recognized as the transform of $j\,f(t) \sin \omega_c t$

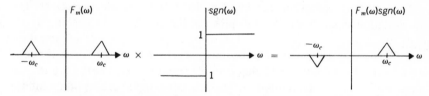

Figure 4.49 Construction of transform of the imaginary part of $z_m(t)$ for Illustrative Example 4.9.

(verify this statement!). The pre-envelope is therefore given by

$$z_m(t) = f(t) \cos \omega_c t + jf(t) \sin \omega_c t \qquad (4.55)$$

The envelope is defined as the magnitude of $z_m(t)$ and is therefore given by

$$\begin{aligned}
|z_m(t)| &= \sqrt{f^2(t) \cos^2 \omega_c t + f^2(t) \sin^2 \omega_c t} \\
&= \sqrt{f^2(t)(\cos^2 \omega_c t + \sin^2 \omega_c t)} \qquad (4.56) \\
&= \sqrt{f^2(t)} = |f(t)|
\end{aligned}$$

This agrees with the result accepted using our intuitive definition. We note that this rigorous definition only requires that $\omega_m \leq \omega_c$. Previously, we required that the maximum frequency component of $f(t)$ be much less than ω_c. The mathematical definition of envelope is less restrictive than the intuitive one.

We therefore have a rigorous definition of the envelope of any function of time as the magnitude of the pre-envelope of the time function.

It would be a worthwhile piece of work if one could analyze the envelope detector circuit in the light of the definition of envelope presented here. Since the envelope detector has not been accurately analyzed, this task would seem to be extremely difficult.

The intuitive definition of envelope which applies when the envelope is slowly varying will be sufficient for all work that we will encounter in amplitude modulation theory. The concept of pre-envelope will prove useful in detection theory (e.g., radar).

4.10 THE SUPERHETERODYNE AM RECEIVER

We are now in a position to fulfill the "dream" of most students when they first start studying communication theory. That is, we can understand the operation of the standard radio receiver. We probably could not repair it (a technician with a familiarity with electronics is needed for this) but we certainly can understand a block diagram of it. This section will analyze the superheterodyne AM receiver from a system's approach standpoint (block diagram). The entire section can be skipped on a first reading without loss of continuity.

Several basic operations can be identified in any broadcast receiver. The first is station separation. We must pick out the one desired signal and reject all of the other signals (stations). The second operation is that of amplification. The signal picked up by a radio antenna is far too weak to drive the cone of a loudspeaker without first being amplified many times. The third, and final operation is that of demodulation. The incoming signal is amplitude modulated, and contains frequencies centered around the carrier frequency. The carrier is usually near one million hertz in frequency (standard broadcast), and the signal received must therefore be demodulated before it can be fed into a speaker.

In an actual receiver, these three operations are sometimes combined, and the actual order in which they are performed is somewhat flexible.

In standard broadcast AM the highest frequency component of the information signal (ω_m) is set at 5,000 Hz. The adjacent carrier signals are separated by 10 kHz (the minimum separation allowed) so that the transform of the signal received by the antenna would look something like that shown in Fig. 4.50.

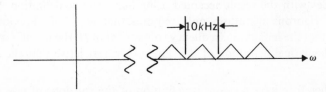

Figure 4.50 A transform of the signal at the input to an AM receiver.

We see that to receive one signal and reject all others requires a very accurate filter with a sharp frequency cutoff characteristic.

Assuming that the listener wanted the capability of choosing any station (a reasonable assumption), this filter would have to be tuneable. That is, the band of frequencies which it passes must be capable of variation.

We have seen how one can build filters which approximate ideal band pass filters to any desired degree of accuracy. However, when we require that the pass band of the filters be capable of moving all over the frequency axis, we have made the practical construction of the filter virtually impossible.

We are rescued from this dilemma by recalling that multiplication of the incoming signal by a sinusoid shifts all frequencies up and down by the frequency of the sinusoid. Because of this, station separation can be accomplished by building a fixed band pass filter and shifting the input frequencies so that the station of interest falls in the pass band of the filter. That is, we sort of construct a viewing "window" on the frequency axis, and shift the desired station so that it sits in the window. This shifting process (multiplication by a sinusoid) is sometimes called *heterodyning*. The receiver which we shall describe is called a "superheterodyne" receiver. Although we will only discuss the AM version of this receiver, the principle of superheterodyning is equally applicable to other forms of modulation (e.g., FM). We shall now analyze the total block diagram of such a receiver. (*See* Fig. 4.51.)

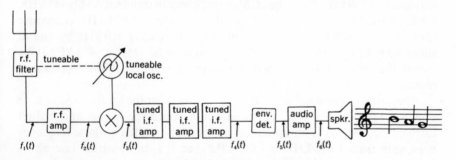

Figure 4.51   The superheterodyne receiver.

Starting at the antenna, a signal consisting of the sum of all broadcasted signals (stations) is received. We then amplify all stations in an r.f. (radio frequency) amplifier. The tuneable r.f. filter shown will be described later. The signal, $f_2(t)$ is then shifted up and down in frequency by multiplying it by a sinusoid generated by a "local" oscillator. The output of this heterodyner is applied to the almost ideal band pass filter (usually part of the i.f. amplifiers). In standard AM, this fixed filter has a bandwidth of 10 kHz, and is centered at 455 kHz, known as the intermediate frequency (i.f.) of the receiver. In most receivers, the filter is made up of three tuned circuits which are often aligned so as to generate the Butterworth filter poles. These three stages of filtering are usually combined with three stages of i.f. amplification. Thus, at $f_4(t)$, we have a modulated signal whose carrier frequency has been shifted to 455 kHz, and which has already been amplified and sifted out from all of the other signals.

The receiver then simply envelope detects $f_4(t)$ and then amplifies once more before applying the signal to a loudspeaker. More expensive receivers may add a bass (low pass filter) or treble (high pass filter) control. Inexpensive radios may eliminate one or more stages of amplification from the above block diagram.

It remains to justify the presence of the tuneable r.f. filter. One problem was not mentioned in the above analysis. The heterodyning process shifts the incoming frequencies both up and down in frequency. Therefore, while the "up-shift" might place the desired signal in the viewing window (pass band) of the i.f. filter, the downshift might place an undesired signal right on top of that deposited during the upshift. This undesired signal is called an "image" signal. Its elimination is not very difficult to perform as we shall presently see.

In order to continue, we will need to know how one chooses the frequency of the local oscillator. In standard broadcast AM, we may wish to listen to any station with carrier frequency between 550 kHz and 1,600 kHz. This is the band of frequencies which the FCC (Federal Communications Commission) has allocated to AM transmission.

For example, suppose that we wished to listen to a station with carrier frequency of 550 kHz. Since the i.f. filter frequency in standard AM is 455 kHz, the local oscillator must be tuned to either 95 kHz or 1,005 kHz. (Convince yourself that multiplication of a sinusoid of frequency 550 kHz by one of either 95 kHz or 1,005 kHz results in a sinusoidal term at 455 kHz.) We choose the higher of the two oscillator frequencies for the following practical reason.

Reception of the standard AM band requires that the local oscillator tune over a range of 1,050 kHz, the difference between the lowest and highest carrier frequency. It is far easier to construct an oscillator whose frequency must vary from 1,005 kHz to 2,055 kHz than it is to construct one whose frequency varies from 95 kHz to 1,145 kHz. This is due to the fact that the bandwidth of the latter oscillator is very large compared to the frequencies involved.

Returning now to the image problem, if the incoming signal is multiplied by a sinusoid of 1,005 kHz, the signal with carrier frequency 550 kHz is shifted into the pass band of the of the i.f. filter (455 kHz). The station with carrier frequency 1,460 kHz is also shifted into the pass band of the filter! This is the image station.

It turns out to be relatively simple to reject this unwanted signal. Had we first placed the incoming r.f. signal through a very sloppy band pass filter that passed the signal around 550 kHz but rejected that around 1,460 kHz, we would have eliminated the image signal.

A general analysis would show that, regardless of what station we wish to tune to, the image station would be separated from the desired station by 910 kHz (twice the i.f. frequency). Therefore, a band pass filter with bandwidth of 1,820 kHz would accomplish image rejection (*see* Fig. 4.52). Since the actual form of this filter is not critical (i.e., we don't really care what it does to signals lying between the desired and image station in frequency) it is not difficult to construct even though it must be tuneable.

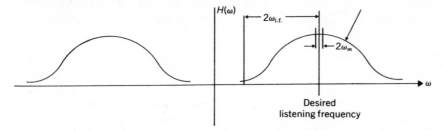

Figure 4.52   The rejection filter need not be ideal.

In practice, when one turns the tuning dial on his receiver, he is tuning this sloppy r.f. rejection filter and also changing the frequency of the local oscillator simultaneously.

4.11   SINGLE SIDEBAND

In the standard AM systems which we have been describing, the range of frequencies required to transmit a signal, $f(t)$, is that band between $\omega_c - \omega_m$ and $\omega_c + \omega_m$. This represents a bandwidth of $2\omega_m$ rad/sec. It was shown that, if we frequency multiplex two signals, their carrier frequencies must be separated by at least $2\omega_m$ rad/sec so that the signals do not have overlapping frequency components. This was necessary to allow separation of the various signals at the receiver.

In addition to smog and atomic fallout, our air is becoming polluted with electromagnetic signals. Considerations relating to antenna design tend to limit the range of usable frequencies for transmission. Therefore, the number of channels which can be frequency multiplexed is limited due to practical considerations.

Certain ranges (bands) of frequencies prove better than others for transmitting over long distances, as in the case of transoceanic transmission. This is due to ionospheric skip conditions. These frequencies are obviously at a premium.

For the above reasons, the Federal Communications Commission regulates the frequencies allocated to each particular class of user. Wouldn't it be nice if each channel required a smaller portion of the frequency band than $2\omega_m$? We could then stack adjacent channels closer together.

Single sideband transmission is a technique which allows adjacent carriers to be closer together than that separation permitted in double sideband transmission (the type of AM which we have been discussing until now).

As in Fig. 4.53 we define that portion of $F_m(\omega)$ which lies in the band of frequencies above the carrier as the "upper sideband." Similarly, that portion which lies in the band of frequencies below the carrier is called the "lower sideband." A double sideband AM suppressed carrier wave is therefore made up of a lower and an upper sideband. A double sideband AM transmitted carrier wave is composed of upper and lower sidebands plus a carrier term.

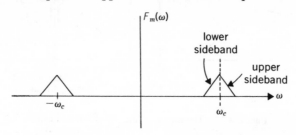

Figure 4.53   Definition of upper and lower sideband.

If the information signal, $f(t)$, is a real function of time, the magnitude of $F(\omega)$ must be even in ω. It should therefore be clear that $F(\omega - \omega_c)$ is symmetrical about ω_c. That is, the upper sideband is a mirror image of the lower sideband. Since the upper sideband can be exactly derived from the lower sideband, and vice versa, all of the information about the modulated waveform is contained in either the lower or the upper sideband. Why should we transmit both sidebands?

In the past, we transmitted both sidebands since the corresponding modulators and demodulators were easy to construct. We automatically got the double sideband version with each modulation scheme we examined. If however, frequency bands are really at a premium in a particular application, the extra work we will find necessary to send only one sideband may prove profitable.

Figure 4.54 shows the Fourier Transforms of the upper, and lower sideband versions of $f_m(t)$, denoted by $f_{usb}(t)$ and $f_{lsb}(t)$, respectively.

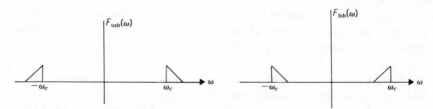

Figure 4.54   Upper and lower sideband transforms.

Since the double sideband suppressed carrier waveform, $f_m(t)$, is composed of upper and lower sidebands, it should be clear that

$$f_m(t) = f_{usb}(t) + f_{lsb}(t) \qquad (4.57)$$

This can be seen from Fig. 4.54 since the sum of the two transforms shown is the suppressed carrier version of $F_m(\omega)$.

SSB Modulators and Demodulators

Since the upper and lower sidebands are separated in frequency, our old standby of using filters to reject signals can be used as a method of generating single sideband signals. We simply form the double sideband suppressed carrier waveform as discussed previously and pass it through a band pass filter. The filter should only transmit the sideband of interest. Figures 4.55 and 4.56 show such systems to generate the lower and the upper sideband

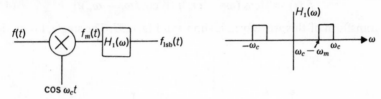

Figure 4.55   Single sideband generator for lower sideband.

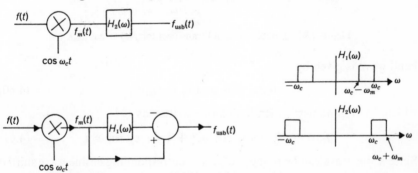

Figure 4.56   Single sideband generators for upper sideband.

signals, respectively. In Fig. 4.56, we either use a band pass filter which passes the upper sideband, or we can take advantage of Eq. (4.57) to write

$$f_{usb}(t) = f_m(t) - f_{lsb}(t) \qquad (4.58)$$

Both of these methods are illustrated in Fig. 4.56.

The systems shown in Figs. 4.55 and 4.56 do not present very satisfying ways in which to generate single sideband signals since they require filters whose band pass characteristics are perfect. For example, in the lower sideband transmitter, just below a frequency of ω_c the filter must transmit unattenuated while just above ω_c it must completely reject or attenuate the signals. This is fine for an ideal filter, but practical filters could never accomplish this task. It would therefore be advantageous to find another method of generating single sideband signals which would not require any perfect devices. We present a system which shifts the problem from that of designing a very good band pass filter to that of designing a very good phase shifter. For certain applications, this is a better approach.

We note that if the information signal, $f(t)$, is a pure sinusoid, $\cos \omega_m t$, the single sideband waveform is easy to derive explicitly. The double sideband wave is given by

$$f_m(t) = f(t) \cos \omega_c t = \cos \omega_m t \cos \omega_c t$$

which, using trigonometric identities, becomes

$$f_m(t) = \tfrac{1}{2}[\cos (\omega_c + \omega_m)t + \cos (\omega_c - \omega_m)t] \qquad (4.59)$$

The transform of this waveform is shown in Fig. 4.57. The single, lower side-

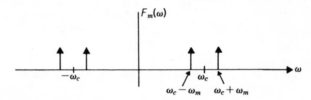

Figure 4.57 Double sideband transform for $f(t) = \cos \omega_m t$.

band wave is given by

$$f_{lsb}(t) = \tfrac{1}{2} \cos (\omega_c - \omega_m)t \qquad (4.60)$$

which, by trigonometric identity, can be written as

$$f_{lsb}(t) = \tfrac{1}{2} \cos \omega_c t \cos \omega_m t + \tfrac{1}{2} \sin \omega_c t \sin \omega_m t \qquad (4.61)$$

Since a sine wave can be thought of as the corresponding cosine wave shifted by $-90°$ in phase, the $f_{lsb}(t)$ given in Eq. (4.61) can be generated by the system shown in Fig. 4.58.

Before examining this system in more detail, we would like to derive an equivalent system which will generate a single sideband signal from any general $f(t)$. That is, must any modifications be made to the system shown in Fig. 4.58 so that it generates a single sideband wave when the input is not a pure sinusoid? The derivation follows.[2]

---

[2]One can wave his hands and say that any $f(t)$ can be expressed as a sum of sinusoids, and then use a linearity argument. We will not do this as this text intends to promote some familiarity and comfort with Fourier Transform analysis techniques.

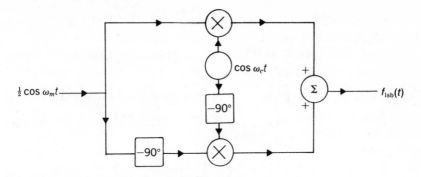

Figure 4.58   lsb generator for $f(t) = \cos \omega_m t$.

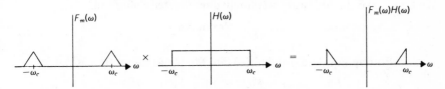

Figure 4.59 $F_{\text{lsb}}(\omega)$ derived from $F_m(\omega)$.

From Fig. 4.59 we see that the Fourier Transform of $f_{\text{lsb}}(t)$ is given by $F_m(\omega)H(\omega)$, where

$$H(\omega) = U(\omega + \omega_c) - U(\omega - \omega_c) \qquad (4.62)$$

Writing $F_m(\omega)$ as

$$F_m(\omega) = \tfrac{1}{2}[F(\omega - \omega_c) + F(\omega + \omega_c)]$$

we recognize that (*see* Fig. 4.60)

$$U(\omega + \omega_c)F(\omega - \omega_c) = F(\omega - \omega_c) \qquad (4.63a)$$

and

$$U(\omega - \omega_c)F(\omega + \omega_c) = 0 \qquad (4.63b)$$

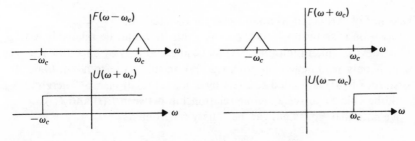

Figure 4.60   Products appearing in Eq. (4.63).

Using these, we find

$$F_{lsb}(\omega) = F_m(\omega)H(\omega),$$
$$= \tfrac{1}{2}F(\omega + \omega_c)U(\omega + \omega_c) - \tfrac{1}{2}F(\omega - \omega_c)U(\omega - \omega_c) \quad (4.64)$$
$$+ \tfrac{1}{2}F(\omega - \omega_c)$$

We note that the unit step function can be rewritten in the following form,

$$U(\omega + \omega_c) = \frac{1 + sgn(\omega + \omega_c)}{2} \quad (4.65a)$$

and

$$1 - U(\omega - \omega_c) = \frac{1 - sgn(\omega - \omega_c)}{2} \quad (4.65b)$$

We now make the following seemingly unmotivated definition,

$$F_h(\omega) \triangleq \frac{F(\omega)\,sgn(\omega)}{j} \quad (4.66)$$

Substituting Eqs. (4.65) and (4.66) into Eq. (4.64) we get the desired result,

$$F_{lsb}(\omega) = \frac{1}{2}\left[\frac{F(\omega + \omega_c) + F(\omega - \omega_c)}{2} + \frac{F_h(\omega - \omega_c) - F_h(\omega + \omega_c)}{2j}\right]$$
$$(4.67)$$

Both of the ratios of Eq. (4.67) look familiar. The first is the form of the Fourier Transform of an AM wave. The positive and negative frequency shift represents a multiplication of $f(t)$ by a cosine in the time domain. If we now let the inverse transform of $F_h(\omega)$ be called $f_h(t)$, whatever that might be, we can find the inverse transform of Eq. (4.67).

$$f_{lsb}(t) = \tfrac{1}{2}f(t)\cos\omega_c t + \tfrac{1}{2}f_h(t)\sin\omega_c t \quad (4.68)$$

The second identity used in Eq. (4.68) is easy to prove. We used it once before in the pre-envelope discussion. That is, if

$$f(t) \longleftrightarrow F(\omega)$$

then

$$f(t)\sin\omega_c t \longleftrightarrow \frac{1}{2j}[F(\omega - \omega_c) - F(\omega + \omega_c)] \quad (4.69)$$

The proof of this is left as an exercise for the student.

Equation (4.68) represents the final result. It would be relatively easy to build a device which multiplies $f(t)$ by cosine and $f_h(t)$ by sine and adds the result, if only we knew what $f_h(t)$ was. We recall that, from Eq. (4.66) the transform of $f_h(t)$ is related to $F(\omega)$ by a simple relationship. Therefore, we should be able to derive a simple relationship between $f_h(t)$ and $f(t)$.

We note that Eq. (4.66) can be simply rewritten as

$$F_h(\omega) = \frac{sgn(\omega)}{j}F(\omega) = G(\omega)F(\omega) \quad (4.70)$$

where we have defined

$$G(\omega) \triangleq \frac{sgn(\omega)}{j}$$

Some careful juggling of complex forms will convince one that (I again say "do it!")

$$G(\omega) = \frac{sgn(\omega)}{j} = -j\,sgn(\omega) = e^{-j\pi sgn(\omega)/2} \qquad (4.71)$$

$G(\omega)$ is a complex function whose magnitude and phase are sketched in Fig. 4.61.

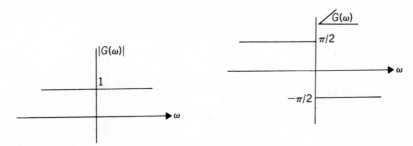

Figure 4.61 Magnitude and phase of $G(\omega)$ of Eq. (4.71).

$G(\omega)$ is the system function of a filter whose output consists of all positive input frequencies shifted by an angle of $-90°$. Therefore, $f_h(t)$ represents $f(t)$ shifted by $-90°$. This operation can be approximately performed over any finite range of frequencies. That is, the filter characteristics shown in Fig. 4.61 are not difficult to approximate. Since $\sin \omega_c t$ can be thought of as a $\cos \omega_c t$ shifted by $-90°$, Eq. (4.68) leads to the expected system block diagram shown in Fig. 4.62. Note that this is the same system we used when $f(t)$ was a pure sinusoid.

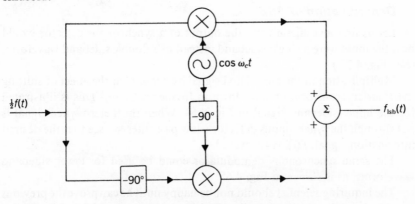

Figure 4.62 A lower sideband SSB generator.

If we wished to build a transmitter to generate the upper sideband instead of the lower sideband, we would simply subtract the lower sideband signal from $f_m(t)$. That is,

$$f_{usb}(t) = f_m(t) - f_{lsb}(t)$$
$$f_{usb}(t) = f(t) \cos \omega_c t - [\tfrac{1}{2}f(t) \cos \omega_c t + \tfrac{1}{2}f_h(t) \sin \omega_c t]$$
$$= \tfrac{1}{2}f(t) \cos \omega_c t - \tfrac{1}{2}f_h(t) \sin \omega_c t \qquad (4.72)$$

The block diagram of a modulator which forms the upper sideband version is shown in Fig. 4.63. We note that it differs from the lower sideband generator by a simple change of sign in the output summer.

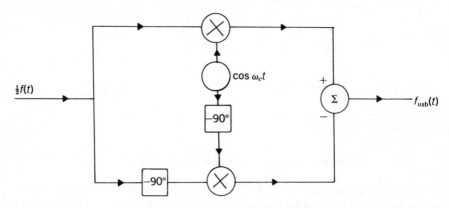

Figure 4.63 An upper sideband SSB generator.

We comment that if one were to sketch the time waveform of a single sideband signal, the sketch would reveal no obvious relationship to $f(t)$. That is, while for double sideband transmission, the envelope was a replica of $f(t)$, the envelope of a SSB signal does not resemble $f(t)$.

Demodulation of SSB

Let us first investigate what the output of a synchronous detector would be if the input were a single sideband instead of a double sideband waveform. (*See* Fig. 4.25.)

Multiplication of the modulated signal by $\cos \omega_c t$ has the effect of shifting the Fourier Transform up and down in frequency by ω_c. This is illustrated for the upper sidegand signal in Fig. 4.64. When this heterodyned signal is put through the synchronous detector's low pass filter, we see that the desired information signal, $f(t)$, is recovered.

The same synchronous demodulator could be used for lower sideband waveforms, as is shown in Fig. 4.65.

The inquiring scientist should not be happy until he can prove the previous graphical results mathematically. Consider the upper sideband waveform treated earlier. We found that its transform could be expressed as the stan-

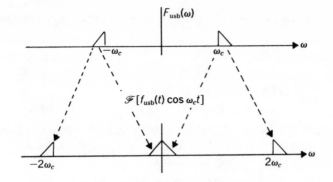

Figure 4.64   Heterodyning of SSB waveform.

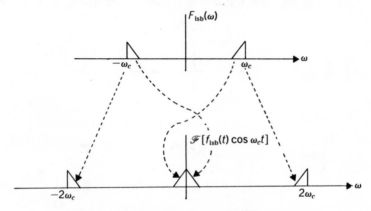

Figure 4.65   Heterodyning of SSB LSB waveform.

dard double sideband transform multiplied by the system function of an ideal band pass filter which passed frequencies between ω_c and $\omega_c + \omega_m$. (We repeat Fig. 4.56 as 4.66.) That is,

$$F_{usb}(\omega) = F_m(\omega)H(\omega) = \tfrac{1}{2}H(\omega)[F(\omega - \omega_c) + F(\omega + \omega_c)] \tag{4.73}$$

where

$$H(\omega) = \begin{cases} 1 & \omega_c < |\omega| < \omega_c + \omega_m \\ 0 & \text{otherwise} \end{cases}$$

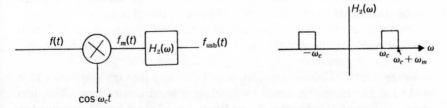

Figure 4.66   One type of SSB generator (repeat of Fig. 4.56).

If we now introduce $f_{usb}(t)$ into a synchronous demodulator, we first multiply it by cos $\omega_c t$.

$$f_{usb}(t)\cos\omega_c t \longleftrightarrow \tfrac{1}{4}\{[H(\omega + \omega_c)][F(\omega - \omega_c + \omega_c) + F(\omega + 2\omega_c)]$$
$$+ [H(\omega - \omega_c)][F(\omega - 2\omega_c) + F(\omega + \omega_c - \omega_c)]\}$$
$$= \tfrac{1}{4}F(\omega)[H(\omega + \omega_c) + H(\omega - \omega_c)] \qquad (4.74)$$
$$+ \tfrac{1}{4}F(\omega + 2\omega_c)H(\omega + \omega_c) + \tfrac{1}{4}F(\omega - 2\omega_c)H(\omega - \omega_c)$$

If we now pass this product, $f_{usb}(t)\cos\omega_c t$, through a low pass filter, the terms centered in frequency about $2\omega_c$ and about $-2\omega_c$ will be rejected, and the output will have the following transform.

$$\tfrac{1}{4}F(\omega)[H(\omega + \omega_c) + H(\omega - \omega_c)] \qquad (4.75)$$

We note from Fig. 4.67 that $[H(\omega + \omega_c) + H(\omega - \omega_c)] = 1$, for ω between

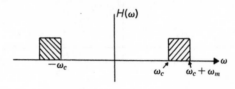

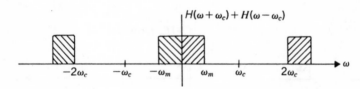

Figure 4.67 $H(\omega + \omega_c) + H(\omega - \omega_c)$ for Eq. (4.75).

$-\omega_m$ and $+\omega_m$. That is, it resembles the transfer function of an ideal low pass filter. Therefore,

$$\tfrac{1}{4}F(\omega)[H(\omega + \omega_c) + H(\omega - \omega_c)] = \tfrac{1}{4}F(\omega) \longleftrightarrow \tfrac{1}{4}f(t) \qquad (4.76)$$

Figure 4.67 illustrates this "low pass filter" characteristic of the sum, $H(\omega + \omega_c) + H(\omega - \omega_c)$. Note that $F(\omega) = 0$ for $|\omega| > \omega_m$. Thus the sections shown in Fig. 4.67 at $2\omega_c$ and at $-2\omega_c$ do not affect the final result since they are multiplied by $F(\omega) = 0$.

The mathematical proof that a synchronous detector can also be used for the lower sideband case is similar to that given above for the upper sideband. The details are left to the exercises at the end of this chapter.

We therefore see that a synchrononus detector will work as well with SSB as it does with double sideband signals. Unfortunately, this is not very well as you may recall.

In the double sideband case, we talked our way through a system which would use the incoming signal to generate a local carrier term. This was desirable since any slight error in the frequency of the sinusoid generated in the receiver proved disastrous. This is also true in the SSB case, although the physical manifestations of this error are not the same as in the double side-band case.

While the distortion in the double sideband case exhibited itself as a beating of the amplitude of the demodulated wave, in the single sideband case it results in total garbling of the message. Anybody who is an amateur radio operator can tell you how difficult it is to "tune in" a SSB station, and of the funny-sounding doubletalk that comes out whenever the local oscillator drifts in frequency.

If we view the transform of a SSB wave, it seems intuitively reasonable that sufficient information does *not* exist to derive the carrier from the SSB signal. The symmetry (redundance) that was present in the AM double sideband wave no longer exists. One might be tempted to say (for the upper sideband case) that ω_c can be found just by looking for the lowest frequency component present in the incoming modulated waveform. The fallacy in this reasoning arises in one of two ways. First, other stations may be occupying the frequency slot just below the signal of interest. There is no way of distinguishing the modulated portion due to the high information frequencies of one channel from that due to the low frequencies of an adjoining channel. (*See* Fig. 4.68(a).) Second, we cannot be sure that the information signal

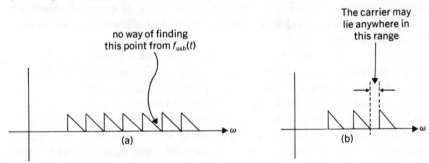

Figure 4.68   Carrier frequency cannot be derived from the incoming SSB waveform.

has frequencies that go right down to zero. It is more likely that the signal cuts off at some non-zero low frequency. We would not want to falsely label this lowest frequency as the carrier (*see* Fig. 4.68(b)). It would therefore appear that we must simply "fish" for the proper carrier frequency and readjust it whenever the oscillator drifts slightly in frequency.

Suppose now, just as in the double sideband case, one decided to transmit a carrier term in the hopes that envelope detection might be used. In the lower sideband case, this would result in the following.

$$f_{lsb}(t) + A \cos \omega_c t = [A + \tfrac{1}{2}f(t)] \cos \omega_c t + \tfrac{1}{2}f_h(t) \sin \omega_c t \qquad (4.77)$$

Performing a trigonometric identity, we find

$$f_{lsb}(t) + A \cos \omega_c t = B \cos (\omega_c t + \sigma) \qquad (4.78)$$

where

$$B = \sqrt{[A + \tfrac{1}{2}f(t)]^2 + [\tfrac{1}{2}f_h(t)]^2}$$

and

$$\sigma = \tan^{-1}\left[\frac{\tfrac{1}{2}f_h(t)}{A + \tfrac{1}{2}f(t)}\right]$$

The output of an envelope detector with this as an input is the term "B." In general, $f_h(t)$ looks completely different from $f(t)$ since every frequency component is shifted by the same phase shift. Recall that distortionless transmission corresponded to a phase shift which was linear with frequency. It would therefore appear that the envelope detector output would not resemble the desired output, $f(t)$. This is indeed the case unless "A" is made very large. In that case, $A + \tfrac{1}{2}f(t)$ is much larger than $\tfrac{1}{2}f_h(t)$, and the output of the envelope detector can be approximated by

$$B = \sqrt{[A + \tfrac{1}{2}f(t)]^2 + [\tfrac{1}{2}f_h(t)]^2} \approx A + \tfrac{1}{2}f(t) \qquad (4.79)$$

from which $f(t)$ is easily extracted. A receiver to accomplish this demodulation would differ only slightly from the standard (DSB) AM receiver. The catch is that the system is horribly inefficient. Large amounts of power are wasted in the sending of this large carrier term, and efficiencies of well under 5% are to be expected.

If power is a very minor consideration, and bandwidth and simple receiver design are major ones, the SSB system with large transmitter carrier and envelope demodulation would be used (television does something like this, as we shall see in the next section). If both bandwidth and power efficiency are serious considerations, but constant and sensitive monitoring of the local oscillator can be tolerated, SSB with synchronous demodulation may be used (we assume that the listener can tell when the signal sounds "right." This could rarely be done in data communications). This system is used in standard Ham radio and for overseas voice communications. For example, overseas news reports on radio probably come via this type of communications. Even though the stations use expensive crystal frequency control, you can occasionally hear the voice garbled due to a slight difference in transmitter and receiver oscillator frequencies.

For all of the above reasons, SSB does not find much application in common broadcast communication systems.

4.12 VESTIGIAL SIDEBAND TRANSMISSION

The only advantage of SSB over DSB is the economy of frequency usage. That is, SSB uses half the corresponding bandwidth required for DSB transmission of the same information signal. The primary disadvantage of SSB is the difficulty in building a transmitter or an effective receiver. If we could eliminate one of these disadvantages, SSB would become more attractive than it now appears.

Vestigial sideband possesses approximately the same frequency bandwidth advantage of SSB without the disadvantage of difficulty in building a modulator.

As the name implies, vestigial sideband (VSB) is a sloppy form of SSB where a vestige, or trace, of the second sideband remains.

We said earlier that one way of generating SSB is by perfect band pass filtering of a double sideband modulated waveform. Suppose that we band pass filtered the DSB wave, but used a non-ideal band pass filter to do this. The result might resemble that shown in Fig. 4.69.

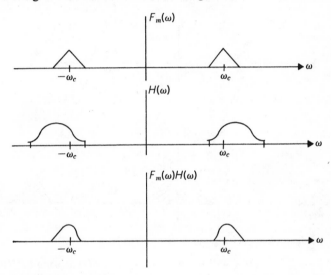

Figure 4.69 Vestigial sideband generated by non-ideal band pass filtering of double sideband.

From the figure, we can see that the band of frequencies occupied by this VSB wave is not much larger than that occupied by the corresponding SSB waveform. If we can show that the demodulation of this wave is essentially the same as that for SSB, we have accomplished our objective. That is, instead of the complicated modulator required for SSB, we can use a modulator that is only slightly more complex than the standard DSB modulator.

The mathematical argument parallels that used to show that synchronous demodulation could be used for single sideband detection. We call $F_v(\omega)$ the Fourier Transform of the VSB wave, $v(t)$. $F_m(\omega)$ is the transform of the double sideband wave, and $H(\omega)$ is the system function of the band pass filter. We therefore have (*see* Fig. 4.69)

$$F_v(\omega) = F_m(\omega)H(\omega) = \tfrac{1}{2}[F(\omega + \omega_c) + F(\omega - \omega_c)]H(\omega) \qquad (4.80)$$

We would now like to find those conditions under which $v(t)$, the VSB waveform, can be demodulated by a synchronous detector. If $v(t)$ is the input to a synchronous demodulator, the output will have the transform $Y(\omega)$, where

$$Y(\omega) = \begin{cases} \frac{1}{2}[F_v(\omega + \omega_c) + F_v(\omega - \omega_c)] & |\omega| < \omega_m \\ 0 & |\omega| > \omega_m \end{cases} \qquad (4.81)$$

In the above, we have taken the synchronous detector's multiplication by a sinusoid and low pass filtering into account. Substituting Eq. (4.80) for $F_v(\omega)$, we see

$$Y(\omega) = \tfrac{1}{4}[F(\omega)][H(\omega + \omega_c) + H(\omega - \omega_c)] \qquad (4.82)$$

The multiplying factor, $[H(\omega + \omega_c) + H(\omega - \omega_c)]$ is sketched in Fig. 4.70 for a typical band pass filter transfer function. We see that $y(t)$, the output of

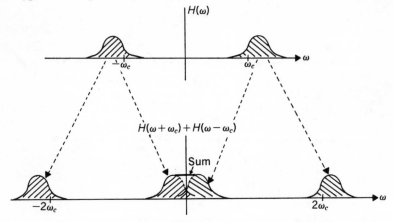

Figure 4.70 $H(\omega + \omega_c) + H(\omega - \omega_c)$ for a band pass filter.

the synchronous detector, will be proportional to $f(t)$ as long as $H(\omega + \omega_c) + H(\omega - \omega_c)$ is equal to a constant in the range of frequencies occupied by $f(t)$. In this case, the combination will resemble an ideal low pass filter. In intuitive terms, the tail of the filter characteristic must be asymmetric about $\omega = \omega_c$. That is, the outer half of the tail must fold over and fill in any difference between the inner half values of the tail and the values for an ideal filter characteristic curve. This is not a very stringent restriction, and a sloppy common practical filter approximately accomplishes this.

With the addition of a strong carrier term, we have a transmitted carrier VSB waveform,

$$v(t) + A \cos \omega_c t \qquad (4.83)$$

If "A" is large, the VSB signal can be approximately demodulated using an envelope detector, just as in the SSB case. The proof of this will not be presented since it requires that we assume some specific form for $H(\omega)$.

In addition to simplified transmitter design, VSB possesses some additional attractive features when compared to SSB. First, since VSB is somewhere in between DSB and SSB, the size of a carrier term (A in Eq. (4.83)) required to permit envelope demodulation is smaller than that required by SSB. Therefore, although efficiency is not as great as that achieved in DSB, it is nevertheless higher than the SSB value. The second attractive feature of VSB arises out of a practical consideration. The filter required to separate one station from another (i.f. filter in the superheterodyne receiver case) need not be as perfect as that required for SSB station separation. The Fourier Transform of the VSB signal does not cut off sharply at the frequency limits, and therefore a less ideal filter can be used without appreciable distortion.

4.13 TELEVISION

Public television had its beginnings in England in 1927. In the United States, it started three years later, in 1930. These early forms used *mechanical* scanning of the picture to be transmitted. That is, a picture was changed into an electrical signal by scanning the entire image along a spiral starting at the center. This scanning was accomplished by means of a rapidly rotating wheel with holes cut in it. As the wheel rotated, light from various parts of the total picture passed through the holes.

During this early period, broadcasts did not follow any regular schedule. Such regular scheduling did not begin until 1939, during the opening of the New York World's Fair.

The concepts of television and picture transmission spread into many exciting areas; facsimile transmission, pictures from outer space, video telephone, and cable TV represent a few examples. Many predict that a cable TV revolution will soon take place; two-way communication links will be common, and the TV will replace the newspaper, supermarket, baby-sitter, theater, and perhaps even the university campus. The theory about to be presented, although geared toward broadcast TV, is applicable to most forms of picture transmission, including video games and video computer terminals.

A Picture Is Worth a Thousand Words?

Nonsense! A picture is equivalent to far more than a thousand words. In fact, as a highly oversimplified example, we could take a picture of 1,001 words of text, thereby having a picture worth more than 1,000 words. But how much information does a picture contain? This is a serious technical question that must be answered before we can talk about specific ways of implementing transmission from one point to another.

While the information content of a message can be rigorously analyzed using concepts from the science of information theory (*see* Chapter 8), we will make a simple adaptation for our present applications. We divide a picture into squares where each square is a certain shade (this will later be extended to color TV). The number of squares in any given area determines the *resolution*. For example, Fig. 4.71(a) shows a picture of the letter "*A*" where 81

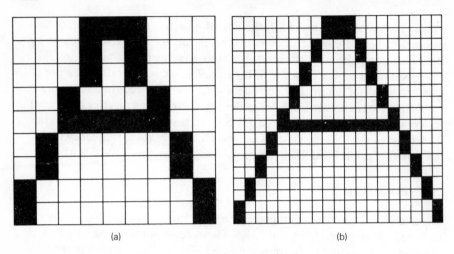

(a)   (b)

Figure 4.71   The letter "*A*" with two different resolutions.

squares have been used to define the picture. In Fig. 4.71(b), the same letter is shown where the number of squares has increased to 324. The result is a "clearer," more highly resolved picture of the letter "A."

The number of squares needed depends upon the type of picture being considered. The less detail in a picture, the fewer squares are needed to adequately describe it.

The number of squares used in U.S. television is prescribed by law. The Federal Communications Commission (FCC) sets the number of boxes (known as *picture elements*) of U.S. TV at 211,000. This is divided into 426 elements in each horizontal line and 495 visible horizontal lines in each picture. All that television transmitters must do is to step through each of these 211,000 picture elements and send an intensity value for each one. The receiver interprets these transmissions and reconstructs the picture from the 211,000 intensity levels.

Now to the details. A TV receiver is not much different from a laboratory oscilloscope. A beam of electrons is shot toward a screen and bent by use of deflection plates. When a negative charge is placed on a plate, the electron beam is repelled. In the oscilloscope, we apply a sawtooth waveform to the horizontal deflection plates to sweep the beam from the left to the right edge of the screen (and then, more rapidly, back to the left). This traces a line. In TV receivers, we add a second dimension to this sweep. While the beam is rapidly sweeping from left to right on the screen, it is less rapidly sweeping from top to bottom on the screen. The net result is a series of (almost) horizontal lines on the screen, as sketched in Fig. 4.72. This is known as the TV *raster*.

The screen has 495 of these horizontal lines in order to comply with the FCC regulation. The beam then returns to the top of the screen, taking the equivalent time of an additional 30 lines to do so. We can therefore think of

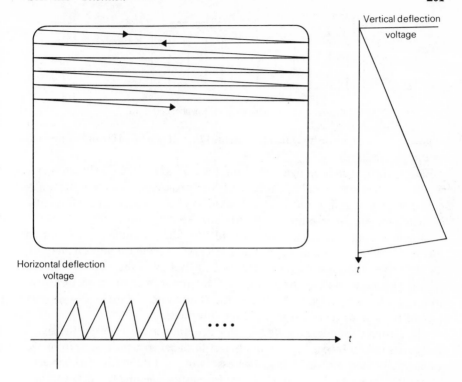

Figure 4.72   Generation of the raster.

the picture as having 525 lines, 30 of which occur during the *vertical retrace* time.

No times have yet been assigned to this process. The human eye requires a certain picture rate (number of times the entire screen is traced each second) to avoid seeing flicker, as in early motion pictures. This number is somewhere near 40 per second. U.S. TV decided to use 60 frames per second. This matches the frequency of household current and is chosen that way to minimize the effects of the video equivalent of 60-Hz hum. At 525 lines/frame and 60 frames/sec, we take the product to get 31,500 lines/sec. The reciprocal of this gives the time per line as 31.75 μsec/line. Of this 31.75 μsec, 5.1 μsec is used for horizontal retrace from right to left. This leaves 26.65 μsec for the visible part of each horizontal line. Dividing this by the 426 elements in a line finally yields the time per element of 0.0625 μsec/element, or 16 million elements/sec. The system would therefore have to be capable of transmitting 16 million different shades (black, white, gray, etc.) per second. In the worst case, one may wish to display a perfect checkerboard design of alternating black and white squares. The system would then jump from the darkest to lightest and back again 16 million times a second.

If we now think of this light intensity information as a signal, we see from Fig. 4.73 that it has a fundamental frequency of 8 MHz. Thus, no matter

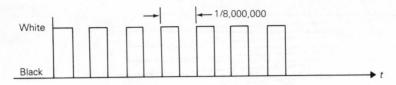

Figure 4.73 Intensity signal for checkerboard.

what scheme we choose by which to transmit this, at least 8 MHz of bandwidth would be required.

Here, the law steps in again, and the FCC says that the maximum bandwidth that the video signal can have is 4.1 MHz. Alas, how can these seemingly contradictory specifications (60 frames/sec to avoid flicker, 211,000 picture elements for proper resolutions, 4.1-MHz maximum bandwidth) be met?

Engineers had observed many decades of the development of motion pictures in which a similar predicament occurs. In standard motion pictures, 24 different pictures are shown each second. But 24 flashes/sec on a screen would appear to flicker considerably. Contemporary motion picture projectors flash each image twice. Thus, the frame rate is 48/sec, though only half of the frames represent new picture information.

Television's founders decided to play a similar trick. They cut the signal frequency in half by cutting the number of lines per second in half, from 525 to 262.5 lines/sec. However, it was necessary to still fill the entire screen each $\frac{1}{60}$ sec to prevent flicker. The technique for cutting the line frequency in half without changing the frame frequency is known as *interlaced scanning*. In the first $\frac{1}{60}$ sec, all of the odd-numbered lines are traced (ending with a $\frac{1}{2}$ line). The beam then returns to the top center of the screen to trace the even-numbered lines in the next $\frac{1}{60}$ sec. Thus, while it now takes $\frac{1}{30}$ sec to send all 211,000 picture elements, during this $\frac{1}{30}$ sec the entire screen is covered twice. The eye fills in the missing rows and detects no flicker. There is, of course, some loss of resolution on very fast-moving objects, but television was never intended to follow such high speeds anyway.

Signal Design and Transmission

We now translate this information into an electrical signal. If we plot light intensity as a function of time, a staircase function results. Figure 4.74 shows an example of the letter "T" in dark black followed by the punctuation "period" in light gray. The associated signal is shown where, for simplicity, interlaced scanning has not been shown and the number of lines has been drastically reduced.

For broadcast TV, the information (video) signal would be a similar staircase (smoothed to reduce bandwidth) function where the minimum step width is 0.125 μsec. In the actual video signal, the voltage corresponds to the light intensity since, in the receiver, this voltage is used to control the electron gun. The higher the voltage applied to a grid placed between the gun and the

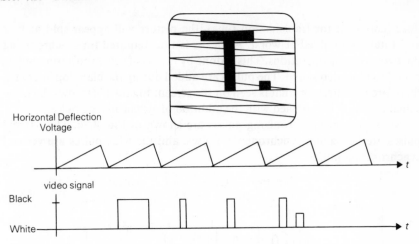

Figure 4.74 Video signal for message "*T*."

screen, the fewer electrons hit the screen and the darker the spot on the screen. As an additional modification, the staircase function is smoothed. The eye cannot tell the difference between a smooth or rapid transition in the signal during 0.125 μsec. After all, this corresponds to only $\frac{1}{426}$ of the width of the TV screen.

There is one additional aspect to this electrical video signal known as *blanking*. While the beam on the cathode ray tube is retracing (either horizontally or vertically), it is desirable that the beam of electrons be "turned off" so that this retrace is not seen as a line on the screen.

Taking all of this into account, the video signal corresponding to the picture shown in Fig. 4.74 is redrawn as Fig. 4.75.

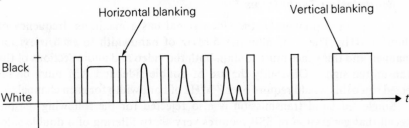

Figure 4.75 Figure 4.74 modified to include smoothing and blanking.

Synchronization

The transmitter at the TV studio is tracing rapidly from left to right and from top to bottom many times a second. It is sending a record of light intensity as a function of time. The receiver must be sure it is placing the transmitted intensity in the same spot on the screen as intended. If the beam in the receiver doesn't start a scan at the same instant that the transmitter does (corrected for transmission time), the picture will appear split at best and totally scrambled at worst. A means is thus required for synchronizing the two sweeping operations. This is done by means of synchronization pulses added to the video signal. The pulses are added during the blanking intervals, therefore not affecting what is seen on the screen. Figure 4.76 shows the same signal as Fig. 4.75 modified with the addition of synchronizing pulses.

Two types of synchronizing pulses are shown in Fig. 4.76. The narrow pulses are horizontal synchronizing pulses, and the wide pulses are vertical synchronizing pulses.

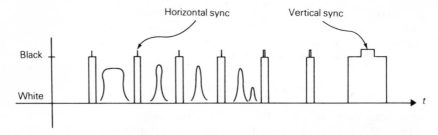

Figure 4.76   Video signal including sync pulses.

The receiver separates these pulses from the remaining signal by means of a threshold circuit. The horizontal and vertical pulses are then separated by means of a simple RC integrator circuit. The integral of the wider vertical sync pulses is higher than that of the narrower horizontal pulses. The separated pulses are then used to synchronize (trigger) the horizontal and vertical oscillators.

Modulation Techniques

As discussed previously, the video signal has a maximum frequency of about 4 MHz. The FCC allocates 6 MHz of bandwidth to each television channel, and this space must contain both the video and audio sections of the transmitted signal. Obviously the use of double sideband AM must be rejected since this would require over 8 MHz of bandwidth for each channel.

Single sideband transmission is also rejected for the following reason. Recall that generation of SSB requires very sharp filtering of a double sideband signal to remove one of the sidebands. As a filter is designed to closely approach the desired amplitude characteristic, it becomes increasingly difficult to control the phase characteristic, which, ideally, should be linear with frequency. That is, in designing practical filters one can approach either the phase or amplitude characteristics as closely as desired, but to achieve both simultaneously is extremely difficult. In audio applications, phase deviations from the ideal characteristic are not very serious. They represent varying delays of the frequency components of the message, and the human ear is not sensitive to such variations. In a video signal, these varying delays would be manifested as position shifts on the screen. These are commonly referred to as *ghost images* and are highly undesirable. Thus, SSB is not used for television transmission.

The video portion of the TV signal is sent using vestigial sideband (VSB). A strong carrier is added to enable the use of the envelope detector for demodulation. The entire upper sideband and a portion of the lower sideband are sent. Figure 4.77 shows the frequency composition of a TV signal. Note

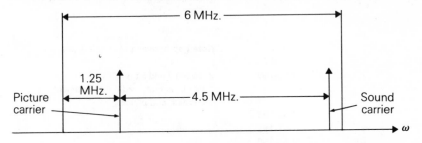

Figure 4.77   Frequency composition of TV signal.

that the audio and video are frequency multiplexed, and their carriers are separated by 4.5 MHz. The audio is sent via wideband FM, which is described in the following chapter.

Functional Block Diagram of TV Receiver

We are now in a position to examine the overall block diagram of a black and white (monochrome) TV receiver. A representative block diagram is shown in Fig. 4.78.

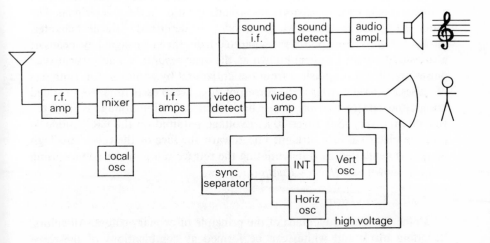

Figure 4.78   TV receiver block diagram.

The composite (video plus audio) signal is received by the antenna and is amplified by an r.f. amplifier. It then enters the tuner which is a mixer and i.f. filter just as in the case of the superheterodyne radio receiver. The frequency band allocated to commercial television is shown in the following table:

Channel	Frequency Range (MHz)	Comments
2	54–60	
3	60–66	
4	66–72	
5	76–82	Note gap between channels 4 and 5
6	82–88	
7	174–180	Between 7 and 6 there is the FM radio band, aircraft radio, government, railroad, and police
8	180–186	
9	186–192	
10	192–198	
11	198–204	
12	204–210	
13	210–216	
14–83	470–890	

Most large cities with many TV stations use channels 2, 4, 5, 7, 9, 11, and 13. Examination of the table of frequency allocations shows that no two of these channels are adjacent to each other in frequency. This fact slightly eases the design requirements on the i.f. filters.

The i.f. frequency used for TV is 40 MHz. After i.f. amplification and filtering, the signal enters the video detector, which is simply an envelope detector.

Since the sound carrier is 4.5 MHz above the picture carrier frequency, a filter (trap) is used to separate the sound signal from the video signal. The sound is FM and is demodulated in a way to be described in the next chapter.

The synchronizing pulses are separated from the video signal by means of a threshold circuit (clipper) known as the *sync separator*. The vertical sync pulses are then distinguished from the horizontal by an integrator. Both sets of pulses are used to trigger the corresponding sweep oscillators. As an added bonus, the retrace portion of the horizontal deflection voltage is used to generate the very high (over 20 kV) voltage required on the CRT anode to pull the electrons in a straight line toward the face of the tube. This high voltage is generated by differentiating the retrace ramp of the voltage using what is known as a *flyback transformer*.

Color Television

Color TV takes advantage of the principle of primary colors. All colors, including black and white, can be formed as combinations of the three primary colors red, blue, and yellow. TV modifies this slightly by using red, blue, and green (a combination of blue and yellow) since phosphors that glow with these colors when excited with an electron beam are commonly available.

The color TV receiver is essentially three separate TVs, one generating red, one generating blue, and one generating green. By mixing these in varying strengths, any color can be formed.

Traditional (this is rapidly changing) color CRTs actually contain three separate electron guns, one for each color. The transmitted signal must therefore generate three separate video signals to control each of these guns.

But, you ask, where can these two additional signals be squeezed? We are already using all of the bandwidth allowed for TV. We need a system to transmit all three video signals without increasing the 6-MHz bandwidth *and*, at the same time, allowing for compatible transmission—a black and white TV receiver should be capable of receiving and reproducing a color signal (in black and white, of course).

Again, the engineers designing TV systems were ingenious. In fact, they used the dual of interlaced scanning.

If we were to examine the Fourier Transform of a black and white video signal, we would find it resembles a train of impulses. An actual signal for one trace of a picture contains 262.5 horizontal traces. Since most pictures contain some form of vertical continuity (that is, one horizontal line is very close to the next horizontal line in content), the video signal is almost periodic with fundamental frequency equal to the line frequency (15,750 lines/sec).

If the video signal were exactly periodic, its Fourier Transform would be a train of impulses at multiples of the fundamental frequency. Since the signal is almost periodic, the Fourier Transform consists of pulses (not impulses) centered around multiples of the line frequency.

Some caution is required. The above argument applies to most real life pictures. The argument fails if TV is pushed to its limits, and a high-resolution detailed picture is sent.

Since the Fourier Transform is essentially zero between multiples of the line frequency, additional information can be placed in these spaces by use of a form of frequency multiplexing or frequency interlacing. The additional information needed for color transmission modulates a carrier which is midway between two multiples of the line frequency. This is assured by using a carrier whose frequency is an odd multiple of half of the line frequency. The figure used is 3.579545 MHz, which is the 455 th harmonic of half of the line frequency.

To make the signal compatible, the three signals are not the three colors. The three parameters sent are brightness, *hue* (position in the color spectrum), and *saturation* (how close is the color to a pure single frequency?). Instead of the word brightness, we use *luminance*. The required bandwidths of these three parameters are 4, 0.4, and 1.3 MHz, respectively. The color subcarrier is amplitude-modulated with the saturation signal and phase-modulated (see the next chapter) with the hue signal. The receiver demodulates this subcarrier using synchronous demodulation. The frequencies are matched at the transmitter and receiver by sending a segment of the exact sinusoidal subcarrier (the *color burst*) during each blanking interval.

4.14 NOISE IN AM SYSTEMS

In the process of communication of a signal $f(t)$ via AM techniques, noise arises in various ways. The information signal $f(t)$ is corrupted by some noise before it even reaches the modulator in the transmitter. This noise occurs because of the presence of electronic devices (e.g., audio amplifiers) and also because of electromagnetic radiation and pickup in the transmitter wires, which act as small antennas. Thus, the signal at the output of the modulator would be of the form

$$\{f(t) + n_1(t)\} \cos \omega_c t$$

where $n_1(t)$ is the noise. There is additional noise due to the fact that the carrier sinusoid is not a pure cosine wave. In fact, it contains harmonic distortion.

The modulated signal experiences multiplicative noise in the process of being transmitted from transmitter to receiver. This type of noise is due to both turbulence in the air and reflections of the signal. The turbulence causes the characteristics of the transmission medium, $H(\omega)$, to change constantly with time. The reflections, when recombined with the main path signal, either enhance the signal strength or subtract from it. Taking the multiplicative noise into account, the signal at the receiving antenna would be of the form

$$A\{1 + n_2(t)\} f_m(t)$$

where $n_2(t)$ is the multiplicative factor of noise.

The AM signal also experiences additive noise during transmission. This noise is generated by a multitude of sources, as mentioned in Chapter 3. For example, passing automobiles, static electricity, lightning, power transmission lines, and sunspots contribute to the overall noise effect. If one could listen to this noise, it would sound like static heard on radio transmission with occasional crackling sounds added. If the transmitted signal is $f_m(t)$, the received signal would be of the form

$$A f_m(t) + n_3(t)$$

where $n_3(t)$ is the additive noise.

An additional source of noise occurs in the receiver. Electronic devices and components are present, thus giving rise to thermal and shot noise. In addition, the wires in the receiver act as small antennas, thereby picking up some transmission noise. For purposes of analysis, this receiver noise can be treated as additive noise and included in $n_3(t)$.

Of the various types of noise introduced, the additive transmission noise is the most annoying type of noise. It normally contains the most power of the types discussed above. This is not to imply that other types of noise are not critical. Multiplicative transmission noise (turbulence) can become a significant factor at certain high frequencies. For example, if visible light frequencies are used for transmission, rain and clouds have a substantial impact upon the ability to receive a signal. To a certain extent, UHF television signals can experience multiplicative noise, which is more detrimental to reception than is additive transmission noise.

Double Sideband Suppressed Carrier

Suppose that a noise-free suppressed carrier AM waveform $f(t) \cos \omega_c t$ is transmitted. In the process of travelling through the air from transmitter to receiver, noise is added to the signal. Thus the waveform received by the antenna can be assumed to be composed of the sum of the desired AM waveform and undesired noise. We shall assume that this additive noise is white and that its power density spectrum is $N_0/2$ for all values of ω.

Assume that a synchronous demodulator is employed in the receiver and that the demodulator is preceded by a band pass filter which filters the desired frequency range from the remainder of energy entering the antenna. The receiver is as shown in Fig. 4.79.

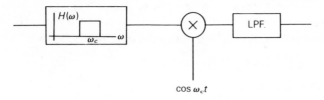

$$\cos \omega_c t$$

Figure 4.79   DSBSC AM receiver.

The signal portion of the output of this system would be $\frac{1}{2} f(t)$. The noise portion can be found by expanding the noise at the input to the multiplier according to Eq. (3.79). That is, since the noise input to the multiplier is filtered white noise, it can be expanded into quadrature components. The noise into the final low pass filter would then be

$$\{x(t) \cos \omega_c t - y(t) \sin \omega_c t\} \cos \omega_c t$$
$$= \tfrac{1}{2}x(t) + \tfrac{1}{2}x(t) \cos 2\omega_c t - \tfrac{1}{2}y(t) \sin 2\omega_c t \tag{4.84}$$

The only term in Eq. (4.84) which is not rejected by the low pass filter is $\frac{1}{2}x(t)$. The other two terms have frequencies centered about $2\omega_c$.

The spectral density of $\frac{1}{2}x(t)$ is given by $N_0/4$ for ω between $-\omega_m$ and $+\omega_m$ (*see* Illustrative Example 3.13).

The average noise power at the output of the demodulator would then be given by

$$\frac{1}{\pi} \int_0^\infty S(\omega) \, d\omega = \frac{1}{\pi} \int_0^{\omega_m} \frac{N_0}{4} \, d\omega = \frac{N_0 \omega_m}{4\pi} \qquad (4.85)$$

Since the signal part of the output is $\frac{1}{2} f(t)$, its power is $\frac{1}{4} P_f$, where P_f is the average power of $f(t)$, the information signal. The ratio of signal to noise power, the so-called *signal to noise ratio*, is then given by

$$\text{S/N} = \frac{\pi}{\omega_m} \frac{P_f}{N_0} \qquad (4.86)$$

This result will be referred back to later for comparison.

Double Sideband Transmitted Carrier

Suppose that a carrier is added and that an envelope detector is used instead of a synchronous demodulator. The receiver would then be of the form shown in Fig. 4.80. We have again included a band pass filter at the

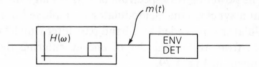

Figure 4.80 Receiver for DSBTC AM.

front end of the receiver. The input to the envelope detector is

$$m(t) = \{A + f(t)\} \cos \omega_c t + x(t) \cos \omega_c t - y(t) \sin \omega_c t \qquad (4.87)$$

This can be written as

$$m(t) = B(t) \{\cos \omega_c t + \theta(t)\} \qquad (4.88)$$

where

$$B(t) = \sqrt{\{A + f(t) + x(t)\}^2 + y^2(t)} \qquad (4.89)$$

The output of the envelope detector would then simply be $B(t)$.

Further analysis requires some assumptions. Let us first assume $A + f(t)$ is very large compared to $y(t)$. This can be restated as a high input (to the demodulator) S/N. In this case, $y^2(t)$ can be ignored in Eq. (4.89), and the output of the envelope detector is $A + f(t) + x(t)$. The signal portion of this output is $f(t)$, and the noise portion is $x(t)$. Both of these are exactly twice the corresponding results for the synchronous demodulator discussed earlier. The output S/N is therefore the same as that given in Eq. (4.86).

If $A + f(t)$ is not much larger than $y(t)$, the analysis is far more complex. It should be somewhat evident from Eq. (4.89) that cross products are present (i.e., signal multiplied by noise). The noise power is higher than in the high-

S/N case, and performance deteriorates. We shall not present a detailed analysis herein. The interested reader is referred to the references.

In summary, the performance of the envelope detector is worse than that of the synchronous demodulator. At high input S/N, the envelope detector performance approaches that of the synchronous demodulator.

Single Sideband

Suppose that single sideband AM were used instead of double sideband AM. The signal transmitted is then of the form (assuming upper sideband is used)

$$f_{usb}(t) = \tfrac{1}{2}f(t)\cos \omega_c t - \tfrac{1}{2}f_h(t)\sin \omega_c t \qquad (4.90)$$

Recall that $f_h(t)$ was a phase-shifted version of $f(t)$ used in the analysis of SSB.

The demodulator of Fig. 4.79 would require only one modification. The initial band pass filter would have a pass band from ω_c to $\omega_c + \omega_m$, with a bandwidth of ω_m instead of $2\omega_m$.

The output signal would be $\tfrac{1}{4}f(t)$, with power $P_f/16$. The noise into the final low pass filter would again be of the form

$$\tfrac{1}{2}x(t) + \tfrac{1}{2}x(t)\cos 2\omega_c t - \tfrac{1}{2}y(t)\sin 2\omega_c t$$

but the spectral density of $\tfrac{1}{2}x(t)$ would now be given by $N_0/8$ for ω between $-\omega_m$ and $+\omega_m$ (*see* Example 3.13). The average noise power at the output would be $N_0\omega_m/8\pi$.

The output signal to noise ratio is now given by

$$S/N = \frac{\pi}{\omega_m}\frac{P_f}{2N_0} \qquad (4.91)$$

Note that the signal to noise ratio for single sideband is one half as great as that for double sideband. One would therefore expect better performance from a double sideband set (i.e., less static).

An intuitive explanation of this result is often useful. In double sideband AM, either sideband contains all of the information in the signal. In the demodulation process, the two sidebands are added together. One might think that this simply doubles the signal and doubles the noise, thus yielding no advantage in signal to noise ratio. In fact, the noise in the lower sideband is not the same as the noise in the upper sideband. When the two noise signals are added together, some cancellation takes place, and the total noise power only doubles instead of being four times the power in either sideband. That is, we are adding two independent random processes. On the other hand, the resulting signal power is four times that in either sideband. Thus the S/N ratio doubles. This principle of signal redundancy is basic to many forms of noise reduction. In its most elementary form, a teacher will repeat a sentence if a loud noise prohibits the class from hearing the entire message. In more sophisticated systems (electronically, we mean), we find antenna arrays instead of a single antenna and frequency diversity (where a signal is sent on several carriers at once to reduce effects of selective interference) instead of a single carrier.

The above comparison of DSB and SSB illustrates the very important concept of redundancy as used to reduce the effects of noise. It is not, however, a fair comparison between SSB and DSB. One must be very careful to make the mathematical model approximate real life as closely as possible.

Suppose you had to choose between SSB and DSB in a real life system. You have no control over the additive noise that the signal experiences between transmitter and receiver. However, you do have control over the transmitted power. A fair way to compare SSB and DSB would seem to be to keep the transmitted power constant for both forms of AM. This was not the case in the above example. $f(t) \cos \omega_c t$ and $\{\frac{1}{2}f(t) \cos \omega_c t - \frac{1}{2}f_h(t) \sin \omega_c t\}$ do not contain the same power. Recall that the SSB equation was derived by chopping one sideband from the DSB signal. In fact, to match the powers, the SSB signal would have to be multiplied by $\sqrt{2}$. The student should not simply accept this. Try to verify the statement for a representative case of a sinusoidal $f(t)$ (i.e., $f(t) = a \cos \omega_m t$).

If the SSB signal is multiplied by $\sqrt{2}$, the signal power at the output of the demodulator becomes $P_f/8$ instead of $P_f/16$. The signal to noise ratio for SSB is therefore *exactly the same* as it was for the DSB case.

PROBLEMS

4.1. Give two reasons for using modulation rather than simply sending an information signal through the air.

4.2. What is the reason for using AM transmitted carrier instead of AM suppressed carrier? Why might suppressed carrier be preferred over transmitted carrier?

4.3. It is decided to amplitude-modulate a carrier of frequency 10^6 rad/sec with the signal shown below.

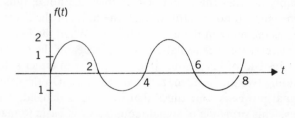

(a) If this modulation is performed as double sideband suppressed carrier (DSBSC), sketch the modulated waveform.

(b) The modulated wave of part (a) is the input to an envelope detector. Sketch the output of the envelope detector.

(c) If a carrier term is now added to the modulated wave to form DSBTC, what is the minimum amplitude of the carrier term such that an envelope detector could be used to recover $f(t)$ from the modulated signal?

(d) For the modulated signal of part (c), sketch the output of an envelope detector.

(e) Draw a block diagram of a synchronous detector that could be used to recover $f(t)$ from the modulated wave of part (a).

(f) Sketch the output of this synchronous detector if the input is the modulated waveform of part (c).

4.4. You are given the voltage signals, $f(t)$ and $\cos \omega_c t$, and you wish to produce the AM wave, $f(t) \cos \omega_c t$. Discuss two practical methods of generating this AM wave. Block diagrams would be useful.

4.5. The signal, $f(t) = (2 \sin t)/t$ is used to amplitude-modulate a carrier of frequency $\omega_c = 100$ rad/sec. The modulated signal is sent through the air. At the same time, a strong signal, $f_s(t) = (\sin 99.5t)/t$ is being fed into a nearby antenna (without modulation). This adds to the desired modulated signal and both are received by the receiver. The receiver contains a synchronous demodulator which is perfectly adjusted. What is the output of the synchronous demodulator?

4.6. Prove that the half-wave rectifier circuit is a non-linear system.

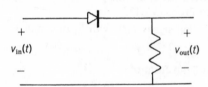

4.7. The waveform, $v_{in}(t)$, shown below is the input to an envelope detector. Sketch the output waveform.

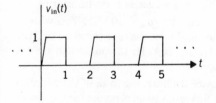
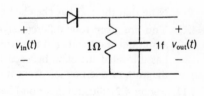

4.8. You are given the system shown below. $h(t)$ is the periodic function shown, and $F(\omega)$ is as sketched. $F(\omega)$, $G(\omega)$, $S(\omega)$, and $R(\omega)$ are the Fourier Transforms of $f(t)$, $g(t)$, $s(t)$, and $r(t)$ respectively. Assume $\omega_c \gg \omega_m$.

(a) *Sketch* $|G(\omega)|$.

(b) *Sketch* $|S(\omega)|$.

(c) *Sketch* $|R(\omega)|$.

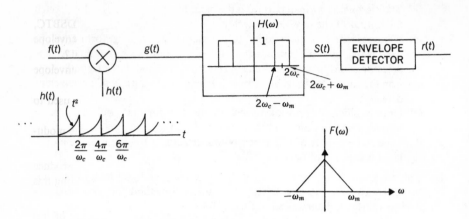

4.9. Given that the input to an envelope detector is

$$f(t) \cos \omega_c t$$

where $f(t)$ is always greater than zero.
(a) What is the output of the envelope detector?
(b) What is the input average power in terms of the average power of $f(t)$?
(c) What is the average power at the output?
(d) Describe any apparent discrepancies.
(Hint: If you have trouble, do this problem for the special case of $f(t) = A$, and refer back to Problem 2.12.)

4.10. Given an information signal, $f(t)$, with $F(\omega)$ complex,

$$F(\omega) = A(\omega)e^{j\theta(\omega)}$$

Find the Fourier Transform of

$$f(t) \cos \omega_c t$$

Also find the Fourier Transform of

$$f(t) \cos (\omega_c t + \tfrac{1}{4}\pi)$$

Note that the text has sketched only the magnitudes of these transforms.

4.11. You are given four different signals, each with $\omega_m = 10$ kHz. You wish to amplitude modulate and then frequency multiplex the four signals. The FCC has allocated the band between 100 kHz and 200 kHz to you. Sketch a block diagram of one possible transmitter scheme.

4.12. Figure 4.21 illustrated a non-linear device. Indicate the form of the output if the input is $x(t) = \cos \omega_c t$. Try to use this result to incorporate the non-linear device in a frequency multiplexing system (similar to that in Problem 4.11). The non-linear device should enable you to build the system using only one carrier oscillator.

4.13. An AMTC signal, $f_m(t)$,

$$f_m(t) = [A + f(t)] \cos (\omega_c t + \theta)$$

is applied to both systems shown below. The maximum frequency of $f(t)$ is ω_m, which is also the cutoff frequency of the low pass filters. Show that the two systems will yield the same output. Also comment upon whether the two low pass filters of system (b) can be replaced by a single filter following the square root operation or preceding the square root operation.

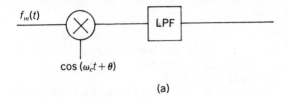

(a)

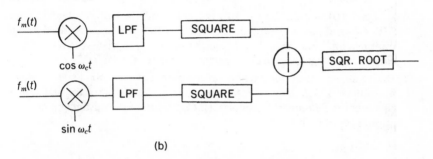

(b)

4.14. Prove that if $f(t)$ is even, then the Hilbert Transform of $f(t)$ is odd. Also prove the reverse. That is, if $f(t)$ is odd, the Hilbert Transform is even.

4.15. Starting with the transform of a lower sideband SSB wave,

$$F_{\text{lsb}}(\omega) = \tfrac{1}{2}H(\omega)[F(\omega - \omega_c) + F(\omega + \omega_c)]$$

where $H(\omega)$ is shown below, prove that a synchronous demodulator can be used to recover $f(t)$ from $f_{\text{lsb}}(t)$.

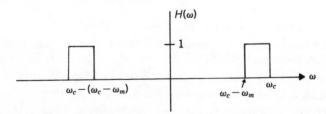

4.16. Discuss the trade-off decisions required if it were necessary to increase the vertical resolution of commercial TV by 10%.

4.17. Refer to the block diagram of Fig. 4.78. Pinpoint the problem block(s) which could result in the following symptoms:
   (a) Horizontal bright line in the middle of the TV screen. Remainder of screen is dark. Sound OK.
   (b) Picture is rolling vertically.
   (c) Vertical line in the center of the screen. Remainder of screen is dark.
   (d) No light on screen. Sound OK.
   (e) Light on screen filling entire area but no sound or picture.

4.18. An information signal $f(t) = 5 \cos 1000t$ is transmitted using DSBSC and demodulated using a synchronous demodulator. Noise with power spectral density $S_n(\omega) = 10^{-4}$ W/rad/sec is added to the signal during transmission. Find the S/N at the output of the receiver and express this in decibels.

4.19. Repeat Problem 4.18 for SSB transmission.

4.20. A signal $f(t) = 20 \cos 500t + 10 \cos 1000t$ is transmitted via SSB. Noise of power spectral density $S_n(\omega) = 10^{-3}$ W/rad/sec is added during transmission.
(a) Sketch a block diagram of the required receiver.
(b) Find the S/N at the output of the receiver.
(c) If a band pass filter is added to the output of the receiver with $H(\omega) = 1$ for $400 < |\omega| < 1,100$, find the signal to noise improvement of this filter.

4.21. A signal $f(t) = 4 \sin(200t + 10°)$ is transmitted via DSBTC, DSBSC, and SSB. Noise of power spectral density $S_n(\omega) = 10^{-2}$ W/rad/sec is added during transmission. Find the S/N at the output of the appropriate receiver for each case (assume that an envelope detector is used for the DSBTC case).

4.22. Derive an expression for the noise at the output of the second system shown in Problem 4.13. Compare this to the noise at the output of the first system (the usual synchronous demodulator). Comment on the advantages and disadvantages of each system.

REFERENCES

BECKMANN, P., *Probability in Communication Engineering*, Harcourt Brace Jovanovich, New York, 1967.

CARLSON, A. B., *Communication Systems*, McGraw-Hill, New York, 1975.

KENNEDY, R. S., *Fading Dispersive Communication Channels*, Wiley, New York, 1969.

LATHI, B. P., *Communication Systems*, Wiley, New York, 1968.

―――― *Random Signals and Communication Theory*, International Textbook Co., Scranton, Pennsylvania, 1968.

McMULLEN, C. W., *Communication Theory Principles*, Macmillan, New York, 1968.

SAKRISON, D. J., *Communication Theory*, Wiley, New York, 1968.

SCHWARTZ, M., *Information Transmission, Modulation and Noise*, McGraw-Hill, New York, 1970.

STREMLER, F. G., *Introduction to Communication Systems*, Addison-Wesley, Reading, Mass., 1977.

TAUB, H., and SCHILLING, D. L., *Principles of Communication Systems*, McGraw-Hill, New York, 1971.

ZIEMER, R. E., and TRANTER, W. H., *Principles of Communications*, Houghton Mifflin, Boston, 1976.

FREQUENCY
AND
PHASE MODULATION
5

As in the type of modulation previously discussed, we start with an unmodulated carrier wave,

$$f_c(t) = A \cos (\omega_c t + \theta) \tag{5.1}$$

If ω_c is varied in accordance with the information we wish to transmit, the carrier is said to be *frequency modulated*. However, when ω_c is varied with time, $f_c(t)$ is no longer a sinusoid, and the definition of frequency which we have used all of our lives must be modified accordingly.

Let us examine three time functions.

$$f_1(t) = A \cos 3t \tag{5.2a}$$
$$f_2(t) = A \cos (3t + 5) \tag{5.2b}$$
$$f_3(t) = A \cos (te^{-t}) \tag{5.2c}$$

The frequencies of $f_1(t)$ and $f_2(t)$ are clearly 3 rad/sec. The frequency of $f_3(t)$ is, at present, undefined. Our old definition of frequency does not apply to this type of waveform. Some thought would indicate that the usual intuitive definition of frequency can be stated as follows: "If $f(t)$ can be put into the form $A \cos (\omega t + \theta)$, where A, ω, and θ are constants, then the frequency is *defined* as ω rad/sec." $f_3(t)$ cannot be put into this form.

As is usually done to get out of a predicament such as this, we will define a new quantity (recall envelope discussion). This new quantity will be called *instantaneous frequency*. We shall define instantaneous frequency in such a way that, in those cases where the old intuitive definition of frequency can be applied, the two definitions will yield the same value of frequency.

Given $f(t) = A \cos \theta(t)$, where A is a constant, the instantaneous frequency of $f(t)$ is defined as the time derivative of $\theta(t)$. Note that at any time, t_0, $\theta(t_0)$ is the argument of the cosine function, and therefore can be thought of as its phase.

Before continuing, we note that this definition is not as restrictive as it might appear. Any time function, $f(t)$, can be put into the form, $f(t) = A \cos \theta(t)$. One does this by setting $\theta(t) = \cos^{-1}[f(t)/A]$, where A is chosen large enough so that the inverse cosine is defined. That is,

$$\left| \frac{f(t)}{A} \right| \leq 1 \qquad \text{for all } t$$

The symbol $\omega_i(t)$ will be used for the instantaneous frequency.

$$\omega_i(t) \triangleq \frac{d\theta}{dt} \tag{5.3}$$

For example, for $f_2(t)$ above, $\theta(t) = 3t + 5$, and $d\theta/dt$ is equal to 3 rad/sec as we found using our old definition. For $f_3(t)$, $\theta(t) = te^{-t}$, and $\omega_i(t) = e^{-t} - te^{-t}$. We have no intuitive answer with which to compare this value, so we must accept it as the definition of the frequency of $f_3(t)$.

Illustrative Example 5.1

Find the instantaneous frequency of the following waveform,

$$f(t) = \begin{cases} \cos t & t < 1 \\ \cos 2t & 1 \leq t \leq 2 \\ \cos 3t & 2 < t \end{cases}$$

Solution

This wave is of the form

$$f(t) = \cos[tg(t)] \tag{5.4}$$

where $g(t)$ is as sketched in Fig. 5.1. Therefore,

$$\omega_i(t) = \frac{d}{dt}[tg(t)] = g(t) + t\frac{dg}{dt}$$

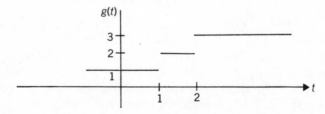

Figure 5.1 $g(t)$ for Illustrative Example 5.1.

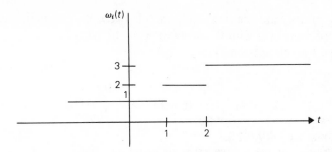

Figure 5.2 $\omega_i(t)$ for Illustrative Example 5.1.

This instantaneous frequency is sketched in Fig. 5.2. We note that the frequency is undefined at $t = 1$ and $t = 2$.

Illustrative Example 5.2

Find the instantaneous frequency of the following function,

$$f(t) = 10 \cos [1000t + \sin 5t] \tag{5.5}$$

Solution

For this waveform,

$$\theta(t) = 1000t + \sin 5t$$

and

$$\omega_i(t) = \frac{d\theta}{dt} = 1000 + 5 \cos 5t$$

This instantaneous frequency is sketched in Fig. 5.3.

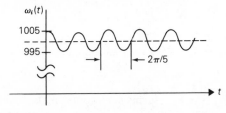

Figure 5.3 $\omega_i(t)$ for Illustrative Example 5.2.

In frequency modulation, we modulate (vary) $\omega_i(t)$ with the signal, $f(t)$, just as in amplitude modulation we modulated A with the signal. Since, intuitively, we wish to shift the frequencies of $f(t)$ up to the vicinity of ω_c for efficient transmission, we will immediately add the constant, ω_c, to $\omega_i(t)$. We will do this instead of first trying $\omega_i(t)$ of the form of $f(t)$. What we are saying here is that, in FM, there is no analogy to the suppressed carrier case in AM. Therefore, we shall set

$$\omega_i(t) = \omega_c + k_f f(t) \tag{5.6}$$

where ω_c and k_f are constants. Given this $\omega_i(t)$, we can derive the total transmitted waveform with this instantaneous frequency.

$$\lambda_{fm}(t) = A \cos \theta(t) \tag{5.7}$$

where

$$\theta(t) = \int_0^t \omega_i(\tau) \, d\tau = \omega_c t + k_f \int_0^t f(\tau) \, d\tau \tag{5.8}$$

We have assumed $\theta(0) = 0$.

Finally, the modulated waveform is given by

$$\lambda_{fm}(t) = A \cos \left[\omega_c t + k_f \int_0^t f(\tau) \, d\tau \right] \tag{5.9}$$

where the constant of integration has been set equal to zero since it represents an arbitrary time delay which is not a form of distortion. Note that $\lambda_{fm}(t)$ is a pure carrier wave if $f(t) = 0$. This would not be true if we hadn't added the constant, ω_c, in Eq. (5.6).

Illustrative Example 5.3

Sketch the FM and AMSC modulated waveforms for the following information signals. (*See* Fig. 5.4.)

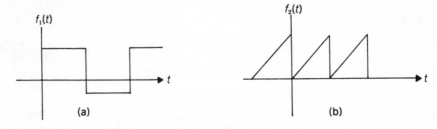

Figure 5.4 Waveforms for Illustrative Example 5.3.

Solution

The AMSC and FM waveforms corresponding to the information signals shown in Fig. 5.4 are illustrated in Fig. 5.5. The student should stare at these results until he is convinced that he has a feel for what is taking place. The concept of a frequency being controlled by the amplitude of the information signal is a difficult one to grasp. Note that as $f(t)$ increases, it is the frequency of the FM wave which increases, not the amplitude, which remains constant.

When FM was first investigated, it was reasoned that the frequency of $\lambda_{fm}(t)$ varied from $\omega_c + k_f[\min_t f(t)]$ to $\omega_c + k_f[\max_t f(t)]$. Therefore, by making k_f arbitrarily small, the frequency of $\lambda_{fm}(t)$ can be kept arbitrarily close to ω_c (i.e., small variations). This would result in a great bandwidth savings without resorting to SSB or other techniques. It was quickly realized

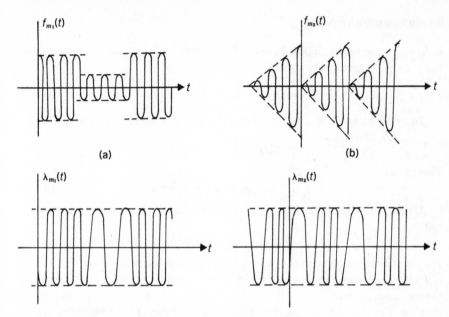

Figure 5.5 AMSC and FM waveforms for Illustrative Example 5.3.

that this reasoning is completely fallacious. That is, given a signal whose instantaneous frequency varies from ω_1 to ω_2, we shall show that the Fourier Transform of the signal is certainly *not* confined to the range of frequencies between ω_1 and ω_2. Clearly, one must be careful not to confuse the concept of instantaneous frequency with that of frequency in the Fourier Transform representation.

Proceeding in the usual manner which we have established for modulation systems' treatment, we must now show that this form of transmitted wave satisfies the three criteria for a modulating scheme. That is, it must be capable of effecting efficient transmission through the air; be in a form that allows multiplexing; and $f(t)$ must be uniquely recoverable from the modulated waveform. The demonstration of these properties is easier said than done. We will be forced to make many approximations. In the end, we will be satisfied with a far less detailed analysis than was possible in the AM wave case.

In order to show that FM is efficient and can be adapted to frequency multiplexing, we must first find the Fourier Transform of the FM wave. We shall find things much simpler if we first divide FM into two classes depending upon the size of the constant, k_f. A relatively simple approximation is possible for very small values of k_f.

For reasons that will become obvious, the small k_f case is known as *narrowband FM*. If k_f is not very small, we have what is called *wideband FM*.

5.1 NARROWBAND FM

If $k_f \ll 1$, the approximate Fourier Transform of $\lambda_{fm}(t)$ can be found. We start with the general form of the FM wave,

$$\lambda_{fm}(t) = A \cos \left[\omega_c t + k_f \int_0^t f(\tau) \, d\tau \right] \qquad (5.10)$$

In order to avoid having to rewrite the integral many times, we define

$$g(t) \triangleq \int_0^t f(\tau) \, d\tau$$

Therefore,

$$\lambda_{fm}(t) = A \cos \left[\omega_c t + k_f g(t) \right] \qquad (5.11)$$

Using the trigonometric cosine sum formula, $\lambda_{fm}(t)$ can be rewritten as follows,

$$\lambda_{fm}(t) = A \cos \omega_c t \cos k_f g(t) - A \sin \omega_c t \sin k_f g(t) \qquad (5.12)$$

If k_f is small enough such that $k_f g(t)$ is always much less than one, the cosine of $k_f g(t)$ is approximately equal to unity, and the sine approximately equal to the radian argument (i.e., we are taking the first term in a Taylor series expansion for the sine and cosine functions). With these approximations, $\lambda_{fm}(t)$ becomes

$$\begin{aligned} \lambda_{fm}(t) &= A \cos \omega_c t \cos k_f g(t) - A \sin \omega_c t \sin k_f g(t) \\ &\approx A \cos \omega_c t - A g(t) k_f \sin \omega_c t \end{aligned} \qquad (5.13)$$

We could easily find the transform of this expression if we only knew the transform of $g(t)$. Recall that $g(t)$ is the integral of $f(t)$. Referring to the properties of the Fourier Transform, we see that,

$$G(\omega) = \frac{F(\omega)}{j\omega}$$

That is, integration in the time domain corresponds to division by $j\omega$ in the frequency domain (compare this to LaPlace Transforms, where integration corresponds to division of the transform by s).

Therefore, if we make the usual assumption that $f(t)$ is band limited to frequencies below ω_m, $g(t)$ must also be band limited to these frequencies. This is true since, for all values of ω that $F(\omega)$ is equal to zero, $F(\omega)/j\omega$ must also be zero. We can now write the transform of $\lambda_{fm}(t)$ in terms of $F(\omega)$. We start by writing $\sin \omega_c t$ in terms of its complex exponential expansion.

$$\lambda_{fm}(t) = A \cos \omega_c t - A k_f g(t) \left[\frac{e^{j\omega_c t} - e^{-j\omega_c t}}{2j} \right]$$

$$\Lambda_{fm}(\omega) = A\pi[\delta(\omega - \omega_c) + \delta(\omega + \omega_c)] - \frac{A k_f}{2j}[G(\omega - \omega_c) - G(\omega + \omega_c)]$$

$$= A\pi[\delta(\omega - \omega_c) + \delta(\omega + \omega_c)] + \frac{A k_f}{2}\left[\frac{F(\omega - \omega_c)}{\omega - \omega_c} - \frac{F(\omega + \omega_c)}{\omega + \omega_c} \right]$$

$$\qquad (5.14)$$

The magnitude of this transform is sketched in Fig. 5.6.

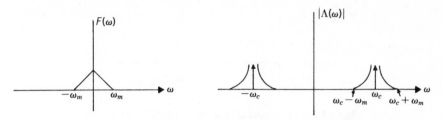

Figure 5.6   Magnitude transform of a narrowband FM waveform.

The fact that the transform, $G(\omega)$, apparently approaches infinity at $\omega = 0$ is not too troublesome. The important consideration is whether the corresponding time function is bounded. Indeed, the impulse function also goes to infinity although the inverse transform is bounded.

In real life situations, $F(\omega)$ can usually be assumed to go to zero as ω approaches zero. For example, standard audio signals are essentially zero below about 15 Hz. Therefore, $G(\omega)$, which is equal to $F(\omega)/j\omega$, would also be zero for frequencies below 15 Hz.

Figure 5.6 illustrates that, as far as narrowband FM is concerned, the first two objectives of modulating are automatically achieved. That is, the frequencies present can be made as high as is necessary for efficient transmission by raising ω_c to any desired value. By using different carrier frequencies separated by at least $2\omega_m$, many signals can be transmitted simultaneously on the same channel. As for the third requirement, being able to recover $f(t)$ from the modulated waveform, we shall defer that consideration to our later study. We do this since the demodulation process does not depend upon the value of k_f. The same detector will be used for both small and large k_f.

We now digress for a moment to present two vector plots for those students familiar with this type of analysis. In later work these will prove useful in comparing AM and FM with respect to noise rejection.

In the AM case, for a sinusoidal information signal, we have

$$f_m(t) = A \cos \omega_c t + \cos \omega_m t \cos \omega_c t \tag{5.15}$$

where we have set $f(t) = \cos \omega_m t$.

This can be written as

$$f_m(t) = Re\{e^{j\omega_c t}[A + \tfrac{1}{2}e^{j\omega_m t} + \tfrac{1}{2}e^{-j\omega_m t}]\} \tag{5.16}$$

where "Re" stands for "real part of."

If we plotted the phasor of this sinusoidal signal (the quantity in the braces in Eq. (5.16)), it would have a large angular term due to the $\omega_c t$ phase factor. Recall that $\omega_c \gg \omega_m$. We will use a standard technique of assuming that a stroboscopic picture of this phasor is taken every $2\pi/\omega_c$ seconds. That is, we will only plot the phasor of the inner bracket in Eq. (5.16). This is shown as Fig. 5.7.

We note that the resultant is in phase with the carrier term, A, as expected since there is no phase variation in AM.

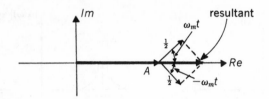

Figure 5.7   "Stroboscopic" phasor plot for AM waveform.

For narrowband FM, with the same information signal we have

$$\omega_i(t) = \omega_c + k_f \cos \omega_m t \tag{5.17}$$

and, from Eq. (5.13),

$$\lambda_{fm}(t) = A \cos \omega_c t - \frac{k_f}{\omega_m} A \sin \omega_m t \sin \omega_c t \tag{5.18}$$

This can be rewritten as

$$\lambda_{fm}(t) = Re\left\{e^{j\omega_c t}\left[A - \frac{Ak_f}{2\omega_m}e^{-j\omega_m t} + \frac{Ak_f}{2\omega_m}e^{+j\omega_m t}\right]\right\} \tag{5.19}$$

The quantity in the inner brackets of Eq. (5.19) is plotted in Fig. 5.8.

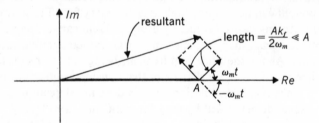

Figure 5.8   "Stroboscopic" phasor plot for narrowband FM waveform.

Note that the resultant has almost the same amplitude as A. The reason that the amplitudes are not exactly the same is that Eq. (5.13) is an approximation. If we had not used an approximation, the amplitude must be constant and only the frequency varies.

We shall defer our discussion of narrowband FM modulators to Section 5.3.

5.2   WIDEBAND FM

If k_f is not small enough to make the approximations for the sine and cosine valid, we have what is known as *wideband FM*. The transmitted signal is of the form found earlier. That is,

$$\lambda_{fm}(t) = A \cos\left[\omega_c t + k_f g(t)\right] \tag{5.20}$$

If we knew $g(t)$, we could find the Fourier Transform of this FM waveform. We would simply evaluate $\Lambda_{fm}(\omega)$ from the defining integral for the transform. However, we do not wish to restrict ourselves to a particular $f(t)$. We only know that $f(t)$ is band limited to frequencies below ω_m. Therefore, the only information we have about $g(t)$ is that it is band limited to frequencies below ω_m. With this information only, it is not possible to find the Fourier Transform of the FM wave. That is, in the AM case the transform of the modulated waveform was very simply related to the transform of the information signal. In the FM case, this is not true. Since we cannot find the Fourier Transform of the FM wave in the general case, we will see how much we can possibly say about the modulated signal.

In order to show that FM can be efficiently transmitted and that several channels can be multiplexed, we must only gain some feeling for what range of frequencies the modulated wave occupies (note that we are now discussing wideband FM since we have already answered all of these questions for the narrowband signal). That is, if we knew that $\Lambda_{fm}(\omega)$ occupied the band of frequencies between ω_1 and ω_2 without knowing the exact form of the Fourier Transform, we would be satisfied. This will, indeed, be the only information we will be able to obtain in general about the transform of the FM wave. The modulation and demodulation processes will be analyzed entirely in the time domain. Let us emphasize this fact. We will only use the Fourier Transform domain to get a rough idea of the range of frequencies occupied by the modulated waveform. This is an extremely significant point.

Even finding this range of frequencies is impossible in general, so we shall begin by restricting ourselves to a specific type of information, or modulating signal. $f(t)$ will be assumed to be a pure sinusoid. Very few people are interested in transmitting a pure sinusoid. It would not constitute a very pleasing piece of music since it would be a flat sounding single tone for all time. We shall, however, make one of our frequent generalizations after analysis of this special case is completed.

Assuming that $f(t) = a \cos \omega_m t$, where "a" is a constant amplitude, we have

$$\begin{aligned} \omega_i(t) &= \omega_c + k_f f(t) \\ &= \omega_c + a k_f \cos \omega_m t \end{aligned} \tag{5.21}$$

and

$$\lambda_{fm}(t) = A \cos \left[\omega_c t + \frac{a k_f}{\omega_m} \sin \omega_m t \right] \tag{5.22}$$

Rather than be forced to use trigonometric sum-difference relationships at this time, we shall use the complex exponential notation, and later take the real part of the result.

$$\lambda_{fm}(t) = Re\{A e^{j[\omega_c t + (a k_f/\omega_m) \sin \omega_m t]}\} \tag{5.23}$$

In order to find the Fourier Transform of this, we recognize that the second part of the exponent,

$$\frac{ak_f}{\omega_m} \sin \omega_m t$$

is a periodic time function. Therefore,

$$e^{j[(ak_f/\omega_m) \sin \omega_m t]}$$

is also a periodic function. That is, the only place where "t" appears is in the term, $\sin \omega_m t$, which is itself periodic. We shall let

$$\beta \triangleq \frac{ak_f}{\omega_m} \tag{5.24}$$

in order to avoid having to write this term many times. Therefore,

$$e^{j\beta \sin \omega_m t}$$

is periodic with fundamental frequency, ω_m. It can be expanded in a Fourier Series to yield

$$e^{j\beta \sin \omega_m t} = \sum_{n=-\infty}^{\infty} c_n e^{jn\omega_m t} \tag{5.25}$$

where the Fourier coefficients are given by

$$c_n = \frac{1}{T} \int_{-T/2}^{T/2} e^{j\beta \sin \omega_m t} e^{-jn\omega_m t} \, dt \tag{5.26}$$

and

$$T = \frac{2\pi}{\omega_m}$$

This integral cannot be evaluated in closed form. It does however converge to some real value (*see* Problem 5.3). Thus the fact that it cannot be expressed in closed form should not bother anybody. One can always approximate the value of the integral to any desired degree of accuracy using numerical techniques and a digital computer. As a last resort, one can build a trough in the shape of the real part of the integral, fill it with water, and see how much it holds. This situation is not basically different from the case of the integral of cos t, which also cannot be expressed in closed form. This latter case arises often enough that it has been tabulated under the name "sin t."

The expression in Eq. (5.26) turns out to be a function of n and β (*see* Problem 5.4). That is, given n and β, one can find a number for the integral evaluation. These numbers are given the name, "Bessel function of the first kind," and are tabulated under the symbol, $J_n(\beta)$. Given any value of n and β, we simply look up the corresponding value of $J_n(\beta)$. Using this new symbol, Eq. (5.26) becomes

$$c_n = J_n(\beta)$$

and Eq. (5.23) becomes

$$\lambda_{fm}(t) = Re\left\{Ae^{j\omega_c t} \sum_{n=-\infty}^{\infty} J_n(\beta)e^{jn\omega_m t}\right\} \tag{5.27}$$

Since $e^{j\omega_c t}$ is not a function of n, we can bring it under the summation sign to get

$$\lambda_{fm}(t) = Re\left\{A \sum_{n=-\infty}^{\infty} J_n(\beta)e^{jt(n\omega_m + \omega_c)}\right\}$$

Finally, taking the real part of the expression,

$$\lambda_{fm}(t) = A \sum_{n=-\infty}^{\infty} J_n(\beta) \cos(n\omega_m + \omega_c)t \tag{5.28}$$

The FM waveform, in this case, is simply a sum of cosines. Its transform will be a train of impulses in the frequency domain. Taking the transform of the FM wave term by term, we get $\Lambda_{fm}(\omega)$ as sketched in Fig. 5.9.

It appears that, unless $J_n(\beta)$ is equal to zero for "n" above a certain value, the Fourier Transform of the FM wave has frequency components spaced

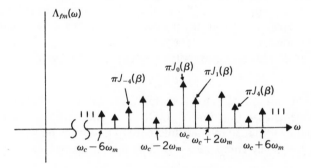

Figure 5.9   Transform of wideband FM wave with sinusoidal modulating signal (positive ω shown).

by ω_m and extending over all frequencies. Unfortunately, the Bessel functions are not equal to zero for large values of n. That is, one FM signal seems to occupy the entire band of frequencies. This is certainly disastrous if we wish to frequency multiplex.

Fortunately, for fixed β, the functions $J_n(\beta)$ approach zero as n increases. That is, if we decide to approximate as zero all $J_n(\beta)$ that are less than some specified value, the FM wave will be essentially band limited. We can find some N for any given β such that, if $n > N$, $J_n(\beta)$ is less than any specified amount.

As Fig. 5.9 indicates, the Bessel functions do not necessarily decrease monotonically. In fact, it is possible to choose β such that $J_0(\beta) = 0$, and therefore the carrier term is eliminated from the FM wave. In the AM case, such an accomplishment would increase power efficiency. In FM, elimination of the carrier does not buy anything since the total power remains constant.

The question now arises as to what level should be chosen below which we shall consider a frequency component to be insignificant. Clearly, no matter how small we choose this level, we will be distorting the waveform by neglecting all components below it. In other words, we are proposing "chopping off the tails of the transform" in order to force the modulated waveform to be band limited. How much of the tail can we chop before the result is too highly distorted?

The answer to this question must lie in the intended application. For high fidelity music, one would have to be more careful than for general purpose voice communications. The only true answer is an operational one. That is, tentatively decide where to chop off the frequencies; put the signal through a band pass filter which chops at these chosen frequencies; compare the output of the filter with the input; decide whether the output is acceptable (whatever that means). At this point, we simply state that any reader who owns an FM receiver can attest to the fact that this approach does indeed work. The stations to which you listen have been artificially band limited in the above manner, and they apparently still sound good to the human ear.

As an example, we shall arbitrarily define as significant, those components whose amplitudes are greater than 1 % of the unmodulated carrier amplitude. That is, given β, the significant components are those for which

$$|J_n(\beta)| \geq 0.01 \qquad (5.29)$$

Armed with a table of Bessel functions, one can now tabulate the significant components as a function of β. Once we know the significant components, the bandwidth is simply the difference in frequency between the highest and lowest frequency components. A brief table is shown in Fig. 5.10.

β	Significant components	Bandwidth
0.01	J_{-1}, J_0, J_1	$2\omega_m$
0.5	$J_{-2}, J_{-1}, J_0, J_1, J_2$	$4\omega_m$
1.0	$J_{-3}, J_{-2}, J_{-1}, J_0, J_1, J_2, J_3$	$6\omega_m$
2.0	$J_{-4}, J_{-3}, \ldots, J_3, J_4$	$8\omega_m$
Large β $\beta \geq 25$	$J_{-\beta}, J_{-(\beta-1)}, \ldots, J_{\beta-1}, J_\beta$	$2\beta\omega_m$

Figure 5.10   Table of value of n for which Eq. (5.29) is satisfied.

It turns out that, for very small or very large β, the significant components do not depend strongly upon the actual definition of significance (Eq. 5.29). That is, for example, choosing $\beta = 25$, one will find the $J_{25}(25)$ is quite significant while $J_{26}(25)$ is small. $J_{26}(\beta)$ is small for $\beta < 26$. Therefore, even if Eq. (5.29) had read

$$|J_n(\beta)| \geq 0.1 \tag{5.30}$$

which is a difference of a factor of 10 in the definition of significance, the significant components and bandwidth would be the same as those in the table for $\beta > 25$. It is only the mid-range values of β for which the definition of significance greatly affects the bandwidth definition. The significant points are that for β very small, the bandwidth is equal to $2\omega_m$, and for β very large, the bandwidth is equal to $2\beta\omega_m$. Recalling that

$$\beta = \frac{ak_f}{\omega_m}$$

we see that very small values of β correspond to small k_f and therefore to the narrowband case. For this case, the bandwidth of $2\omega_m$ agrees with that previously found by trigonometric approximations.

It is convenient to have a general rule of thumb to determine the bandwidth of an FM wave as a function of β and the frequency of the information signal, ω_m. Since the bandwidth is a function of the definition of significant components, any function which goes smoothly between the two limiting cases could be deemed acceptable. One rule of thumb which was advanced by John Carson (one of the first to investigate FM in 1922), and which has gained wide acceptance since it works, is that the bandwidth is approximately equal to the following,

$$BW \approx 2(\beta\omega_m + \omega_m) \tag{5.31}$$

We see that for large β, this is approximately equal to $2\beta\omega_m$, and for small β, $2\omega_m$ as desired. The slight disagreement with Fig. 5.10 for values in between these limits indicates that Carson used a different definition of significant components than the 1 % definition that we used to develop the table.

Recalling that $\beta = ak_f/\omega_m$, Eq. (5.31) becomes

$$BW \approx 2(ak_f + \omega_m) \tag{5.32}$$

We have accomplished our objective for the sinusoidal information signal case. That is, given a, k_f, and ω_m, we can find the approximate bandwidth, and we therefore know that the Fourier Transform occupies that portion of the frequency axis which is shaded in Fig. 5.11.

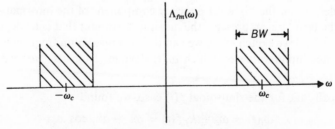

Figure 5.11   Band of frequencies occupied by FM wave.

Illustrative Example 5.4

Find the approximate band of frequencies occupied by an FM wave with carrier frequency 5,000 rad/sec, $k_f = 10$, and

(a) $f(t) = 10 \cos 5t$.

(b) $f(t) = 5 \cos 10t$.

(c) $f(t) = 100 \cos 1,000t$.

Solution

The bandwidth in each case is approximately given by

(a) BW $\approx 2(ak_f + \omega_m) = 2[10(10) + 5] = 210$ rad/sec.

(b) BW $\approx 2(ak_f + \omega_m) = 2[5(10) + 10] = 120$ rad/sec.

(c) BW $\approx 2(ak_f + \omega_m) = 2[100(10) + 1,000] = 4,000$ rad/sec.

The band of frequencies occupied is therefore

(a) 4,895 rad/sec to 5,105 rad/sec.

(b) 4,940 rad/sec to 5,060 rad/sec.

(c) 3,000 rad/sec to 7,000 rad/sec.

Unfortunately, Eq. (5.32) does not tell us too much about the general case when the modulating signal is not a pure sine wave. You might be tempted to say that since any $f(t)$ can be approximated by a sum of sinusoids we need only sum the individual sinusoidal transforms. This is not true since

$$\cos [\omega_c t + k_f(g_1(t) + g_2(t))]$$

is *not* equal to

$$\cos [\omega_c t + k_f g_1(t)] + \cos [\omega_c t + k_f g_2(t)]$$

That is, the FM wave due to the sum of two information signals is not the sum of the two corresponding FM waveforms. People sometimes call this property "non-linear modulation." (Recall the discussion concerning AM and linear time varying systems.)

As promised, it is now time for a rather bold generalization. What we desire is an equation similar to Eq. (5.32) that applies to a general modulating information signal. We therefore ask what each term in Eq. (5.32) represents in terms of the information signal. ω_m is clearly the highest frequency component present. In the single sinusoid case, it is the only frequency present. Indeed, one would expect the bandwidth of the modulated waveform to be dependent upon the highest frequency component of the information signal, $f(t)$. The term, ak_f, represents the maximum amount that $\omega_i(t)$ deviates from ω_c. That is, if we sketch $\omega_i(t)$, we can define a maximum frequency deviation as the maximum value by which $\omega_i(t)$ deviates from its "*dc*" value of ω_c. (*See* Fig. 5.12.)

Recall that for the sinusoidal $f(t)$ case, we found

$$\omega_i(t) = \omega_c + k_f f(t) = \omega_c + ak_f \cos \omega_m t$$

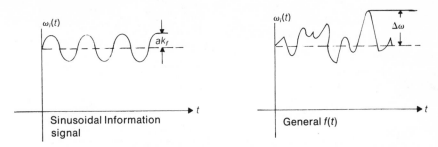

Figure 5.12 Definition of maximum frequency deviation.

The maximum frequency deviation is given the symbol $\Delta\omega$ and is defined as

$$\Delta\omega \triangleq \max_t [\omega_i(t) - \omega_c] = \max_t [\omega_c + k_f f(t) - \omega_c] = \max_t [k_f f(t)] \quad (5.33)$$

which is clearly defined for any general signal, $f(t)$.

We can now make an educated guess that Eq. (5.32) can be modified to read

$$BW \approx 2(\Delta\omega + \omega_m) \quad (5.34)$$

for any general $f(t)$ information signal.

One can give an extremely intuitive interpretation to Eq. (5.34). If $\Delta\omega$ is much larger than ω_m (wideband FM), the frequency of the carrier is varying by a large amount, but at a relatively slow rate. That is, the instantaneous frequency of the carrier is going from $\omega_c - \Delta\omega$ to $\omega_c + \Delta\omega$ very slowly. It therefore approximates a pure sine wave over any reasonable length of time. We can almost think of it as a sum of many sine waves with frequencies between these two limits. The transform is therefore approximately a super-position of the transform of each of these many sinusoids, all lying between the frequency limits. It is therefore reasonable to assume that its bandwidth is approximately the width of this frequency interval, or $2\Delta\omega$. On the other hand, for very small $\Delta\omega$, we have a carrier which is varying over a very small range of frequencies, but doing this relatively rapidly. One can almost think of this as being caused by two oscillators, one at a frequency of $\omega_c - \Delta\omega$, and the other at $\omega_c + \Delta\omega$, each being on for half of the total time, and alternating with the other. Thus each can be considered as being multiplied by a gating function at a frequency of $\Delta\omega$. (*See* Fig. 5.13.)

From an analysis similar to that of the gated modulator studied in Chapter 4, we would find that the band of frequencies occupied is approximately that from $\omega_c - \Delta\omega - \omega_m$ to $\omega_c + \Delta\omega + \omega_m$. For small $\Delta\omega$, this gives a bandwidth of $2\omega_m$, thus verifying Eq. (5.34). The details of this derivation are left to Problem 5.12.

The previous analysis was highly intuitive. It is doubtful that one could suggest such plausibility arguments if he did not previously know the result he desired.

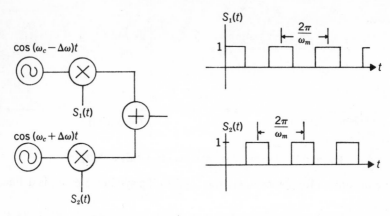

Figure 5.13 Visualization of narrowband FM.

It should be clear that the bandwidth of an FM wave increases with increasing k_f. Therefore at this point there appears to be no reason to use other than very small values of k_f (narrowband FM) and thus require the minimum bandwidth. We state without elaboration at this time that the advantages of FM over AM lie in its ability to reject unwanted noise. This ability to reject noise increases with increasing k_f, and therefore, with increasing bandwidth.

Illustrative Example 5.5

A 10-MHz carrier is frequency modulated by a sinusoidal signal of 5-kHz frequency such that the maximum frequency deviation of the FM wave is 500 kHz. Find the approximate band of frequencies occupied by the FM waveform.

Solution

We must first find the approximate bandwidth. This is given by

$$BW \approx 2(\Delta\omega + \omega_m)$$

For this case, we are told that $\Delta\omega$ is 500 kHz, and the maximum frequency component of the information is 5 kHz. In this case this is the *only* frequency component of the information, and therefore, certainly is the maximum. Therefore the bandwidth is approximately

$$BW \approx 2(500 \text{ kHz} + 5 \text{ kHz}) = 1{,}010 \text{ kHz}. \tag{5.35}$$

Thus the band of frequencies occupied is centered around the carrier frequency, and ranges from 9,495 kHz to 10,505 kHz. Note that while "ω" usually represents radian frequency, we can use it as cycles per second (hertz) in our bandwidth formula as long as we are consistent throughout. The FM signal of this example represents a wideband FM signal. If it were narrowband, the bandwidth would only be 10 kHz as compared with 1,010 kHz found above.

Illustrative Example 5.6

A 100-MHz carrier is frequency modulated by a sinusoidal signal of unit amplitude. k_f is set at 100 Hz/volt. Find the approximate bandwidth of the FM waveform if the modulating signal has a frequency of 10 kHz.

Solution

Again we use the only formula which is available. That is,

$$BW \approx 2(\Delta\omega + \omega_m)$$

Since the information, $f(t)$, has unit amplitude, the maximum frequency deviation, $\Delta\omega$, is given by k_f, or 100 Hz. ω_m is simply equal to 10 kHz. Therefore,

$$BW \approx 2(100 + 10 \text{ kHz}) = 20{,}200 \text{ Hz} \tag{5.36}$$

Since ω_m is much greater than $\Delta\omega$ in this case, the above signal would be a narrowband signal. The bandwidth necessary to transmit the same information waveform using double sideband AM techniques would be 20 kHz. This is almost the same amount required in this example.

Illustrative Example 5.7

An FM waveform is described by

$$\lambda_{fm}(t) = 10 \cos [2 \times 10^7 \pi t + 20 \cos 1{,}000\pi t] \tag{5.37}$$

Find the approximate bandwidth of this waveform.

Solution

ω_m is clearly equal to 1,000π rad/sec. In order to compute $\Delta\omega$, we first find the instantaneous frequency $\omega_i(t)$.

$$\begin{aligned}
\omega_i(t) &= \frac{d}{dt}(2 \times 10^7 \pi t + 20 \cos 1{,}000\pi t) \\
&= 2 \times 10^7 \pi - 20{,}000\pi \sin 1{,}000\pi t
\end{aligned} \tag{5.38}$$

The maximum frequency deviation is the maximum value of $20{,}000\pi \sin 1{,}000\pi t$ which is simply

$$\Delta\omega = 20{,}000\pi$$

The approximate bandwidth is therefore given by

$$\begin{aligned}
BW &\approx 2(20{,}000\pi + 1{,}000\pi) = 42{,}000\pi \text{ rad/sec} \\
&= 21{,}000 \text{ Hz}
\end{aligned} \tag{5.39}$$

This is clearly a wideband waveform since $\Delta\omega$ is much greater than ω_m.

5.3   MODULATORS

We have shown that in the case of narrowband or wideband FM waveforms the transform of the modulated waveform is band limited to some range of frequencies around ω_c, the carrier frequency. The first two criteria of a useful modulation system are therefore satisfied. We can transmit efficiently by choosing ω_c high enough. We can frequency multiplex many separate signals by making sure that the adjacent carrier frequencies are separated by a sufficient amount such that the transforms of the FM waveforms do not overlap in frequency.

We need now only show a few ways to generate an FM wave and to recover the information signal from the received waveform.

In the narrowband case, a simple trigonometric identity allows us to rewrite the FM waveform in the following form (we repeat Eq. (5.13)),

$$\lambda_{fm}(t) = A \cos \omega_c t - Ag(t)k_f \sin \omega_c t \qquad (5.40)$$

where $g(t)$ was the integral of the information signal. This immediately leads to the block diagram of the narrowband FM modulator shown in Fig. 5.14.

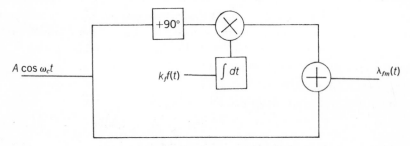

Figure 5.14   Narrowband FM modulator.

We now make the general observation that if the narrowband FM output of this modulator is put through a frequency multiplier, it can be made into a wideband waveform. A frequency multiplier is a device which multiplies all frequency components of its input by a constant.

We start with

$$\lambda_{fm}(t) = \cos\left[\omega_c t + k_f g(t)\right] \qquad (5.41)$$

which has an instantaneous frequency given by

$$\omega_i(t) = \omega_c + k_f f(t)$$

If we multiply all frequencies by a constant, C, we get a new FM wave with an instantaneous frequency of $\omega_i(t)$.

$$\omega_i(t) = C\omega_c + Ck_f f(t) \qquad (5.42)$$

The maximum frequency deviation of this new waveform is C times the maximum frequency deviation of the old waveform. The maximum frequency of the information signal (which can be thought of as being $Cf(t)$ for the modified waveform) is still ω_m. That is, although the frequency is covering a much wider band, it is changing just as rapidly as it did for the narrowband case. Therefore, if C is chosen large enough, the frequency deviation may end up being large enough to cause the resulting wave to be labelled "wideband."

This can be seen quite easily in the frequency domain. If the original wave is narrowband, it occupies a bandwidth of approximately $2\omega_m$ around ω_c as shown in Fig. 5.15. If we now multiply all frequencies by C, the new wave-

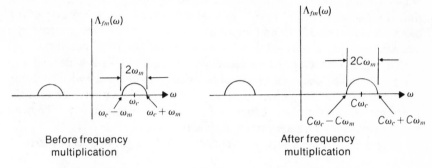

| Before frequency multiplication | After frequency multiplication |

Figure 5.15 Frequency multiplication to change narrowband FM into wideband FM.

form occupies frequencies from $C(\omega_c - \omega_m)$ to $C(\omega_c + \omega_m)$ as shown. Its bandwidth has increased accordingly. If the bandwidth is much larger than $2\omega_m$ (i.e., C is much large than 1), then the new wave must be classified as wideband. Thus we have a technique for generating the wideband waveform from the narrowband signal. This yields the FM modulator shown in Fig. 5.16.

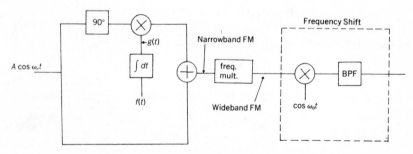

Figure 5.16 Modulator for wideband FM.

In practice, a squarer can be used as a frequency doubler. This is possible due to the trigonometric identity,

$$\cos^2 x = \tfrac{1}{2}(1 + \cos 2x). \tag{5.43}$$

A diode acts as a squarer over a limited range of inputs. If a frequency multiplication factor of 4 is desired, two squarers can be used.

We note that while the frequency multiplier increases the bandwidth of the FM wave, it also increases the carrier, or center frequency. If for some reason we were not happy with the new carrier frequency, $C\omega_c$, we could easily use the standard technique of multiplying by a cosine wave and then band pass filtering in order to shift the entire transform to any desired section of the frequency axis. This shift would not change the bandwidth of the wave which would therefore remain wideband. The shifting process is shown in the dotted section of Fig. 5.16. The essential difference between shifting all frequencies via heterodyning and multiplying of all frequency components by a constant should be clarified in the reader's mind before he continues.

There are also more direct techniques for generating a wideband FM wave. A standard electronic oscillator has an output frequency which depends upon the resonant frequency of a tuned L-C circuit. If either L or C of the tuned circuit is a function of time, the output of the oscillator will be some type of sinusoid with time-varying instantaneous frequency.

Figure 5.17 shows a highly simplified block diagram of a feedback oscillator.

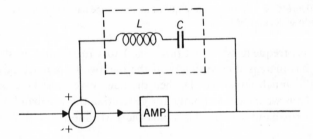

Figure 5.17   A feedback oscillator.

The output frequency would be

$$\omega_i = \frac{1}{\sqrt{LC}}$$

If either L or C is varied, the output frequency will also vary. That is,

$$\omega_i(t) = \frac{1}{\sqrt{L(t)C(t)}} \tag{5.44}$$

If FM is our goal, the trick becomes varying L or C such that

$$\omega_i(t) = \omega_c + k_f f(t)$$

Due to the presence of the square root relationship, achieving this particular $\omega_i(t)$ requires a complicated control over L or C. However, since the required variations are relatively small (i.e., $\Delta\omega \ll \omega_c$), the square root can be approximated by a linear term. Thus if C or L varies with $f(t)$ such that

$$C = C_0 + k_1 f(t) \tag{5.45a}$$

or

$$L = L_0 + k_2 f(t) \tag{5.45b}$$

and the constants, k_1 or k_2 are small, the output will be an FM wave.

The question now is whether or not we can vary a capacitance (or inductance) in this manner. It certainly isn't practical to physically move the knob of a variable capacitor back and forth.

Electronic devices do exist which exhibit a capacitance which is dependent upon an input voltage. We could feed $f(t)$ into one of these devices (e.g., from a microphone) and immediately get the desired results.

Two such devices are the *reactance tube* and the *varactor diode*. The interested student is referred to a textbook on electronics for details.

5.4   DEMODULATORS

The problem of FM demodulation can be stated as follows. Given $\lambda_{fm}(t)$ in the form

$$\lambda_{fm}(t) = A \cos\left[\omega_c t + k_f \int_0^t f(\tau)\, d\tau\right]$$

recover the information signal, $f(t)$. We note that the entire analysis to follow will be performed in the time domain since we do not know the Fourier Transform of the FM wave. We only have some feeling for the range of frequencies occupied by the transform.

With apparently no motivation, we differentiate the FM waveform to get

$$\frac{d\lambda}{dt} = -A[\omega_c + k_f f(t)] \sin\left[\omega_c t + k_f \int_0^t f(\tau)\, d\tau\right] \qquad (5.46)$$

This derivative is sketched in Fig. 5.18 for a typical $f(t)$.

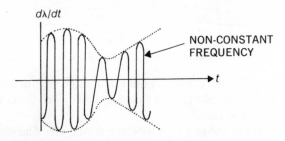

Figure 5.18   Derivative of FM waveform.

If we now assume that the instantaneous frequency of the sinusoidal part in Eq. (5.46) is always much greater than ω_m (a fair assumption in real life), this carrier term fills in the area between the amplitude and its mirror image. That is, we have exaggerated the carrier in sketching Fig. 5.18. Actually, the area between the upper and lower outline should be shaded in due to the extremely high carrier frequency. Thus even though the carrier frequency is not constant, the envelope of the waveform is still clearly defined by

$$|A[\omega_c + k_f f(t)]| \qquad (5.47)$$

The slight variation in the frequency of the carrier would not even be noticed by an envelope detector.

Illustrative Example 5.8

Use the concept of pre-envelope to prove that the envelope of

$$r(t) = -A[\omega_c + k_f f(t)] \sin\left[\omega_c t + k_f \int_0^t f(\tau)\, d\tau\right] \qquad (5.48)$$

is given by

$$|A[\omega_c + k_f f(t)]|$$

Solution

Using techniques similar to those of Illustrative Example 4.9 we find that the pre-envelope of $r(t)$ is

$$
\begin{aligned}
z(t) = &-A[\omega_c + k_f f(t)] \sin\left[k_f \int_0^t f(\tau)\, d\tau\right] \\
&-jA[\omega_c + k_f f(t)] \cos\left[k_f \int_0^t f(\tau)\, d\tau\right]
\end{aligned}
\qquad (5.49)
$$

The envelope of $r(t)$ is the magnitude of $z(t)$,

$$
\begin{aligned}
|z(t)| &= \sqrt{A^2[\omega_c + k_f f(t)]^2\left\{\sin^2 k_f \int_0^t f(\tau)\, d\tau + \cos^2 k_f \int_0^t f(\tau)\, d\tau\right\}} \\
&= |A[\omega_c + k_f f(t)]|
\end{aligned}
\qquad (5.50)
$$

A differentiator followed by an envelope detector can therefore be used to recover $A[\omega_c + k_f f(t)]$ from $\lambda_{fm}(t)$. This is shown in Fig. 5.19. Since ω_c is always much greater than $k_f f(t)$, we have removed the absolute value sign.

Figure 5.19 An FM demodulator.

We note that the above analysis did not assume anything about the size of k_f. The demodulator of Fig. 5.19 therefore works well for either wideband or narrowband FM signals.

The occurrence of the envelope detector should give one clue that AM is somehow rearing its head again. Indeed this is the case as the following analysis will reveal.

If we examine the transfer function of a differentiator, the operation of the FM demodulator will become immediately clear. The system function of a differentiator is

$$H(\omega) = j\omega \qquad (5.51)$$

with magnitude characteristic sketched in Fig. 5.20.

The magnitude of the output of the differentiator is linearly related to the frequency of its input. The differentiator therefore changes FM into AM!! When a differentiator is used in this manner, it is often called a *discriminator*.

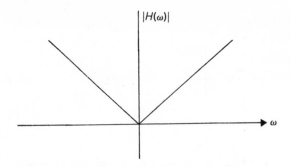

Figure 5.20 Magnitude characteristic of differentiator.

In practice, the differentiator need not be constructed. Any system which has a system function that is approximately linear with frequency in the range of frequencies of interest will change the FM into AM. Thus, even a sloppy band pass filter will work as a discriminator if we operate on the "up slope" (or the "down slope" as a similar argument would show). The filter would have to be followed by an envelope detector. (*See* Fig. 5.21.)

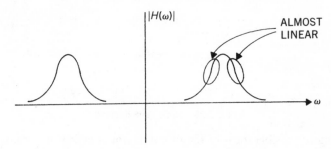

Figure 5.21 A band pass filter as a discriminator.

If quality of reproduction is not a critical factor, one can actually use an AM table radio to receive FM signals. A heterodyner would first have to be constructed to shift the desired FM signal into the range of frequencies receivable on the AM receiver. The AM radio is then tuned to a frequency slightly (several kHz) above or below the shifted FM carrier frequency. Since the image rejection filter in the AM receiver is not ideal, it will change the FM signal into an AM signal (i.e., discriminate). The receiver's envelope detector will then complete the demodulation. This technique is actually used in inexpensive converters in order to listen to police and emergency calls. These are often transmitted via narrowband FM.

Most of the non-linearity in the slope of the band pass filter characteristic of Fig. 5.21 can be eliminated if the characteristic is subtracted from a shifted version of itself as shown in Fig. 5.22. That is, we take the difference between the output of two band pass filters whose center frequencies are separated as shown.

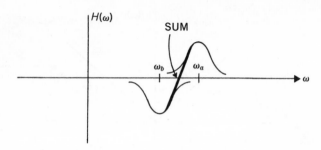

Figure 5.22 Improving the linearity of the discriminator.

Figure 5.23 shows a circuit diagram of a demodulator which uses the principle shown in Fig. 5.22. The tuned circuit consisting of the upper half of the output winding of the transformer and C_1 is tuned to ω_a (*see* Fig. 5.22) and the tuned circuit consisting of the other half of the output winding and C_2 is tuned to ω_b.

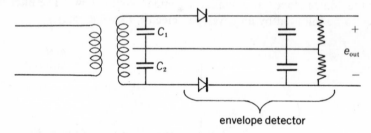

Figure 5.23 Slope demodulator.

One more possible form of the discriminator is called the "phase shift" detector. This is shown in block diagram form as Fig. 5.24. If t_0 is made small enough, the delay and subtraction forms an approximation to the derivative operation.

$$f(t) - f(t - t_0) \approx t_0 \frac{df}{dt} \tag{5.52}$$

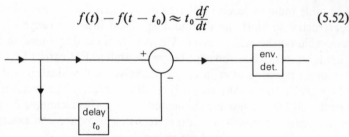

Figure 5.24 Phase shift demodulator.

There are other possible forms of the FM demodulator. Many quality FM receivers severely clip the incoming FM waveform in order to eliminate variations in the amplitude caused by noise. Recall that the noise free FM wave has constant amplitude. After severe clipping the waveform resembles a square wave. Several types of demodulators take advantage of this fact in order to essentially time the spacings between zero crossings of the clipped signal. We shall not consider these here.

5.5 PHASE MODULATION

There is no basic difference between phase modulation and frequency modulation. In spite of this fact, we will begin our study of phase modulation by treating it as an independent technique. We will continue this approach until a point is reached where the similarity between the two techniques cannot be overlooked.

We start with the ummodulated carrier signal

$$f_c(t) = A \cos (\omega_c t + \theta)$$

If θ is varied in accordance with the information we wish to transmit, the carrier is said to be *phase modulated*. We will therefore let θ vary in the following manner.

$$\theta = \theta(t) = k_p f(t) \tag{5.53}$$

where k_p is a constant factor associated with phase modulation. We use the subscript "p" to distinguish this constant from k_f used in FM (if for no other reason than the fact that the two constants have different units). The total phase modulated waveform will therefore be of the form

$$\lambda_{pm}(t) = A \cos [\omega_c t + k_p f(t)] \tag{5.54}$$

with corresponding instantaneous frequency

$$\omega_i(t) = \omega_c + k_p \frac{df}{dt} \tag{5.55}$$

This form does not look basically different from that of the FM signal. If the information signal used to frequency modulate a carrier were df/dt, the resulting FM waveform would look exactly the same as that of a carrier phase modulated with $f(t)$. If a signal which is either a phase or frequency modulated carrier is received, there is no way of telling whether it is an FM or PM wave without knowing $f(t)$ beforehand. Of course knowledge of $f(t)$ is absurd. If it were known, why waste time transmitting it?

The techniques of generating and receiving phase modulated carriers are the same as those for FM with the addition of an integrator or differentiator in the proper places. Figure 5.25 shows a modulator and a demodulator. The student should compare these with the block diagrams of the corresponding FM systems.

There is a definition applied to phase modulation which is analogous to that of maximum frequency deviation for FM. That is, the *maximum phase deviation*, $\Delta\theta$, is defined as the maximum amount by which the phase deviates from that of an unmodulated carrier wave. It is therefore equal to the maximum value of $k_p f(t)$.

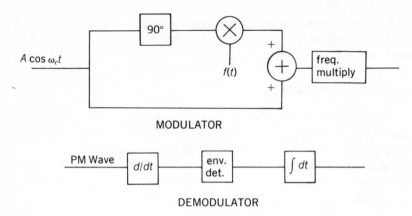

MODULATOR

DEMODULATOR

Figure 5.25 A phase modulator and demodulator.

The bandwidth of the phase modulated waveform is found by forgetting the fact that it is not an FM wave. We write the form of the modulated carrier, find ω_m and $\Delta\omega$, and plug these values into the "rule of thumb" formula previously developed.

Illustrative Example 5.9

The following waveform is received,

$$g(t) = 10 \cos (10^6 t + 200 \sin 500t)$$

(a) What is the approximate bandwidth of the waveform?

(b) If this represents an FM wave, what is $f(t)$, the information signal?

(c) If this represents a phase modulated wave, what is $f(t)$, the information signal?

Solution

(a) The approximate bandwidth of this signal is given by

$$BW \approx 2(\Delta\omega + \omega_m)$$

The instantaneous frequency of the waveform is given by

$$\omega_i(t) = 10^6 + \frac{d}{dt}(200 \sin 500t) \tag{5.56}$$
$$= 10^6 + 10^5 \cos 500t$$

ω_m is the maximum frequency component of $\cos 500t$ which is simply 500 rad/sec. $\Delta\omega$ is the maximum value of $10^5 \cos 500t$ which is simply 10^5 rad/sec. The bandwidth is therefore given by

$$BW \approx 2(10^5 + 500) \approx 2 \times 10^5 \text{ rad/sec} \tag{5.57}$$

Since the bandwidth for the narrowband case would be only 1,000 rad/sec, this is clearly a very wideband waveform. We have not yet needed to know whether the wave is FM or PM.

(b) If this is an FM wave, we would view the instantaneous frequency,

$$\omega_i(t) = \omega_c + k_f f(t)$$
$$= 10^6 + 10^5 \cos 500t \tag{5.58}$$

The information signal is easily identified as

$$f(t) = \cos 500t$$

We note that the assumption was made that $k_f = 10^5$. Our result for $f(t)$ might therefore be wrong by a constant if 10^5 is not the actual value of k_f used. Since a constant multiplier does not represent distortion of the waveform, it does not concern us here.

(c) If this were a PM wave, we would view the instantaneous phase,

$$\theta(t) = \omega_c t + k_p f(t) = 10^6 t + 200 \sin 500t \tag{5.59}$$

It is again easy to identify the information signal as

$$f(t) = \sin 500t \tag{5.60}$$

Here we have assumed that $k_p = 200$. The argument of part (b) concerning the constant multiplier applies again here.

While the difference between $\cos 500t$ and $\sin 500t$ may not appear significant, this will not be the case if $f(t)$ is a more complicated waveform. That is, while differentiation of a pure sinusoid does not alter its basic shape, differentiation of a general time function certainly represents a severe form of distortion. It is therefore important to know what form of modulation was used before one attempts demodulation of an incoming signal.

Since phase modulation is so similar to frequency modulation, a logical question at this time is, "why bother with phase modulation?" It would appear that there is no reason for using one form in preference to the other.

Actually one can see one practical difference by viewing the block diagrams of the modulators and demodulators (Figs. 5.14, 5.19, and 5.25). FM requires an integrator in the modulator. PM requires the integrator in the demodulator. If integrators are expensive, and many receivers are required while only one transmitter is used, FM would prove cheaper than PM.

We also mention at this time that integration of the information signal *before* transmission through the air may make it less susceptible to noise interference.

5.6 BROADCAST FM AND FM STEREO

If you pick up any FM program guide, you will realize that adjacent stations are usually separated by 200 kHz. The FCC has assigned carrier frequencies of the type

$$101.1 \text{ MHz}, \quad 101.3 \text{ MHz}, \quad 101.5 \text{ MHz}, \quad \ldots$$

to the various transmitting stations. Thus, while the bandwidth allocated to each station in AM was 10 kHz, it is an impressive 200 kHz for FM stations. This gives each broadcaster lots of room to work with.

The maximum frequency of the information signal is fixed at 15 kHz. Compare this with the 5-kHz figure for ω_m in the AM broadcast case. It should now be obvious why quality music is reserved for FM.

To transmit this signal via narrowband FM would require a bandwidth of $2\omega_m$, or 30 kHz. Since 200 kHz is available, we see that broadcast FM can afford to be wideband.[1]

Using the approximate formula developed for the bandwidth of an FM signal, we see that a maximum frequency deviation of about 85 kHz is possible. The actual figure used is 75 kHz.

What about FM stereo? Where could we possibly squeeze the second channel?

We have already answered these questions. If narrowband FM had been used instead of wideband, we would have only required 30 kHz out of the 200 kHz allocated to one station.

We begin with a system similar to that described in Illustrative Example 4.5. We take the two audio signals representing the left and right channels, and frequency multiplex them. We choose a carrier frequency of 38 kHz in order to yield an 8-kHz safety band between the two transforms. The system, and the corresponding transforms are sketched in Fig. 5.26.

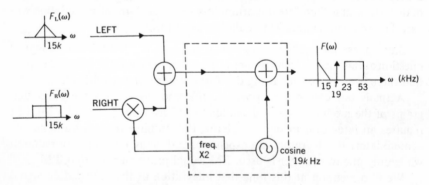

Figure 5.26 Multiplexing stereo signals.

This composite signal represents a perfectly legitimate time function with an upper frequency of 53 kHz. We can therefore transmit it via narrowband FM and be well within the allotted 200-kHz bandwidth. We would actually use only 106 kHz of the band.

---

[1]There is no exact point at which narrowband FM becomes wideband. Thus, if the narrowband bandwidth is 30 kHz, and we actually use, for example, 50 kHz, most people would be at a loss to clearly say which type of FM this represents. We will use the term "wideband" for any modulated signal with a bandwidth greater than $2\omega_m$.

At the receiver we would simply FM demodulate to recover the multiplexed signal. The remainder of the receiver is identical to that described in Illustrative Example 4.5 (*see* Section 4.4). We separate the two signals using a low pass filter and then synchronously demodulate the upper channel using a 38-kHz sinusoid. We must use synchronous demodulation since the carrier frequency (38 kHz) is not much higher than the highest frequency component of the information signal (15 kHz). Therefore envelope detection could not be used, even if a carrier term were added. To avoid the problems inherent in synchronous demodulation, we play a trick. The original multiplexed signal generated by the system of Fig 5.26 is added to a sinusoid of frequency 19 kHz before the frequency modulation (this is taking place at the transmitter). The 38-kHz carrier is formed by doubling the frequency (squaring) so that the two sinusoids are of the same phase and have frequencies which are exactly in the ratio of 2 × 1. This is illustrated in the dashed box in Fig. 5.26.

At the receiver, this 19-kHz sinusoid is separated after FM demodulation. This is easy to accomplish due to the safety band mentioned earlier. The sinusoid is then doubled in frequency and used in the synchronous demodulator. The frequency doubling is again performed by a squarer. Incidentally, it is this 19-kHz sinusoid that is used to light the "stereo" indicator light on the receiver panel.

The entire receiver is illustrated in Fig. 5.27.

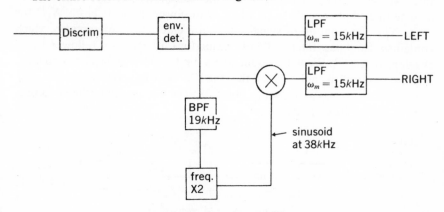

Figure 5.27   FM stereo receiver.

A non-stereo, single channel receiver consists of discriminator, envelope detector, and low pass filter. It would therefore recover the left signal when used in the above system. In order to make this system compatible, one slight modification is made. Instead of sending left and right channels, one sends sum $(L + R)$ and difference $(L - R)$ signals. The monaural receiver only recovers the sum channel. The stereo receiver of Fig. 5.27 recovers the sum and difference. If they are added together in a simple resistive circuit, the result is the left channel signal. If the difference is taken, again in a resistive circuit, the result is the right channel.

The unused portion of the allotted band between 53 kHz and 100 kHz deviation from ω_c is sometimes used for so-called SCA (Subsidiary Communications Authorization) signals. These include the commercial-free music heard in some restaurants, transmitted on a subcarrier of 67 kHz.

If four-channel FM stereo becomes popular, the third and fourth channels will occupy this region. They will probably be transmitted over subcarriers (the 38 kHz used in FM stereo is sometimes called a "subcarrier") of 72 kHz and 92 kHz with a maximum audio frequency of 8 kHz. The reduction from an ω_m of 15 kHz is being considered since the two additional channels will only carry the "fill-in" rear sounds.

This discussion illustrates the flexibility of FM broadcast systems. This flexibility is due to the fact that the bandwidth of an FM signal depends upon more than just the bandwidth of the information (or audio in this case) signal.

5.7 NOISE IN FM SYSTEMS

Let us assume that a noise-free FM waveform is transmitted:

$$\lambda_{fm}(t) = A \cos \{\omega_c t + k_f g(t)\}$$

where $g(t)$ is the time integral of the information signal. Noise is added to this FM signal along the path between the transmitter and receiver. We shall assume this additive noise to be white, with power spectral density $N_0/2$ for all ω. The combination of signal plus noise enters the FM receiver, which contains a band pass filter and a discriminator (differentiator followed by an envelope detector). We will also include a limiter in the receiver to clip large impulse peaks of the noise (recall that the FM wave has constant amplitude). The entire receiver is as shown in Fig. 5.28. The bandwidth of the band pass filter is $BW \simeq 2(\Delta\omega + \omega_m)$.

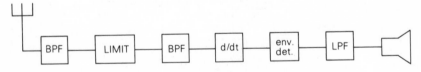

Figure 5.28 FM receiver.

Before proceeding with the mathematics, let's take an intuitive look at the effects of noise in FM.

The FM wave can be represented by a phasor (as in Fig. 5.8) whose length is "A" and whose angle is "$k_f f(t)$." This is a phasor which wiggles around according to $k_f f(t)$. We can add noise directly on the phasor diagram by recalling that the noise can be written as

$$n(t) = x(t) \cos \omega_c t - y(t) \sin \omega_c t$$

where $x(t)$ and $y(t)$ are random processes. We have assumed narrowband noise, which is certainly the case following the band pass filter in the receiver. With the noise broken into these components, we can easily add it to the signal directly on the phasor diagram. Figure 5.29 illustrates the signal and

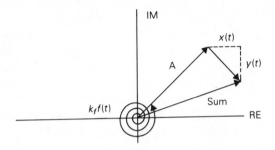

Figure 5.29   Phasor diagram for FM signal and noise.

noise phasors and a representative sum. Since we are considering FM transmission, the angle of the phasor contains all of the useful information, and we disregard the length (the limiter in the receiver will eliminate any magnitude variation). As long as the length of the noise phasor is less than A, the maximum angular variation which the noise can cause is $\pm 45°$ regardless of the magnitude of the noise angle. If the signal angle $k_f f(t)$ is much larger than $45°$ (and it can be many times 2π for large k_f), the effect of the noise is highly reduced.

The observant reader will note that if the noise amplitude gets larger than A, the game rules reverse, and the noise, rather than the signal, is enhanced. This is known as *noise capture* and manifests itself as a sudden replacement of the signal with a rushing noise.

In order to proceed with a mathematical analysis of noise in FM, one usually simplifies the case by setting $f(t)$ equal to zero. That is, we consider sending the carrier only. The signal plus noise entering the limiter is then given by

$$A \cos \omega_c t + x(t) \cos \omega_c t - y(t) \sin \omega_c t$$

Using a trigonometric identity, this can be rewritten as

$$\sqrt{\{A + x(t)\}^2 + y^2(t)} \cos \left[\omega_c t + \tan^{-1} \frac{y(t)}{\{A + x(t)\}} \right]$$

The output of the limiter would have constant amplitude and therefore be of the form

$$\cos \left[\omega_c t + \frac{\tan^{-1} y(t)}{\{A + x(t)\}} \right] + \text{Higher harmonics}$$

The higher harmonics arise when the peaks are clipped by the limiter.

Neglecting higher order terms, the output of the discriminator is the derivative of the inverse tangent shown above. This is given by

$$\frac{d\theta}{dt} = \frac{\{x(t) + A\}\, dy/dt - y(t)\, dx/dt}{y^2(t) + \{x(t) + A\}^2} \tag{5.61}$$

The signal of Eq. (5.61) must then be put through a low pass filter, as shown in Fig. 5.28. In order to perform this operation, we must first find the auto-correlation function of the signal and then the power spectral density and, finally, arrive at the power spectral density of the output of the filter. Following all of these operations, we could evaluate the power of the output noise and thereby find the S/N at the output of the FM receiver. These mathematical operations are somewhat involved, and one usually makes a simplifying assumption at this point. We assume a high carrier to noise ratio (CNR). That is, we assume that

$$A \gg x(t) \qquad A \gg y(t)$$

With this assumption, Eq. (5.61) can be reduced to

$$\frac{d\theta}{dt} = \frac{dy/dt}{A} \tag{5.62}$$

Note that in both Eq. (5.61) and Eq. (5.62) we omit the constant ω_c, as is done in all FM receivers.

Our goal is to find the noise power, so, at this point, we shall investigate the power spectral density of the output noise as given by Eq. (5.62). From Illustrative Example 3.13, we have that the power spectral density of $y(t)$ is N_0 for ω between $-BW/2$ and $+BW/2$, where BW is the bandwidth of the receiver input filter. We can find the power spectral density of dy/dt as in Problem 2.18. That is, since the differentiator has transfer function $H(\omega) = j\omega$, the power spectral density of $d\theta/dt$ is given by ω^2/A^2 multiplied by the density of $y(t)$. This density is shown in Fig. 5.30. Since the discriminator

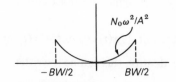

Figure 5.30 Spectral density of $d\theta/dt$.

output is of interest only for frequencies between $-\omega_m$ and $+\omega_m$ (i.e., the audio is low-pass-filtered before entering the speakers), the output noise power can be explicitly calculated as

$$P_N = \frac{1}{\pi} \int_0^{\omega_m} S_{N_0}(\omega)\, d\omega = \frac{1}{3\pi}\frac{N_0}{A^2}\omega_m^3 \tag{5.63}$$

Since the output signal is $k_f f(t)$, the output signal power is $k_f^2 P_f$, where P_f is the power of the information signal. The output signal to noise ratio is finally given by

$$S/N = \frac{3A^2 k_f^2 \pi P_f}{\omega_m^3 N_0} \tag{5.64}$$

Note from this result that as the carrier power increases, the output S/N increases. This is not the case in AM. Note also that as k_f increases, the S/N increases. Thus, as FM becomes more wideband, the noise becomes less disrupting.

Illustrative Example 5.10

Assume that $f(t) = a \cos \omega_m t$. This information signal frequency-modulates a carrier and experiences additive white noise of spectral density $N_0/2$. Find the signal to noise ratio at the output of the receiver.

Solution

From Eq. (5.64), with $P_f = a^2/2$, we have

$$S/N = \frac{a^2 k_f^2 3 A^2 \pi}{2 N_0 \omega_m^3} \tag{5.65}$$

We wish to make one observation from this result. Recall that for a sinusoidal information signal the FM bandwidth is given by $2(ak_f + \omega_m)$. If ak_f is much larger than ω_m (wideband FM), this bandwidth can be approximated by $2ak_f$. Therefore, the signal to noise ratio of Eq. (5.65) can be rewritten as

$$S/N = \frac{3A^2 \pi BW^2}{8 N_0 \omega_m \omega_m^2} \tag{5.66}$$

Equation (5.66) clearly shows the dependence between the S/N and the bandwidth of the FM wave.

Pre-emphasis

An even greater noise advantage can be realized by observing that the noise at the output of the FM receiver is not white even though the input noise is assumed to be white. Figure 5.30 illustrates that the output noise increases parabolically with increasing frequency. Therefore, the higher information frequencies are more severely affected by noise than the lower frequencies. A more efficient use of the band of frequencies would occur if these higher signal frequencies were emphasized and the lower frequencies were de-emphasized (so that the total power remains constant). The signal could therefore be shaped by a filter with $H(\omega)$ as shown in Fig. 5.31. This pre-shaping would be done prior to transmission. All signal frequencies would then be equally affected by the noise. The filter, of course, represents a signal distortion. The filtering operation must therefore be "undone" at the receiver

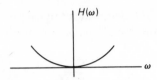

Figure 5.31 $H(\omega)$ of pre-emphasis filter.

by an inverse filter known as the *de-emphasis filter*. This filter would have transfer function $1/H(\omega)$. The composite filter transfer function of the two filters is unity.

Commercial broadcast FM uses pre-emphasis. In the interest of economy, the exact function shown in Fig. 5.31 is not used. A simple RC low pass filter is used in the receiver for de-emphasis. The inverse filter is used at the transmitting station for pre-emphasis.

PROBLEMS

5.1. Find the instantaneous frequency of $f(t)$, where

$$f(t) = 10[\cos(10t)\cos(30t^2) - \sin(10t)\sin(30t^2)]$$

5.2. Find the instantaneous frequency, $\omega_i(t)$, of $f(t)$, where

$$f(t) = 2e^{-t}U(t)$$

5.3. In Eq. (5.26), c_n was found to be

$$c_n = \frac{1}{T}\int_{-T/2}^{T/2} e^{j\beta \sin \omega_m t}e^{-jn\omega_m t}\, dt$$

where $T = 2\pi/\omega_m$. Show that the imaginary part of c_n is zero. That is, show that c_n is real.

5.4. Show that c_n as given in Eq. (5.26) (*see* Problem 5.3) is not a function of ω_m or T. That is, show that c_n only depends upon β and n. (Hint: Let $x = \omega_m t$ and make a change of variables.)

5.5. Find the approximate Fourier Transform of $f(t)$, where

$$f(t) = \cos\left[\frac{ak_f}{\omega_m}\sin \omega_m t\right]$$

Do this by expanding $\cos(x)$ in a Taylor series expansion and retaining the first three terms.

5.6. Repeat Problem 5.5 for $f(t)$ given by

$$f(t) = \sin\left[\frac{ak_f}{\omega_m}\sin \omega_m t\right]$$

5.7. Use the results of Problems 5.5 and 5.6 to find the approximate Fourier Transform of the FM waveform

$$\lambda_{fm}(t) = \cos\left[\omega_c t + \frac{ak_f}{\omega_m}\sin \omega_m t\right]$$

5.8. Use a table of Bessel functions in order to verify the entries in Fig. 5.10.

5.9. Prepare a new table equivalent to Fig. 5.10 but change the criterion of "significance" to

$$|J_n(\beta)| \geq 0.1$$

Compare the results for large and small values of β with the corresponding results in the text.

5.10. Find the approximate band of frequencies occupied by an FM waveform of carrier frequency 2 MHz, $k_f = 100$ Hz/volt, and
(a) $f(t) = 100 \cos 150t$ volts.
(b) $f(t) = 200 \cos 300t$ volts.

5.11. Find the approximate band of frequencies occupied by an FM waveform of carrier frequency 2 MHz, $k_f = 100$ Hz/volt, and

$$f(t) = 100 \cos 150t + 200 \cos 300t \text{ volts}$$

Compare this with your answers to Problem 5.10.

5.12. Consider the system shown below where the output is alternately switched between a source at frequency $\omega_c - \Delta\omega$ and a source at frequency $\omega_c + \Delta\omega$. This switching is done at a frequency ω_m. Find and sketch the Fourier Transform of the output, $y(t)$. You may assume that the switch takes zero time to go from one position to another. (Hint: *See* Fig. 5.13.)

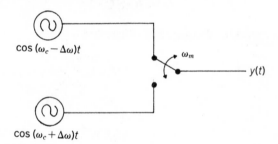

$\cos(\omega_c - \Delta\omega)t$

ω_m

$y(t)$

$\cos(\omega_c + \Delta\omega)t$

5.13. A 10-MHz carrier is frequency modulated by a sinusoid of unit amplitude and $k_f = 10$ Hz/volt. Find the approximate bandwidth of the FM signal if the modulating signal has a frequency of 10 kHz.

5.14. A 100-MHz carrier is frequency modulated by a sinusoid of frequency 75 kHz such that $\Delta\omega = 500$ kHz. Find the approximate band of frequencies occupied by the FM waveform.

5.15. Find the approximate band of frequencies occupied by the waveform

$$\lambda(t) = 100 \cos [2\pi \times 10^5 t + 35 \cos 100\pi t]$$

5.16. Find the approximate pre-envelope and envelope of

$$f(t) \cos \left[\omega_c t + k_f \int_0^t f(\sigma) \, d\sigma\right]$$

where the maximum frequency of $f(t)$ is much less than ω_c.

5.17. An angle modulated wave is described by

$$\lambda(t) = 50 \cos [2\pi \times 10^6 t + 0.001 \cos 2\pi \times 500t]$$

(a) What is $\omega_i(t)$, the instantaneous frequency of $\lambda(t)$?

(b) What is the approximate bandwidth of $\lambda(t)$?

(c) Is this a narrowband or wideband angle modulated signal?

(d) If this represents an FM signal, what is $f(t)$, the information signal?

(e) If this represents a PM Signal, what is $f(t)$?

5.18. You are given the sum $(L + R)$ and difference $(L - R)$ signals in a stereo receiver system. Design a simple resistive circuit that could be used to produce the individual left and right signals from the sum and difference waveforms.

5.19. An information signal $f(t) = 5 \cos 1{,}000t$ frequency-modulates a carrier of frequency 10^6 rad/sec. $A = 100$ and $k_f = 100$ rps/volt. Noise of power spectral density 10^{-3} W/rps is added during transmission.

(a) Find the S/N at the output of an FM receiver, and express this in decibels.

(b) If the output is put through a band pass filter with $H(\omega) = 1$ for $900 < |\omega| < 1{,}100$, find the signal to noise improvement of the filter.

(c) Repeat parts (a) and (b) for AM DSBTC. Compare the results.

5.20. Repeat Problem 5.19 for $f(t) = 5 \cos 1{,}000t + 3 \cos 2{,}000t$.

5.21. Repeat Problem 5.20 for $f(t) = (100 \sin 1{,}000t)/t$.

REFERENCES

See the references listed at the end of Chapter 4.

PULSE
MODULATION

6

Prior to the present chapter we have been discussing analog continuous forms of communication. An analog signal can be used to represent a quantity which can take on a continuum of values. For example, the pressure wave emitted by a pair of vocal chords when a person speaks constitutes an analog continuous waveform.

A *discrete* signal is a signal which is not continuous in time. That is, it only occurs over disconnected sections of the time axis. As an example, a pulse train would be considered as a discrete signal.

6.1 SENDING ANALOG SIGNALS BY DISCRETE TECHNIQUES

The sampling theorem suggests one technique for changing a band limited analog signal, $f(t)$, into a discrete time signal. We need only sample the continuous signal at discrete points in time. For example, a list of numbers could be specified, $f(0), f(T), f(2T), \ldots$, where $T < \pi/\omega_m$. In order to transmit this discrete sampled signal the list of numbers could be read over a telephone or written on a piece of paper and sent through the U.S. mail. Neither one of these two suggested techniques is very desirable for obvious reasons.

A more attractive technique would be to modulate some parameter of a carrier signal in accordance with the list of numbers. This modulated signal could then be transmitted over wires or, if the band of frequencies it occupies allows, through the air.

What type of carrier signal should be used in this case? Since the information is in discrete form, it would be sufficient to use a discrete carrier signal as opposed to a continuous sine wave used previously.

We therefore choose a periodic pulse train as a carrier. The parameters which can be varied are the height, width and the position of each pulse. Varying one of these three leads to pulse amplitude modulation (PAM), pulse width modulation (PWM), and pulse position modulation (PPM), respectively.

Pulse amplitude modulation is the most commonly used technique and is the only one of the three which is easy to analyze mathematically.

6.2 PAM—GENERAL DESCRIPTION

Figure 6.1 illustrates an unmodulated carrier wave, $f_c(t)$, a representative information signal, $f(t)$, and the resulting pulse amplitude modulated signal, $f_m(t)$.

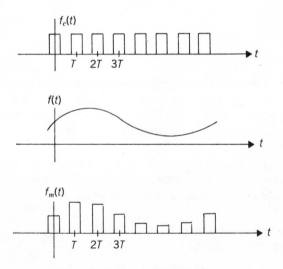

Figure 6.1 Pulse amplitude modulation.

In this form, PAM allows only the amplitude of the carrier pulses to vary. Their shape remains unchanged. We therefore note that $f_m(t)$ is *not* the product of $f(t)$ with $f_c(t)$. If this were the case, the heights of the various pulses would not be constant, but would follow the curve of $f(t)$ as shown in Fig. 6.2.

$f_m(t)$ certainly contains sufficient information to enable recovery of $f(t)$. This is simply a restatement of the sampling theorem.

Figure 6.2 Product of $f(t)$ and $f_c(t)$ from Fig. 6.1.

In order to assess transmission system requirements, we will now find the Fourier Transform of $f_m(t)$. A simple observation makes this calculation almost trivial.

We observe that if $f(t)$ is first sampled with a train of impulses, then the sampled wave consists of impulses whose areas (strengths) are the sample values. If we now put the impulse sampled wave through a filter which changes each impulse into a rectangular pulse, the output of the filter is $f_m(t)$. This is illustrated in Fig. 6.3. The fact that the filter is non-causal will be of no consequence since we will never actually construct this system.

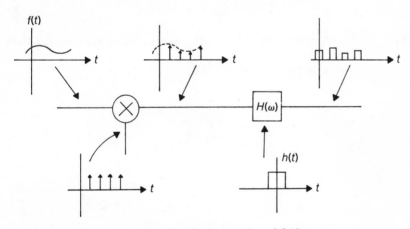

Figure 6.3 Idealized generation of $f_m(t)$.

The input to this filter is

$$\sum_{n=-\infty}^{\infty} f(nT)\delta(t - nT) \tag{6.1}$$

The output of the filter is therefore a superposition of delayed impulse responses. This is given by

$$\sum_{n=-\infty}^{\infty} f(nT)h(t - nT) \tag{6.2}$$

which is seen to be $f_m(t)$. We note that if the discrete carrier, $f_c(t)$, were composed of anything other than rectangular pulses (e.g., triangular) we would simply have to modify $h(t)$ accordingly (*see* Illustrative Example 6.1).

In practice, one would certainly not use a system such as that shown in Fig. 6.3 to generate $f_m(t)$. If one could generate the impulse sampled waveform he would send it without modification. The reasons for this will become clear when we speak about multiplexing in a later section. The system of Fig. 6.3 will simply be used as an analysis tool in order to find the Fourier Transform of the PAM wave.

Proceeding to calculate this transform, we know that the transform of the ideal impulse sampled wave is given by

$$\sum_{n=-\infty}^{\infty} f(nT)\delta(t - nT) \leftrightarrow \frac{1}{T} \sum_{n=-\infty}^{\infty} F(\omega - n\omega_0) \tag{6.3}$$

where

$$\omega_0 \triangleq \frac{2\pi}{T}$$

This was derived in Section 1.10.

Therefore, the transform of the output of the system of Fig. 6.3 is given by

$$\frac{1}{T}H(\omega) \sum_{n=-\infty}^{\infty} F(\omega - n\omega_0) \tag{6.4}$$

where $H(\omega)$ is the transform of $h(t)$ as sketched in Fig. 6.4. Figure 6.5 shows

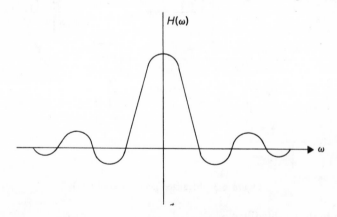

Figure 6.4   Transform of $h(t)$ given in Fig. 6.3.

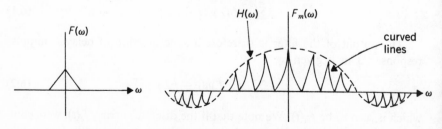

Figure 6.5   Transform of output of system in Fig. 6.3.

the resulting transform of the PAM waveform. We have sketched a general low frequency limited $F(\omega)$ as we have many times before.

Illustrative Example 6.1

An information signal is of the form

$$f(t) = \frac{\sin \pi t}{\pi t} \tag{6.5}$$

It is transmitted using PAM. The carrier waveform is a periodic train of triangular pulses as shown in Fig. 6.6. Find the Fourier Transform of the modulated carrier.

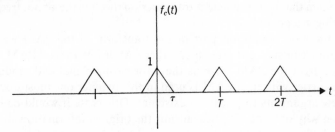

Figure 6.6 Carrier for Illustrative Example 6.1.

Solution

We again refer to the system shown in Fig. 6.3. The output of the ideal impulse sampler has a transform given by

$$F_s(\omega) = \frac{1}{T} \sum_{n=-\infty}^{\infty} F(\omega - n\omega_0) \tag{6.6}$$

where $F(\omega)$ would be the transform of $\sin \pi t/\pi t$. This transform is a pulse, and is shown in Fig. 6.7.

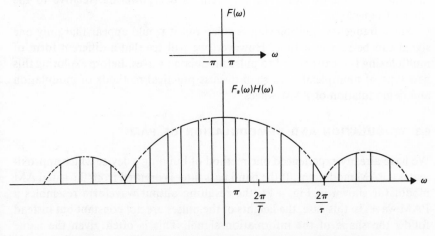

Figure 6.7 Transform of PAM wave of Illustrative Example 6.1.

In this case, the filter would have to change each impulse into a triangular pulse. Its impulse response is therefore a single triangular pulse which has as its transform (*see* Appendix II)

$$H(\omega) = \frac{4 \sin^2 (\omega\tau/2)}{\tau\omega^2} \tag{6.7}$$

Finally, the transform of the PAM waveform is given by the product of $F_s(\omega)$ with $H(\omega)$ as shown in Fig. 6.7.

The significant general observation to make about the transform of a PAM wave is that it occupies all frequencies from zero to infinity. It would therefore seem that neither efficient transmission through the air nor frequency multiplexing are possible.

Since the most significant part of the transform of the PAM wave lies around zero frequency, one often uses either AM or FM to send PAM waves. That is, we treat the PAM wave as the information signal and modulate a sine wave carrier with it. If PAM is used in this manner one must obviously invoke new arguments to justify its existence. Otherwise it would be logical to question why one doesn't just transmit the original information signal by AM or FM. The answer to such a question should be obvious if the original information is in discrete form. If it is in continuous analog form one can only justify the "double modulation" on the basis of processing advantages which depend upon the specific application.

After either amplitude or frequency modulation with the PAM wave the bandwidth is huge (actually it is infinity). For this reason pulse modulation combined with AM or FM is usually not transmitted in the same way as our other forms of modulated signals are. It is often sent over wires which are capable of transmitting a wide range of frequencies efficiently. Sometimes it is sent through the air at microwave frequencies. These are high enough so that the large bandwidth does not seem as overpowering (relative to the carrier frequency).

Since frequency multiplexing is ruled out it would appear that only one signal can be sent at a time. However, we will see that a different form of multiplexing is ideally suited to pulse modulated waves. Before exploring this new type of multiplexing we shall discuss practical methods of modulation and demodulation of PAM waves.

6.3 MODULATION AND DEMODULATION OF PAM

We have already experienced one method of building a device which approximates a PAM modulator. If the band pass filter is omitted from the gated AM modulator shown in Fig. 4.14, the resulting output waveform resembles a PAM wave. In this case, the heights of the pulses are not constant but instead follow the shape of the information signal. This is often given the name *natural sampled PAM*.

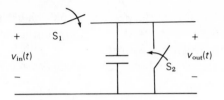

Figure 6.8   Idealized PAM modulator.

A modulator which produces the flat top PAM pulses is idealized in Fig. 6.8.

This is a variation of the *sample and hold* circuit. Switch S_1 closes instantaneously at the sample instants. This charges the capacitor to the sample values. S_2 closes periodically at a fixed short interval after each sample point, thus discharging the capacitor and returning its voltage to zero. With both switches open, the capacitor has no discharge path and remains charged at the previous sample value until S_2 closes.

In the practical realization of this system, S_1 would have to be preceded by some electronics in order to supply the energy to charge the capacitor very rapidly. The capacitor would have to be followed by electronics to prevent the following circuitry from providing a discharge path for the capacitor while both switches are open.

One type of PAM *demodulator* takes advantage of a simple observation of Fig. 6.7, the Fourier Transform of the PAM waveform. From this figure, one can observe that the low frequency section of $F_m(\omega)$ is of the form $F(\omega)H(\omega)$. Therefore, $f(t)$ can be recovered from $f_m(t)$ by first injecting $f_m(t)$ into a filter with transfer function $1/H(\omega)$ and then passing the result through a low pass filter. The first of these two filters is known as an *equalizer* since it cancels the effects of the pulse shape. These two series filters can be combined into a single filter with system function as shown in Fig. 6.9.

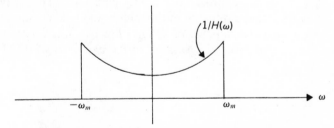

Figure 6.9   System function of PAM demodulator.

To arrive at a second type of PAM demodulator, one often approximates the low pass demodulator filter by a sample and hold circuit similar to the one previously described for use as a modulator. The output of such a circuit consists of periodic samples of the input. For time between the sample points, the output remains constant at the previous sample value. The output of the sample and hold circuit with PAM as the input will therefore be a step approximation to the original information signal. This is illustrated in Fig. 6.10. One form of a sample and hold circuit which accomplishes this (with no

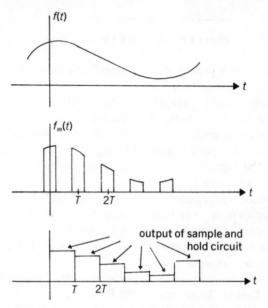

Figure 6.10   Sample and hold circuit used as PAM demodulator.

load applied) consists of a switch that closes momentarily at the sample instants, and a capacitor which charges up and remains charged during the time that the switch is opened (*see* Fig. 6.11).

A semi-mathematical handwaving argument is sometimes given to show that this type of circuit approximates a low pass filter. This is done by approximating the impulse response of the sample and hold circuit and comparing it with the impulse response of an ideal low pass filter. Some thought will show that the principles of operation of the two types of PAM demo-

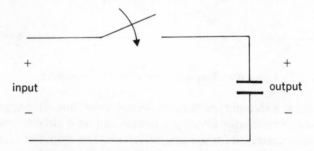

Figure 6.11   Simple form of a sample and hold circuit.

dulators are completely different from each other. The sample and hold circuit only accepts the values of the input which occur at the sampling instants. Therefore the argument relating its operation to that of a low pass filter is not relevant. (Recall that we were tempted to make a similar false comparison in the case of the "rectifier low pass filter" AM demodulator vs. the envelope detector.)

We shall see that the sample and hold circuit is easily adaptable to the type of multiplexing used in pulse systems. For this reason it is often preferred to the low pass filter as a demodulation system.

6.4   TIME DIVISION MULTIPLEXING

We can easily illustrate time division multiplexing (TDM) by returning to our original example of transmitting a signal by reading the list of sample values over a telephone.

Suppose that the reader is capable of reading two numbers every second, but the maximum frequency of $f(t)$ is only $\frac{1}{2}$ Hz. The sampling theorem tells us that at least one sample must be sent every second. If the reader sends this one sample per second, he is idle for one half of every second. During this half second, he could send a sample of another signal. Alternating between samples of the two signals is called time division multiplexing. The person at the receiver will realize that every other number is a sample of the second signal. He can therefore easily separate the two signals.

This operation can be described schematically with two switches, one at the transmitter and one at the receiver. The switches alternate between each of two positions making sure to take no longer than one sampling period to complete the entire cycle. (*See* Fig. 6.12)

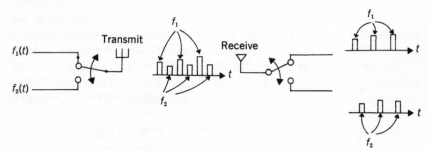

Figure 6.12   TDM of two signals.

If we now increase to 10 channels, we get a system as shown in Fig. 6.13.

The switch must make one complete rotation fast enough so that it arrives at channel # 1 in time for the second sample. The receiver switch must rotate in synchronism with that at the transmitter. In practice, this synchronization is not trivial to effect. If we knew exactly what was being sent as # 1,

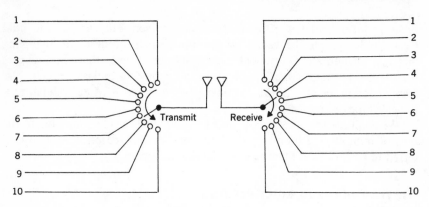

Figure 6.13 TDM of 10 signals.

we could identify its samples at the receiver. Indeed, a common method of synchronization is to sacrifice one of the channels and send a known synchronizing signal in its place.

The only thing that limits how fast the switch can rotate, and therefore how many channels can be multiplexed, is the fraction of time required for each PAM signal. This fraction is the ratio of the width of each pulse to the spacing between adjacent samples of the sample channel. We can now finally appreciate the fact that ideal impulse samples would be beautiful. If this were possible, an infinite number of channels could be multiplexed.

It looks as if we are getting something for nothing. Apparently the only limit to how many channels can be time division multiplexed is placed by the width of each sample pulse.

The minimum width of each sample pulse, and therefore the maximum number of channels which can be multiplexed, is determined by the bandwidth of the communication channel. Examination of Illustrative Example 6.1 (Fig. 6.7) will show that the bandwidth of the PAM wave increases as the width of each pulse decreases. Note that we are assuming some practical definition of bandwidth is used since the ideal bandwidth is infinity.

Problem 6.8 at the end of this chapter indicates that the bandwidth of a multiplexed signal is independent of the number of channels multiplexed, provided that the width of each pulse remains unchanged. Therefore once the pulse width is chosen, system potential would be wasted unless the maximum possible number of non-overlapping signals is squeezed together.

6.5 PULSE WIDTH (DURATION) MODULATION

We again start with a carrier which is a periodic train of pulses as in Fig. 6.1. Figure 6.14 shows an unmodulated carrier, a representative information signal $f(t)$, and the resulting pulse width modulated waveform.

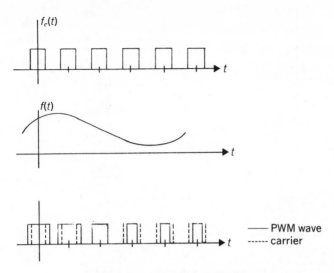

Figure 6.14 Pulse width modulation.

The width of each pulse has been varied in accordance with the instantaneous sample value of $f(t)$. The larger the sample value, the wider the corresponding pulse.

In attempting to find the Fourier Transform of this modulated waveform we run into severe difficulties. These difficulties are not unlike those which we encountered in trying to analyze frequency modulation.

Pulse width modulation is a non-linear form of modulation. A simple example will illustrate this. If the information signal is a constant, say $f(t) = 1$, the PWM wave will consist of equal width pulses. This is true since each sample value is equal to every other sample value. If we now transmit $f(t) = 2$ via PWM, we again get a pulse train of equal width pulses, but the pulses would be wider than those used to transmit $f(t) = 1$. The principle of linearity dictates that if the modulation is indeed linear, the second modulated waveform should be twice the first. This is not the case as is illustrated in Fig. 6.15.

If one assumes that the information signal is slowly varying then adjacent pulses will have almost the same width. Under this condition an approximate analysis of the modulated waveform can be performed. In so doing, a PWM waveform can be expanded in an approximate Fourier series type expansion. Each term in this series is actually an FM wave instead of being a pure sine wave. We will not carry out this analysis here.

In our study of FM, we found that the easiest way to demodulate a signal was to first convert it to AM. We shall take a similar approach to PWM. That is, we will try to show how one would go about converting one form of pulse modulation into another.

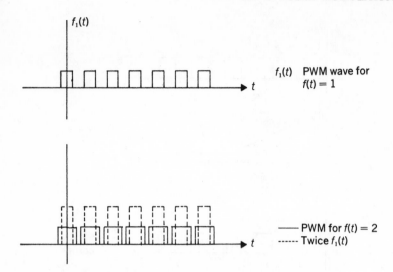

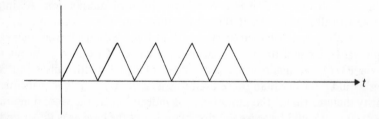

Figure 6.15   Example to illustrate that PWM is non-linear.

The significant difference between PAM and PWM is that PAM displays the sample values as the pulse amplitudes, and PWM displays these values as the "time amplitudes." We therefore search for a technique to convert back and forth between time and amplitude. We require something whose amplitude depends linearly upon time. A sawtooth waveform as sketched in Fig. 6.16 answers this need.

Figure 6.16   A sawtooth waveform.

We shall now describe one possible use of the sawtooth to generate PWM. This is not necessarily the simplest system.

We start with an information signal, $f(t)$. This is put through a sample and hold circuit. The sawtooth is then subtracted, and the resulting difference forms the input to a threshold circuit. The output of the threshold circuit is a narrow pulse each time the input goes through zero. Each of these zero crossings then triggers a bistable multivibrator.

The entire system is sketched in Fig. 6.17. Typical waveforms are shown in Fig. 6.18.

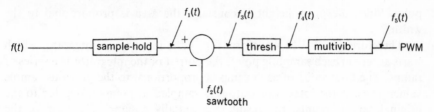

Figure 6.17   One way of generating PWM.

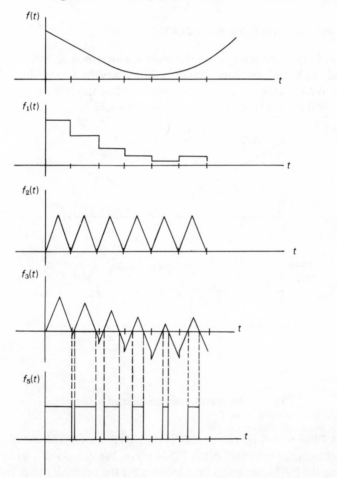

Figure 6.18   Typical waveforms in the system of Fig. 6.17.

A technique similar to the ramp (sawtooth) can be used to change PWM to PAM, and, therefore, to demodulate PWM. If each pulse in a PWM wave-form is integrated, the result is a ramp whose final value is the area of the pulse. Since the pulse height is constant, the area is proportional to the width.

Therefore, if a PWM waveform forms the input to an integrator which starts at zero at each sampling point, the output of the integrator is a series of ramps. The final value of each ramp is proportional to the previous sample value. Thus, if the integrator output is sampled at points just prior to the original sample points, the result is essentially a sampled version of the original waveform.

6.6 PULSE POSITION MODULATION

We will now very briefly consider pulse position modulation. An information signal, $f(t)$, and the corresponding PPM waveform are illustrated in Fig. 6.19. We see that the larger the sample value, the more the corresponding pulse will deviate from its unmodulated position.

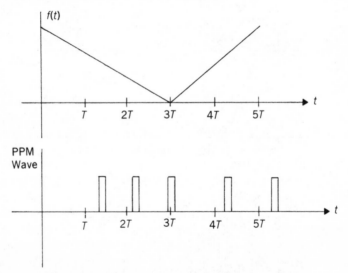

Figure 6.19 PPM—the unmodulated pulses would fall at multiples of T.

A PPM waveform can be simply derived from a PWM waveform. We first detect each "downslope" in the PWM wave. We can do this by first differentiating the PWM wave and then looking for the negative pulses in this derivative (ideally they will be impulses). If we were to now place a pulse at each of these points, the result would be a PPM wave. This process is illustrated in Fig. 6.20.

At this time it would be reasonable to ask what the purposes of the present section were, other than to confuse the student. We have probably said much less about this very complicated area of communication theory

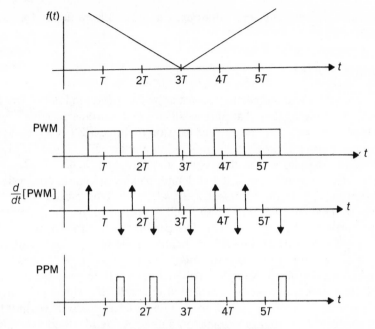

Figure 6.20   PPM derived from PWM.

than we have about some of the much simpler areas. Part of the justification for this is that PWM and PPM are sufficiently complex that not a great deal can be said in general. As such, we have only defined the two types of modulation and have attempted to show possible ways to produce these waveforms. Beyond this, we will close by mentioning that PWM and PPM afford a noise advantage over PAM. This is true since the sample values are not contained in the pulse amplitudes. These amplitudes are easily distorted by noise. This noise advantage is similar to that afforded by FM when compared to AM. In both cases we see that going to the more complicated form of modulation leads to certain disadvantages. In FM we required more space on the frequency axis. In pulse modulation, the two non-linear forms will require more space on the time axis. One must be sure that adjacent sample pulses do not overlap. If pulses are free to shift around or to get wider, as they are in PPM and PWM, one cannot simply insert other pulses in the spaces (i.e., multiplex) and be confident that no interaction will occur. Sufficient spacing must be maintained to allow for the largest possible sample value.

Besides taking more space, these non-linear forms of modulation are difficult, if not impossible, to analyze. The systems necessary to modulate and demodulate are more complicated than those required for linear forms of modulation.

In continuous forms of modulation, if one wanted a noise improvement, he had little choice but to turn to FM. In discrete pulse modulation, an alternative to PWM and PPM does exist as we shall see in the following chapter.

6.7 CROSS TALK

We now consider one source of error in a noise-free pulse modulation system. As an illustration, assume that phone calls were being transmitted via PAM and that 20 channels were being time division multiplexed. If we look at the transmitted pulse train, we can identify pulses numbered 1, 21, 41, 61, 81, . . . as being the samples of the first signal and pulses 2, 22, 42, 62, 82, . . . as samples of the second signal. If the demultiplexing system is not perfectly synchronized, a portion of pulses 2, 22, 42, . . . will be picked up by the receiver for channel 1. That is, pulse 1 will be corrupted by pulse 2, pulse 21 will be corrupted by pulse 22, etc. The person trying to listen to the first channel will hear part of the second channel in the background. This is an example of an error known as *cross talk*.

The above simplified example attributed cross talk to poor synchronization. In fact, this is not the only source of crosstalk. If a single pulse in the train is traced through the system from transmitter to receiver, we find that it will change shape from one point to another. As an idealized example, if the original pulse were a perfect square, it would have rounded corners after transmission. This is due to the fact that the system function of any transmission medium must approach zero as the frequency approaches infinity. The sharp corners of the pulse represent infinite frequencies and will therefore be rounded by the low pass filter effect. Figure 6.21 shows a perfect

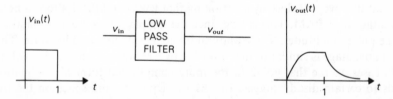

Figure 6.21 Effect of low pass filter on pulse.

square pulse and then the result of passing this pulse through a simple low pass filter comprised of a resistor and capacitor combination. It should be clear that the rounded pulse occupies a wider portion of the time axis than the original pulse. It is therefore possible for the received pulses in a train to overlap even though the transmitted pulses do not overlap. This pulse widening and overlap due to the transmission characteristics are the cause of cross talk. It has its exact dual in a frequency multiplexed system. For example, in broadcast AM radio, if one station were to spread beyond its allocated 5 KHz of bandwidth, a receiver tuned to an adjacent station would hear the undesired signal in the background. The frequency spreading would occur if the signal were gated in time, just as the time spreading in TDM pulse systems results from gating in frequency (the low pass filter characteristic of the transmission system is a frequency gate).

The above discussion of cross talk applies to any multiplexed pulse transmission system, whether it be PAM, PWM, PPM, or the types to be discussed in the next chapter.

PROBLEMS

6.1. Sketch the schematic for a circuit that will produce the product, $f(t)f_c(t)$, where $f(t)$ and $f_c(t)$ are as shown in Fig. 6.1.

6.2. Given the product, $f(t)f_c(t)$ as in Problem 6.1, sketch a system that could be used to recover $f(t)$. Verify that this system works.

6.3. You are given a low frequency band limited signal, $f(t)$. This signal is multiplied by the pulse train, $f_c(t)$ as shown below. Find the transform of the product, $f(t)f_c(t)$. What restrictions must be imposed so that $f(t)$ can be uniquely recovered from the product waveform?

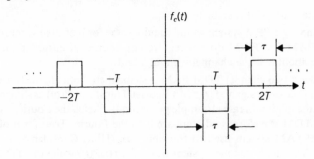

6.4. An information signal is of the form

$$f(t) = \frac{\sin \pi t}{\pi t}$$

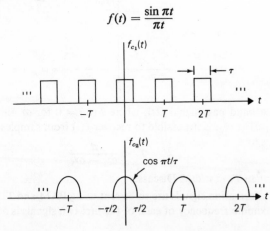

Find the Fourier Transform of the waveform resulting if each of the two carrier waveforms shown is PAM with this signal.

6.5. Sketch a block diagram of a typical pulse amplitude modulator and demodulator.

6.6. Consider the system shown below. We wish to compare $y(t)$ to $x(t)$ in order to evaluate the sample and hold circuit as a PAM demodulator. The comparison between $y(t)$ and $x(t)$ is performed by defining an error, e, as follows,

$$e \triangleq \int_0^{2\pi} [y(t) - x(t)]^2 \, dt$$

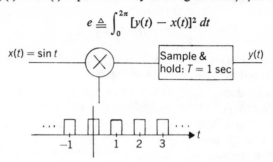

Find the value of this error term.

6.7. Explain why a PPM system would require some form of time synchronization while PAM and PWM do not require any additional information. (We are talking about a single channel without TDM.)

6.8. Consider a two channel TDM PAM system where both channels are used to transmit the same signal, $f(t)$, with Fourier Transform $F(\omega)$ as shown. Sample $f(t)$ at the minimum rate (1 sample per second). Sketch the Fourier Transform of the TDM waveform and compare it to the Fourier Transform of a single channel PAM system used to transmit $f(t)$. (Hint: Consider $S(t)$ as shown below. The single channel system transmits $f(t)S(t)$, while the TDM system can be thought of as transmitting $f(t)\hat{S}(t)$ where $\hat{S}(t) = S(t) \times S(t - \frac{1}{2})$.)

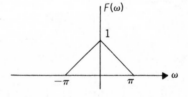

6.9. Given a band pass signal $f(t)$, where $F(\omega) = 0$ for ω outside of the range $\omega_1 < |\omega| < \omega_2$, is it possible to recover $f(t)$ from samples spaced at

$$T_s < \frac{\pi}{\omega_2 - \omega_1}$$

rather than $T_s < \pi/\omega_2$? Discuss.

6.10. Three information signals are to be sent using PAM and TDM. Suppose that the maximum frequency of each of the first two signals is 5 kHz and that the maximum frequency of the third signal is 10 kHz. Can these signals be time division multiplexed? Is it possible to time division multiplex these signals while sampling each at its Nyquist rate? Sketch the block diagram of a system that might accomplish this.

REFERENCES

See the references listed at the end of Chapter 4.

DIGITAL
COMMUNICATION
7

The earlier portions of this text dealt with transmission of a continuous time signal. The vehicle for transmission was either a continuous time function or a discrete (pulsed) time function.

The present chapter adds another dimension to the techniques for transmission. The actual signal values transmitted will become members of a discrete set. Thus, instead of dealing with the transmission of a continuum of voltage values, we concentrate upon a finite set of discrete values. The advantages of this approach will shortly become evident.

As was the case with pulse modulation, we first consider transmission of a list of numbers. This could result from a sampling operation on a time function, or the original information could be in the form of a list. The essential difference between digital communication and pulse modulation is that the numbers in the list can only take on discrete values. Such forms of transmission are termed *digital* as distinct from *analog*.

Many signals are already in the form of a list of numbers drawn from a finite set. The output from a digital computer is in this format. Other common examples include the time of day (if we round to the nearest second or minute), quantities of a particular item manufactured each hour (e.g., number of cars off an assembly line), and Teletype messages (where only the alphanumeric characters are permitted). In contrast with these digital signals, the audio signal representing someone speaking is an important example of an analog signal. If digital transmission techniques are to be used for the latter signal, we must somehow transform to a digital format. The transformation from an analog to a digital format is performed, quite logically, by an analog to digital converter.

7.1 A/D AND D/A CONVERTERS

The first step in changing an analog continuous time signal into a digital form is to convert the signal into a list of numbers. This is accomplished by sampling the time function. The resulting list of numbers will represent a continuum of values. That is, although a particular sample may be stated as a rounded-off number (e.g., 5.758 volts (V)), in actuality it should be continued as an infinite decimal. The list of numbers must therefore be *coded* into discrete *code words*. The first suggestion one might have as a way to accomplish this is to round off each number in the list. Thus, for example, if the samples ranged from 0 to 10 V, each sample could be rounded to the nearest integer. This would result in code words drawn from the 11 integers between 0 and 10.

In a majority of digital communication systems, the actual form chosen for code words is a binary number composed of 1's and 0's. The reasons for this choice will become clear when we discuss specific transmission techniques. With this binary restriction, we can envision one simple form of analog to digital converter for our "0- to 10-V" example (not necessarily the best). The converter would operate on the 0- to 10-V samples by first rounding each sample value to the nearest volt. It would then convert the resulting integer into a 4-bit binary number.

While practical A/D converters perform an operation similar to that outlined above, an explanation of this operation is best introduced by starting from scratch. The conversion of a list of sample numbers into binary code words is usually accomplished by subdividing the range of numbers into regions. A binary number is associated with each of these regions. For the example of the list of voltage values ranging from 0 to 10 V, one might associate the code word "1" with any sample value above 5 V and "0" with any sample value below 5 V. Thus, for example, 2.35 V would be coded as "0" and 6.7 V would be coded as "1." If the coding were done in this manner, a significant amount of information would be lost. A code of 0 would represent any sample voltage between 0 and 5 V. If one therefore looks at the digital signal and attempts to convert back to the analog signal, errors of up to $2\frac{1}{2}$ V/sample are possible. If a finer resolution is required, one might use two binary digits (bits) to represent any one of the sample values. This could be accomplished by dividing the range into four regions. A sample value of

0 to 2.5 V would be coded as "00," 2.5 to 5 V as "01," 5 to 7.5 V as "10," and 7.5 to 10 V as "11." One can continue to increase the length of the code words to increase the resolution. For 3-bit words, the range is divided into eight regions. Each additional bit doubles the number of regions.

We are now in a position to discuss the complete system that performs the conversion. Such systems are known as *analog to digital converters* (A/D or ADC). We shall illustrate several forms of the converter. Our illustrations will be for binary coding. In some cases, other forms of coding are required, but the modifications necessary to the systems presented should be relatively straightforward.

Figure 7.1 illustrates one form of ADC. The diamond-shaped boxes are comparators. They compare the input to some fixed value and give one output

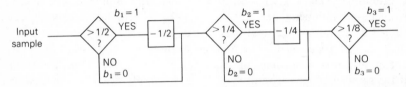

Figure 7.1   A/D converter.

if the input exceeds that fixed value and another output if the reverse is true. The block diagram indicates these two possiblilities as two possible output paths labelled YES and NO. The figure is shown for 3-bit code words and for a range of input values between 0 and 1 V. If the input range of signal sample values were not 0 to 1, the signal could be scaled (amplified or attenuated) to achieve values within this range. If more (or fewer) bits are required, the appropriate comparison blocks can be added (or removed).

Note that b_1 is the first bit of the coded sample value, sometimes known as the *most significant bit* or msb. b_3 is the third and final bit of the coded sample value, known as the *least significant bit* or lsb. The reason for this terminology is that the weight associated with b_1 is 2^2 or 4, while the weight associated with b_3 is 2^0 or 1. The ADC of Fig. 7.1 is performing the exact operation described previously. It performs a successive approximation to the sample value. First we ask whether the sample value is in the upper or lower half of the range. Then we investigate the half-interval to see if the sample is in the upper or lower half *of that half*. Thus, each successive bit (moving from left to right in the binary number) divides the previous region in half. This is true since each bit in a binary number is weighted by a power of 2, and as we move to the right in a binary number, the weight of each bit is one half of the previous weight.

Illustrative Example 7.1

Illustrate the operation of the system of Fig. 7.1 for the following two input sample values: 0.2 and 0.8 V.

Solution

For 0.2 V, the first comparison with $\frac{1}{2}$ would yield a NO answer. Therefore $b_1 = 0$. The second comparison with $\frac{1}{4}$ would yield a NO answer. Therefore, $b_2 = 0$. The third comparison with $\frac{1}{8}$ would yield a YES answer. Therefore, $b_3 = 1$, and the binary coding of 0.2 into 3 bits would be 001.

For the 0.8-V input, the first comparison with $\frac{1}{2}$ would yield a "YES" answer. Therefore $b_1 = 1$, and we subtract $\frac{1}{2}$, leaving 0.3. The second comparison with $\frac{1}{4}$ would yield a YES answer. Therefore, $b_2 = 1$, and $\frac{1}{4}$ is subtracted, leaving 0.05. The third comparison with $\frac{1}{8}$ would yield a NO answer. Therefore, $b_3 = 0$, and the binary coding of 0.8 into 3 bits would be 110.

A simplified system can be realized if, at the output of the block marked "$-\frac{1}{2}$" in Fig. 7.1, a multiplication by 2 is performed and the result is fed back into the comparison with $\frac{1}{2}$. All blocks to the right can then be eliminated, as in Fig. 7.2. The signal sample can be cycled through as many times as desired to achieve any number of bits of code word length.

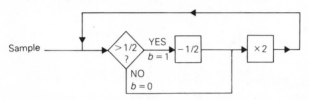

Figure 7.2 Simplified ADC of Fig. 7.1.

Figure 7.3 shows a representative $f(t)$ and the resulting digital form of the signal for both 2-bit and 3-bit ADC.

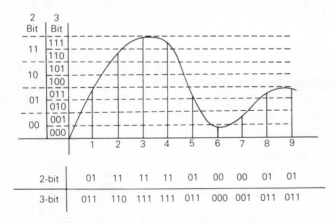

Figure 7.3 Result of ADC for sample $f(t)$.

A second type of ADC is illustrated in Fig. 7.4. A ramp voltage is developed starting at the periodic sampling points. The length of time it takes for

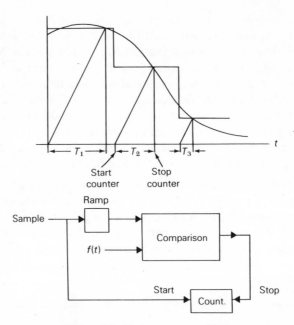

Figure 7.4 Ramp generator ADC.

this ramp voltage to reach the function is proportional to the value of the function at the intersection point. A binary counter is started as the ramp generator leaves the zero point. The counter stops when the ramp voltage intersects the signal value. The binary number then on the counter is a coded version of the signal value. The ramp and counter must be fast enough to accomplish this entire process in the time devoted to taking one sample value.

The time axis has been greatly exaggerated in Fig. 7.4. The ramp must be fast enough so that the value of the function at the point of intersection is very close to the actual sample value. The speed of the counter is set so that a ramp will attain the maximum possible value in the time it takes for the counter to reach its maximum value. Thus, if 2 bits are used, the counter should reach a count of "11" in the time it takes the ramp to go from zero to the maximum amplitude value.

Illustrative Example 7.2

It is desired to build a ramp generator ADC to convert $f(t) = \sin 2\pi t$ into a 4-bit digital signal. Choose appropriate parameters for the ADC.

Solution

To comply with the sampling theorem, the sampling rate must be greater than 2 samples/sec. We can therefore assume a sampling rate of 2.5 samples/ sec (there are many factors which go into choosing the rate, and a rate several times the minimum is often chosen). It is now necessary to set the slope of the ramp generator. The maximum slope of the signal is 2π. The slope of the ramp must be considerably greater than this value to assure that the intersection points will be close to the sample points. For this example, let us choose a ramp slope of 100 V/sec. Since the maximum value of the signal is unity, it will take the ramp function 0.01 sec to reach this maximum. The counter should therefore count from 0000 to 1111 in 0.01 sec. This requires a counting rate (clock) of 1 count/0.000625 sec, or 1,600 counts/sec.

We now shift our attention to the conversion of a digital signal to an analog signal. This is performed by a *digital to analog converter* (D/A converter or DAC). To perform this conversion, we need simply associate a value with each binary code word. If the code word represents a region of sample values, the actual value chosen for the conversion is usually the center point of the region. If the A/D conversion is performed as previously described, the reverse operation is equivalent to assigning a weight to each bit position. In Illustrative Example 7.2, the interval was divided into 16 regions. Therefore, for example, the lower limit of the region corresponding to the code word "0011" is $\frac{3}{16}$ V. The lower limit of the region corresponding to the code word "1111" is $\frac{15}{16}$. To reach the midpoint of the region, we add $\frac{1}{32}$ to each of these lower limits.

We see that conversion to the analog sample value is accomplished by converting the binary number to decimal, dividing by 16, and adding $\frac{1}{32}$. The conversion of the binary number to the decimal equivalent is accomplished by recognizing that each column in a binary number has associated with it a power of 2. Thus, a "1" in the leftmost position indicates 2^3, or 8. A "1" in the next position indicates 2^2, or 4. An analog voltage can be reconstructed from a digital code word by using the 1's and 0's to switch appropriate sources and adding the results together. If a 1 appears in the first position, a $\frac{1}{2}$-V battery is switched in. The second position controls a $\frac{1}{4}$-V battery, and the third position, a $\frac{1}{8}$-V battery. An idealized circuit that accomplishes this conversion is shown in Fig. 7.5. The addition of $\frac{1}{32}$ V to arrive at the midpoint of the region is included in the figure.

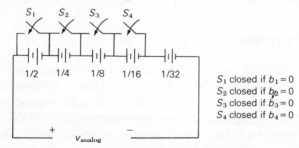

Figure 7.5   Idealized D/A converter—4-bit.

Practical DACs use linear circuits for the addition and one source with dividing networks to generate the various voltages. One practical realization of a DAC using an operational amplifier is shown in Fig. 7.6.

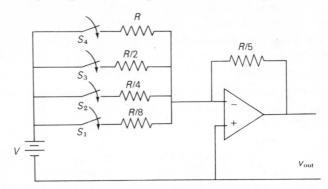

Figure 7.6   DAC using operational amplifier.

The gain of the operational amplifier shown in Fig. 7.6 is $R/5R_{in}$, where R_{in} is the parallel combination of the resistors included in the input circuit (i.e., those with associated switches closed). In terms of input conductance, the gain is $RG_{in}/5$, where G_{in} is the sum of the conductances associated with closed switches. Therefore, closing S_1 contributes $\frac{8}{5}$ to the gain. Closing S_2 contributes $\frac{4}{5}$ to the gain; S_3 contributes $\frac{2}{5}$; S_4 contributes $\frac{1}{5}$. In order to calculate the gain due to more than one switch being closed, we simply add the associated gains. Therefore if the input voltage source, V, is made equal to 5 V, we see that S_1 contributes 8 V to the output; S_2, 4 V; S_3, 2 V; and S_4, 1 V. The output voltage is an integer between 0 and 15 corresponding to the decimal equivalent of the binary number controlling the switches. To convert this system into a DAC analogous to that of Fig. 7.5, we need simply divide the output by 16 and add $\frac{1}{32}$ V. The division can be performed by scaling V or by scaling the resistor values. The addition of $\frac{1}{32}$ V can be accomplished by adding a fifth resistor (unswitched) to the input. The value of this resistor would be $2R$.

7.2   CODED COMMUNICATION

The previous section described techniques for converting a continuous analog signal into a digital signal. The digital signal was composed of a list of numbers, where the numbers could take on only a finite number of values. We saw that the list of numbers did not exactly equal the original sample values but represented rounded-off versions of these values. Therefore in addition to the necessity of doing extra work in converting an analog signal into a digital signal, the digital signal cannot be used to perfectly reconstruct the original analog signal. Why, then, might somebody wish to change an analog signal into a digital signal? The present section seeks to answer this question.

The earliest forms of communication were by telegraph. A series of dots and dashes were transmitted to spell out each word. This was done since the complex audio signals could not be distinguished from the background static along a line, while a long and short duration of a tone could be distinguished.

In early telegraph, the coding and decoding of the signals were done by a human operator. The operator read (or listened to) the message and changed each letter into its Morse coded version. In receiving a message, the operator reversed this procedure. The speed of transmission had to be very carefully controlled so that it did not exceed the speed of the keyer (operator). Although the telegraph transmission line, by virtue of its bandwidth limitations, did place a maximum limit on the speed of transmission, this was never a serious consideration. The human operator never approached this speed limit even though the early telegraph lines were highly inefficient by today's standards.

After many years of favoring non-coded communication, we are rapidly returning to the coded forms for two major reasons. The first is related to the original (telegraph) reason. The air is being polluted with so many electrical signals that noise-free communication is virtually impossible. Signals are constantly corrupted by static and other forms of interfering signal.

Unless a great deal is known about the kind of signal expected at the receiver, no effective techniques of correcting the noise corruption are known. That is, when the signal is changed by noise and the altered signal is received, there must exist some way of separating the noise from the signal. This requires that the signal have some distinguishing form. Such a form is provided through the use of coding.

The second reason for the re-emphasis upon digital coded communication is the changing format of information signals. For many years the only type of information signal encountered was the audio signal limited to frequencies below the cutoff for human hearing. Today we find machines communicating data to other machines using information signals that are considerably different from audio waveforms. The demands placed upon a communication system are much more complicated than those resulting from voice communication.

In a coded communication system, the signal can take on only one of several discrete forms. In most cases when the set of discrete forms is small, an acceptable signal will not change into another acceptable signal when corrupted by noise.

Using more acceptable terminology, we transmit a word from a *dictionary* of acceptable message words. The received word will usually not exactly match any entry in our dictionary.

If the noise corruption is not too great, we can make an educated guess as to which dictionary word was sent. This is the essence of coded communication.

7.3 PULSE CODE MODULATION (PCM)

Pulse code modulation (PCM) is a direct application of the analog to digital converter discussed in Section 7.1. Rather than simply present it as such, we will rederive the results using a slightly different approach. We shall begin this derivation with a pulse modulation system in order to illustrate some physical concepts and to develop some of the relevant terminology.

Suppose that in a PAM system the amplitude of each pulse is rounded off to one of several possible levels. Alternatively, suppose the original continuous time function is first rounded off to several possible levels (yielding a staircase-type function) and then this rounded-off function is sampled for PAM transmission. An example of this rounding-off operation is shown in Fig. 7.7.

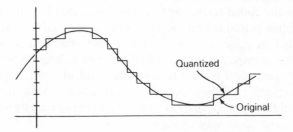

Figure 7.7   Example of rounding off to discrete levels.

The rounding-off process is known as *quantization*, and its performance introduces an error known as *quantization noise*. That is, the staircase approximation is not identical to the original function, and the difference between the two is an error signal. This will be discussed in detail in Section 7.6.

The dictionary of possible signals (pulse heights) has been reduced to include only those quantization levels used. A received pulse is compared to the possible transmitted pulses, and it is decoded into the dictionary entry which most closely resembles the received signal. In this way, small errors (up to one half of the quantization step size) can be corrected.

This principle of error correction is the major reason for the popularity of PCM. Suppose, for example, it were desired to transmit a signal from California to New York over coaxial cable. If the signal were transmitted via PAM, the resulting noise would make the received signal unintelligible. That is, noise would be added along the entire transmission path, with additional noise being added in each amplifier. Many amplifiers would be required to cancel attenuation along the way.

If this same signal were now coded using PCM, under certain conditions most errors could be corrected. If the amplifiers were spaced in such a way that the noise introduced between any two amplifiers is, on the average, less than one half of the quantization step size, each amplifier could restore the function to its original form before amplifying and sending it on its way. That is, each amplifier would round the received pulses to the nearest acceptable level and then retransmit, thereby essentially eliminating all but the strongest noise. In such applications, the amplifiers are known as *repeaters* since they are regenerating the original PCM signal.

The more quantization round-off levels used, the closer the staircase function resembles the desired signal. The number of levels then determines the signal *resolution*, that is, how small a change in signal level can be detected by looking at the quantized version of the signal.

If high resolution is required, the number of quantizing levels must increase. At the same time, the spacing between levels decreases. As the dictionary words get closer and closer together, the noise advantages of this type of transmission decrease.

If resolution could be improved without increasing the dictionary size (i.e., without putting dictionary words closer together), the error correcting advantages could be maintained. PCM is a method of accomplishing this.

In a PCM system, the dictionary of possible transmitted signals contains only two entries, a "0" and a "1." The quantization levels are coded into binary numbers. Thus, if there were eight quantization levels, the values could be coded into 3-bit binary numbers. Three pulses would be required to send each quantization value. Each of the three pulses would represent either a 0 or a 1. Note that we do not necessarily send a 0-V pulse to indicate a coded "0" and 1-V pulse for a coded "1." The actual techniques used for sending the binary digits are discussed in Section 7.5.

When the above description of PCM is compared with the operation of the ADC as described in Section 7.1, we see that if the digital output of the ADC is transmitted, we have PCM. Therefore, a binary ADC is actually a PCM modulator.

Figure 7.8 repeats the representative $f(t)$ of Fig. 7.3 and shows the 2-bit and 3-bit PCM forms of the encoded waveform. For illustrative purposes, a 1-V pulse is used to represent a "1" and a 0-V pulse represents a "0." Again, we emphasize that the PCM signal is not necessarily transmitted in this form.

In a similar manner, a PCM demodulator is simply a binary digital to analog converter, as discussed in Section 7.1. The modulator and demodulator are available as LSI chips and are given the name *CODEC* (for coder-decoder).

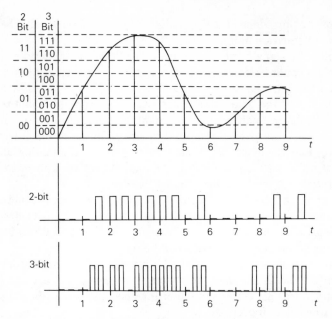

Figure 7.8 Example of 2-bit and 3-bit PCM.

More than one channel of PCM information can be transmitted simultaneously through use of time division multiplexing (TDM). The principles are identical to those described earlier for non-coded forms of pulse modulation. We shall illustrate the application of TDM to PCM with a practical example.

Example of PCM System—Bell System T-1

The Bell Telephone T-1 system of transmission is a popular example of a practical application of PCM. This system is used primarily for short distance transmission, from about 15 to 65 kilometers (km). The spacing of repeaters is less than 2,000 meters (m).

T-1 is a multiplex voice system permitting 24 voice channels of information to be transmitted. Each signal is sampled at a rate of 8,000 samples/sec. We can therefore assume a maximum audio frequency of 4 kHz, which is consistent with telephone voice transmission. The sample values are then encoded into a 7-bit binary code. The amplitudes are therefore quantized into 1 of 128 (2^7) levels. This is more than enough to permit transmission of intelligible speech. In fact, at only two levels of quantization (1 bit) one can still understand speech. An eighth bit is added to transmit signalling information used for such bookkeeping functions as billing.

Since each sample requires 8 pulses and 24 channels are being time division multiplexed, 8×24 or 192 bits must be transmitted during each sampling interval. In fact, a 193rd bit is added to provide synchronization of the time division multiplex commutator. This is known as *framing*. Since there are 8,000 samples, or frames, per second, the rate of bit transmission must be 1,544,000 bits/sec. This allows 0.65 μsec/bit. If pulses are used to transmit these bits, the width of each pulse is one half of the total space available, or 0.325 μsec (i.e., a 50% duty cycle is used).

The only detail missing from the above description is the operation of *companding*. We will reserve discussion of this and present it in the context of quantization noise analysis in Section 7.6.

When PCM signals are time division multiplexed, cross talk becomes a problem. The cross talk occurs in exactly the same manner as with any form of pulse modulation. The reader is therefore referred to the earlier discussion in Chapter 6.

7.4 DELTA MODULATION (DM)

PCM is an excellent method to send data or analog signals when correction for noise errors is critical. However, PCM requires somewhat complex coders decoders, and, as resolution is increased, the number of bits per sample increases. This may severely limit the number of channels that can be time division multiplexed. A continuing search is therefore taking place to devise coded transmission systems which possess noise immunity without having to send as many bits per second as is required by PCM. *Delta modulation* is one such discovery.

In developing PCM, we took each sample value and coded it into a series of binary pulses. Each set of binary pulses gave sufficient information to enable evaluation of the corresponding quantized sample value.

In delta modulation, the knowledge of past information is used to simplify the coding technique and the resulting signal format. The signal is first quantized into discrete levels, but the size of each step in the staircase approximation to the original function is kept constant. That is, the quantized signal is constrained to move by only one quantization level at each transition instant. The location of the steps is controlled to correspond to the sampling instants. Thus, at each sampling instant the quantized waveform must either increase by the standard step size or decrease by this amount. The quantized signal *must* change at each sampling point. It cannot remain constant since this would result in three possible actions, thus negating the possibility of using binary communication techniques. An example of this quantizing process is shown in Fig. 7.9.

Once the quantization is performed, transmission of the quantized signal becomes trivial. We simply transmit a string of 1's and 0's. A 1 can indicate a positive transition, and a 0 indicates a negative transition. The transmitted

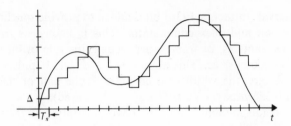

Figure 7.9 Example of quantized signal for delta modulation.

bit train for the example of Fig. 7.9 would then be

1 1 1 1 1 0 0 0 0 1 1 1 1 1 1 1 0 0 0 0 0

A delta modulator (encoder) can be constructed using a staircase generator. At each sampling instant, the generator output is compared to the input waveform. If the input is greater than the staircase function, a positive step is initiated. If the input is smaller than the staircase, a negative step results.

Figure 7.10 shows a block diagram of the modulator.

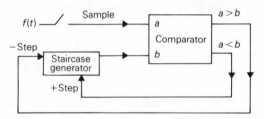

Figure 7.10 Delta modulator.

The demodulator for a delta modulated waveform is simply a staircase generator. If a 1 is received, the staircase increments positively. If a 0 is received, the staircase increments negatively.

The key to effective use of delta modulation is the intelligent choice of the two parameters step size and sampling period. These must be chosen such that the signal cannot possibly change too fast for the steps to accurately follow. Since the signal has a definable upper frequency cutoff, we know the fastest rate at which it can change. However, to account for the fastest possible change in the signal, the sampling frequency and/or the step size must be increased. Increasing the sampling frequency results in the delta modulated waveform requiring a larger bandwidth. Increasing the step size increases the quantization error. That is, the step approximation to the function becomes poorer as the step size increases. This is most obvious during periods when the function is almost constant. Figure 7.11 shows the consequences of a poor choice of step size.

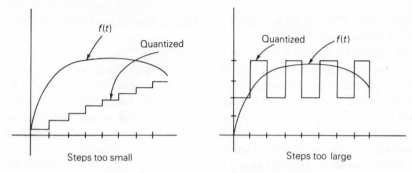

Figure 7.11   Importance of proper step size.

Some improvement is possible over the basic delta modulation system without increasing bandwidth or increasing quantization noise. Such improvement is realized if the step size *adapts* to the situation rather than remaining constant. Thus, during periods when the signal is changing slowly, the step size decreases, thus reducing the quantization error. During periods when the signal is changing rapidly, the step size increases. Such a system is known as *adaptive delta modulation* (ADM). It is, of course, necessary for the receiver to be able to reconstruct the staircase function being transmitted, including the knowledge of the size of each step. Therefore the factor controlling the step size must be derivable both at the transmitter and at the receiver.

Expanding upon this last point, consider one possible adaptive system. The step size is made exactly equal to the difference between the step function and the actual signal. A representative example is shown in Fig. 7.12. This is

Figure 7.12   Adaptive quantization example.

seen to be the sample and hold system and can be implemented with the switch and capacitor as shown. Transmission of the staircase approximation is accomplished by sending a 1 if a positive step occurs and a 0 if a negative step occurs.

For the example shown, the transmitted bit sequence is

$$1\ 1\ 1\ 0\ 0\ 0\ 1$$

This system is not workable for the following reason. The receiver sees the train of 0's and 1's and has no way of extracting the information regarding the step size. There is no way of reconstructing the original staircase function at the receiver.

We require a system whose step sizes depend only upon the sequence of binary numbers being transmitted. We observe that during periods of slow change, the sequence of transmitted bits tends to oscillate between 0 and 1 as the quantized waveform continually compensates for overshoot. During periods of rapid change, the staircase function is trying to catch up. Therefore the transmitted bit train contains strings of either 0's or 1's depending upon whether the original function is decreasing or increasing.

The above observation is the key to a workable adaptive system. If a given length string of bits contains an almost equal number of 1's and 0's, the step size is decreased. If the string contains many more 1's than 0's (or vice versa), the step size increases. Since the step size is dependent only upon the sequence of bits, the exact same adaptive operation can be reproduced at the receiver.

The actual step size control is performed by an integrator. The integrator sums the bits (0's and 1's) over some fixed period. If the sum deviates from one half of the number of bits in the period, the step size is increased. As the sum approaches one half of the number of bits, the step size approaches zero. In practice, the bit sum is translated into a voltage which is then fed into a variable gain amplifier. The amplification is a minimum when the input voltage corresponds to an equal number of 1's and 0's in the period. The amplifier controls the step size.

With the use of adaptive delta modulation, acceptable telephone transmission has been achieved using 32-kilobit/sec transmission. This compares to the 64 kilobits/sec (8 bits/sample × 8,000 samples/sec) required for one channel of telephone transmission using PCM. This nominal two-to-one reduction in bit rate is the reason for considerable attention being given to delta modulation.

7.5 TRANSMISSION TECHNIQUES FOR DIGITAL SIGNALS

The first thought one might have regarding transmission of the 0's and 1's resulting from coded pulse modulators is to send pulses of height 0 V and height 1 V. There are several reasons for rejecting this technique.

Clearly it is desirable to make the difference between a transmitted "1" and a transmitted "0" as large as possible. This will reduce the chance of mistaking a 0 for a 1 at the receiver. If the transmission medium is capable of accepting simple pulses, a better choice than 0 V and 1 V exists. In particular, it would be better to use pulses of height $+V$ and $-V$, where V is the highest transmittable voltage (due to transmitter power limitations, cable breakdown characteristics, or FCC restrictions).

If air is the transmission medium being used, pulses are not the best choice for transmission of digital information. As we saw in the section analyzing pulse transmission, the Fourier Transform of a pulse train covers a very wide range of frequencies. The pulse train could therefore not be efficiently transmitted through the air.

The most commonly used systems for transmitting binary information through the air are *frequency shift keying* (FSK) and *phase shift keying* (PSK). A variation of PSK known as *differential phase shift keying* (DPSK) is also often used.

In the case of FSK, two oscillators operating at different frequencies are employed (or, alternatively, one variable frequency oscillator). Thus, to transmit a binary "0," we transmit a burst of $\cos \omega_0 t$. To transmit the binary "1", we transmit a burst of $\cos \omega_1 t$. This can be seen to be a form of frequency modulation (FM) where the information signal is always one of two possible values.

Since only two possible frequencies are used, the modulator and demodulator need not be so sophisticated as those used for FM. One simple realization of the demodulator can be thought of as consisting of two band pass filters, one tuned to ω_0 and the other tuned to ω_1. As might be expected, the farther apart the two frequencies are, the easier it is to distinguish between them, even in background noise. However, the bandwidth of the coded signal increases as the frequencies are separated, so a trade-off analysis is required in the design of an FSK system.

Figure 7.13 shows two forms of the FSK receiver. The second form substitutes heterodyners for the filters and can be thought of as consisting of two synchronous demodulators. The output of the synchronous demodulator matched to the frequency being transmitted is a constant (actually a pulse since a burst is transmitted), while the output of the other demodulator is a sinusoid whose frequency is equal to the spacing between the two transmitted frequencies. The integral of the first output is therefore a ramp, while the

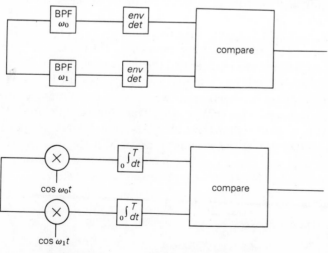

Figure 7.13 Two types of FSK demodulator.

integral of the second output is a sinusoid whose amplitude is inversely proportional to the difference frequency. Therefore, a comparison between the
two integrated outputs should indicate the correct frequency being transmitted. We shall analyze errors in the next section.

Just as FSK corresponds to FM, a second form of binary transmission
corresponds to phase modulation. This technique is known as PSK (phase
shift keying). In this case, both the 0 and 1 are transmitted using oscillations
of the same frequency. Only the phase is varied. For example, the "0" can be
sent by a burst of $\cos \omega_0 t$, and the "1" by a burst of $\cos (\omega_0 t + \theta)$, where θ
is an angle between 0 and 2π. Again, the more different the signals, the easier
it is to distinguish between them at the receiver. For that reason, $\theta = \pi$ is
commonly used. In this case, the two signals are the mirror image of each
other (i.e., $\cos (\omega_0 t + \pi) = -\cos \omega_0 t$).

In order to demodulate PSK, a synchronous demodulator is required.
The demodulator is shown in Fig. 7.14. If the received signal is either

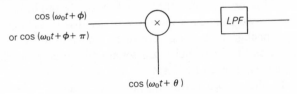

Figure 7.14 PSK demodulators.

$\cos (\omega_0 t + \phi)$ or $\cos (\omega_0 t + \phi + \pi)$, the output of the synchronous demodulator will vary between $\frac{1}{2} \cos (\phi - \theta)$ and $-\frac{1}{2} \cos (\phi - \theta)$ as the transmitted
binary train varies. Note that both of these outputs are constants. The output
waveform would therefore vary between a positive constant and its mirror
image. Without a knowledge of the exact phase of the received signal, it is
not possible to associate 1 and 0 with the two voltage levels. That is, the
positive voltage at the output of the demodulator can correspond to either a
transmitted 0 or a transmitted 1 depending upon the phase difference between
the local oscillator in the receiver and the received waveform. To get around
this ambiguity, *differential phase shift keying* is often used. Here, instead of
the instantaneous phase determining which bit is transmitted, it is the change
in phase which carries the information. Thus, for example, if a "1" is to be
transmitted, the phase of the transmitted sinusoidal signal changes by π from
that transmitted during the previous interval. If a "0" is to be transmitted,
the phase does not change. By examining the phase transitions, one can
decode the received waveform. An example is shown below:

Transmitted bit		1	0	1	0	0	0	1	0
Phase of carrier	0	π	π	0	0	0	0	π	π

Two choices for initial state

A demodulator for DPSK is shown in Fig. 7.15. The received waveform is delayed by one sampling period. The delayed waveform multiplies the non-delayed version. If the phase has not shifted in one sampling period, the

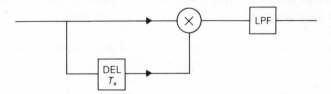

Figure 7.15   DPSK demodulator.

delayed and non-delayed signals are identical, and the result of multiplication and low pass filtering is a positive constant. If the phase has shifted by π during the sampling period, the result of multiplying and low pass filtering is a negative constant. Thus, the positive constant at the output of the low pass filter would correspond to a "0," and the negative constant would correspond to a "1."

Bit synchronization is an important consideration in transmission schemes. It is important that the receiver be able to partition the received signal into time segments corresponding to the sampling points.

Suppose first that we were lucky enough to have the transmitted signal be an alternating train of 0's and 1's. The receiver would then see alternating frequencies or phases, with a transition occurring at each sampling point. In such a case, bit synchronization is relatively simple to perform.

We are not always so lucky to have a transition at each sampling instant. If a particular signal does not have sufficient numbers of transitions to allow accurate synchronization, alternate techniques are required. For example, in a TDM system, one channel could be reserved for use in synchronization. This channel would transmit a very distinctive signal such as one composed of alternating 0's and 1's. This, of course, sacrifices one entire channel. It would be preferable to provide a form of "self-synchronization" of the signal.

To aid in synchronization, one may vary the basic technique of sending a 0 or a 1. Three schemes are commonly used to control the sinusoidal generators in FSK or PSK. These are known as *return to zero* (RZ), *non-return to zero* (NRZ), and *Manchester* coding.

In NRZ, the sinusoidal generator stays on until the next bit is ready for transmission, at which time the new value immediately takes over. Synchronization of this form of transmission relies upon the signal having characteristics which result in a sufficient number of transitions. Since the generator stays constant for one entire sampling period, this form of transmission has the narrowest possible bandwidth.

In RZ, the sinusoid corresponding to a particular bit returns to the "0" state before reaching the end of the sampling interval. Since the pulses are not so wide as in NRZ, the bandwidth is wider with this form of transmission. The advantage is that in the case of a long string of 1's, transitions do occur during each sampling interval to aid in synchronization.

For the Manchester code, a "0" is transmitted by sending one of the two frequencies (or phases) during the first half of the interval and the second frequency (or phase) during the second half. To transmit a "1," the reverse order is used. The advantage of the Manchester code is that a transition occurs at the center of every sampling period, thus aiding synchronization. In addition, the transmitter is in any one of the two states exactly one half of the time. This eliminates certain types of biasing errors.

The three types of transmission are illustrated in Fig. 7.16 using pulses of height 0 and 1. The sequence 11000101 is illustrated.

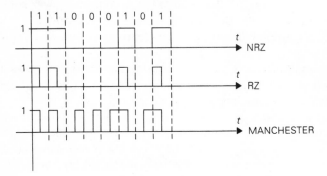

Figure 7.16   Three pulse transmission schemes.

7.6 SOURCES OF ERROR IN DIGITAL COMMUNICATION SYSTEMS

There are two distinct sources of error in a digital communication system. The first is due to inherent characteristics of the system and is independent of any noise added during transmission. The second type is the result of noise being added to the modulated signal during transmission.

When an analog signal is converted into digital format, it is necessary first to quantize the signal to discrete levels. Examination of Fig. 7.7 shows that a difference exists between the quantized waveform and the actual waveform. In PCM, it is the quantized waveform that is coded and transmitted. There is no way to ever recover the original waveform once it has been quantized. Therefore, the quantization operation can be thought of as contributing a noise factor which is defined as the difference between the original function and the quantized waveform. This difference is, quite logically, known as *quantization noise*. It is illustrated for one example in Fig. 7.17.

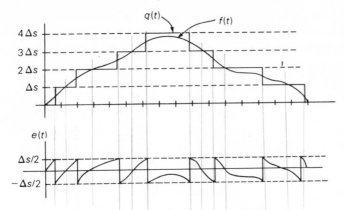

Figure 7.17 Quantization noise in PCM.

Note that the magnitude of the quantization noise is always less than one half of the step size.

The specific form of this noise depends upon the specific form of the original signal, the size and placement of the quantization levels, and the sampling frequency. While it is possible to describe this noise as a precise equation (i.e., nothing in the process is random), it is usually sufficient to think in terms of average quantities and performance. We shall therefore find the mean-square value of this noise, that is, the average noise power.

Looking back at Fig. 7.7, let us designate the quantized levels as $S_1, S_2,$ $\ldots, S_8$. If the original function is designated $f(t)$, then the error within any sampling interval is given by

$$e(t) = f(t) - S_k \tag{7.1}$$

where S_k is the quantized level closest to the sample value for that time interval.

Suppose now that a probability density function is developed for $f(t)$. This is done by observing what fraction of time the signal spends between any two levels. We denote this density as $p(f)$, where $p(f)df$ is the fraction of time (probability) that $f(t)$ is between f and $f + df$. The average power of the quantization noise is then given by

$$\overline{e^2} = \sum_{k=1}^{8} \int_{S_k - \Delta S/2}^{S_k + \Delta S/2} p(f)(f - S_k)^2 \, df \tag{7.2}$$

where ΔS is the spacing between quantization levels. This is simply a weighted average of the quantization noise power in each interval.

In order to simplify this expression, we make use of the mean-value theorem regarding integrals. That is,

$$\int_{t_1}^{t_2} f(t)g(t) \, dt = f(t_0) \int_{t_1}^{t_2} g(t) \, dt \tag{7.3}$$

where t_0 is a point within the range of integration $t_1 < t_0 < t_2$. Using this theorem to remove $p(f)$ from the integral of Eq. (7.2), we denote the values of this function as $p_1, p_2, \ldots, p_8$. Equation (7.2) then reduces to

$$\overline{e^2} = \sum_{k=1}^{8} p_k \int_{S_k - \Delta S/2}^{S_k + \Delta S/2} (f - S_k)^2 \, df \qquad (7.4)$$

The integrals can now be evaluated to yield

$$\int_{S_k - \Delta S/2}^{S_k + \Delta S/2} (f - S_k)^2 \, df = \frac{(\Delta S)^3}{12} \qquad \text{for all } k \qquad (7.5)$$

Therefore,

$$\overline{e^2} = \sum_{k=1}^{8} p_k \frac{(\Delta S)^3}{12} \qquad (7.6)$$

Now, $p_k \Delta S$ is approximately equal to the probability that $f(t)$ is between $S_k - \Delta S/2$ and $S_k + \Delta S/2$. As ΔS approaches zero, this approximation becomes more precise. Therefore,

$$\sum_{k=1}^{8} p_k \Delta S \simeq 1$$

and

$$\overline{e^2} \simeq \frac{\Delta S^2}{12} \qquad (7.7)$$

Equation (7.7) is the desired result.

Quantization noise exists with delta modulation as well as with PCM. It is again defined as the difference between the actual waveform and the quantized (in this case, staircase) approximation to the waveform. In the case of delta modulation, the magnitude of the quantization noise will range up to the full step size (ΔS). This assumes that the sampling period and step size are chosen in such a way that the staircase can respond to the fastest possible change in the signal. In order to carry this analysis further, an assumption is often made that the quantization noise has a uniform probability density function. That is, the probability of any value between plus and minus the step size is the same as the probability of any other value. This assumption is perfect if the original signal is a ramp and the step is just the correct size to "keep up" with the function. Using this uniform probability assumption, it is easy to show that the mean-square value of the quantization noise is $(\Delta S)^2/3$. (The student should prove this as an exercise.) Note that this mean-square noise is four times as great as that for PCM. Adaptive delta modulation reduces this considerably.

We now wish to evaluate the effects of additive transmission noise in digital communication systems. The effect of the noise will depend upon the method of transmission.

Let us first assume that pulses of $+V$ and $-V$ volts are transmitted to represent 1's and 0's, respectively, and that the receiver samples the received signal. We decide that a $+V$ was sent if the sample is positive and that a $-V$ was sent if the sample is negative.

Analysis of this situation is an application of basic discrete probability. We shall illustrate the analysis by means of a specific example.

Illustrative Example 7.3

Assume that in a digital communication system pulses of either $+1$ or -1 V are transmitted to represent a binary 1 and binary 0, respectively. Gaussian noise with zero mean and unit variance is added to the signal during transmission. We wish to find the probability of making a bit error, known as the *bit error rate*, or BER.

Solution

We assume that the receiver samples the received signal and compares the sample value to 0. If the sample is greater than 0, the receiver says that a "1" was transmitted. If the sample is less than 0, a "0" is decoded.

If a "1" is transmitted, the sum of the signal plus noise will be Gaussian with mean 1 and variance 1. The probability of this being erroneously demodulated as a "0" is simply the probability that this Gaussian variable is negative. This is given by

$$\int_{-\infty}^{0} p(s)\, ds = \frac{1}{\sqrt{2\pi}} \int_{-\infty}^{0} e^{-(s-1)^2/2}\, ds \tag{7.8}$$
$$= 0.159$$

The final result in Eq. (7.8) is found by referring to a table of error functions since the integral cannot be evaluated in closed form (*see* Appendix III).

Suppose now that we consider an FSK transmission scheme with the synchronous demodulator shown in Fig. 7.13. We again illustrate the effects of additive noise by means of an example.

Consider an FSK transmission system where the two frequencies are ω_0 and ω_1. The received signal has amplitude A and duration T. White Gaussian noise of power spectral density $N_0/2$ is added during transmission. We shall find the bit error rate.

We repeat Fig. 7.13 as Fig. 7.18.

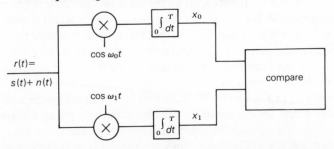

Figure 7.18   FSK demodulator with additive noise.

If we denote the received signal as $r(t) = s(t) + n(t)$, the two inputs to the comparator are given by

$$x_0 = \int_0^T r(t) \cos \omega_0 t \, dt = \int_0^T \{s(t) + n(t)\} \cos \omega_0 t \, dt$$
$$x_1 = \int_0^T r(t) \cos \omega_1 t \, dt = \int_0^T \{s(t) + n(t)\} \cos \omega_1 t \, dt \tag{7.9}$$

The comparator compares x_0 and x_1 and chooses the signal (0 or 1) based upon which of the inputs is larger. This is the same as taking the difference of the inputs and comparing this difference to zero. That is, if x_0 is greater than x_1, we say that a "0" was transmitted. If x_1 is greater than x_0, we say that a "1" was transmitted. If we define the difference as x_2, we have

$$x_2 = \int_0^T r(t) \{\cos \omega_1 t - \cos \omega_0 t\} \, dt \tag{7.10}$$

If x_2 is greater than zero, we say a "1" was transmitted, and if x_2 is less than zero, we say a "0" was transmitted.

In order to find the bit error rate, we must assume that either "0" or "1" was transmitted and find the probability that the erroneous choice was made at the receiver. Thus, we assume that

$$r(t) = A \cos \omega_0 t + n(t) \tag{7.11}$$

which corresponds to "0" being transmitted. With this $r(t)$, we calculate the probability that x_2 is greater than zero, that is, the receiver decodes a "1." Examining x_2, we see that

$$\begin{aligned}
x_2 &= \int_0^T \{A \cos \omega_0 t + n(t)\} \{\cos \omega_1 t - \cos \omega_0 t\} \, dt \\
&= \int_0^T \{A \cos \omega_0 t \cos \omega_1 t - A \cos^2 \omega_0 t + n(t) \cos \omega_1 t \\
&\quad - n(t) \cos \omega_0 t\} \, dt \\
&= \int_0^T n(t) \{\cos \omega_1 t - \cos \omega_0 t\} \, dt - \frac{AT}{2}
\end{aligned} \tag{7.12}$$

In the last line of Eq. (7.12), we have assumed that the integral of the product of the two cosines is zero over the interval T. We have further assumed that the integral of the cosine at $2\omega_0$ (resulting from the squared cosine) is zero. These are standard assumptions and require that the various frequencies be rationally related over the period T. Alternatively, as T gets very large compared to the period of the sinusoids, the integrals in question become negligible compared to the $AT/2$ term.

We now examine x_2 in Eq. (7.12). We note that $n(t)$ is a random process and that all other terms in the equation are deterministic. Since we are integrating over t, x_2 is a random variable. In addition to the noise being white, we asssumed that it was Gaussian. Since an integral is an infinite sum,

x_2 is infinite sum of Gaussian variables and is, therefore, itself Gaussian. We can find its distribution by simply evaluating the mean and variance of this variable.

The mean value of x_2 is $-AT/2$. This is true since the noise is assumed to have zero mean value. The variance of x_2 is found by first evaluating the second moment as in

$$E\{x_2^2\} = E\left\{\int_0^T \int_0^T n(t)n(\tau) (\cos \omega_1 t - \cos \omega_0 t)(\cos \omega_1 \tau - \cos \omega_0 \tau) \, dt \, d\tau \right.$$

$$-\frac{AT}{2}\int_0^T n(t) (\cos \omega_1 t - \cos \omega_0 t) \, dt \qquad (7.13)$$

$$\left. -\frac{AT}{2}\int_0^T n(\tau) (\cos \omega_1 \tau - \cos \omega_0 \tau) \, d\tau + \frac{A^2 T^2}{4}\right\}$$

The second and third integrals in Eq. (7.13) have zero expected value since $n(t)$ has zero expected value. We therefore concentrate upon the first integral:

$$E\{x_2^2\} = \int_0^T \int_0^T E\{n(t)n(\tau)\} (\cos \omega_1 t - \cos \omega_0 t)$$

$$\cdot (\cos \omega_1 \tau - \cos \omega_0 \tau) \, dt \, d\tau + \frac{A^2 T^2}{4}$$

$$= \int_0^T \int_0^T \frac{N_0}{2} \delta(t - \tau) (\cos \omega_1 t - \cos \omega_0 t) \qquad (7.14)$$

$$\cdot (\cos \omega_1 \tau - \cos \omega_0 \tau) \, dt \, d\tau + \frac{A^2 T^2}{4}$$

$$= \frac{N_0}{2} \int_0^T (\cos \omega_1 t - \cos \omega_0 t)^2 \, dt + \frac{A^2 T^2}{4}$$

In Eq. (7.14), we have first moved the expected value into the integral and applied it to the only random quantities present. We then recognized that the expected value was simply the autocorrelation of white noise and replaced it by an impulse. Finally, we used the sifting property of the impulse to reduce the double integral to a single integral. We evaluate the integral as before to obtain

$$E\{x_2^2\} = \frac{N_0 T}{2} + \frac{A^2 T^2}{4} \qquad (7.15)$$

Finally, we find the variance by subtracting the square of the mean from the second moment to get

$$\sigma_{x_2}^2 = \frac{N_0 T}{2} \qquad (7.16)$$

It is an appropriate time to recap what we have been doing. We have assumed that a "0" is being transmitted via FSK. The receiver will erroneously decode this as a "1" if x_2 is greater than 0. x_2 is a Gaussian random variable with mean $-AT/2$ and variance $N_0T/2$. The probability of this bit error is then given by the integral of

$$Pr\{\text{decoding "0" as "1"}\} = \frac{1}{\sqrt{\pi N_0 T}} \int_0^\infty e^{-(x_2 + AT/2)^2/N_0 T} \, dx_2 \qquad (7.17)$$

By a change of variables, $y = (x_2 + AT/2)/\sqrt{N_0 T/2}$, Eq. (7.17) can be simplified to

$$Pr\{\text{error}) = \frac{1}{\sqrt{2\pi}} \int_{\sqrt{A^2 T/2N_0}}^\infty e^{-y^2/2} \, dy \qquad (7.18)$$

Evaluation of the integral in Eq. (7.18) requires reference to a table of error functions. Note that the quantity in the lower limit of the integral, $A^2T/2N_0$, is the signal to noise ratio.

The student should convince himself that the probability of mistaking a transmitted "1" for a "0" is identical to the probability of mistaking a transmitted "0" for a "1".

The resulting bit error rate is plotted in Fig. 7.19 as a function of the signal to noise ratio in decibels. A second curve in this figure shows the result for phase shift keying. We shall now proceed to derive this second result.

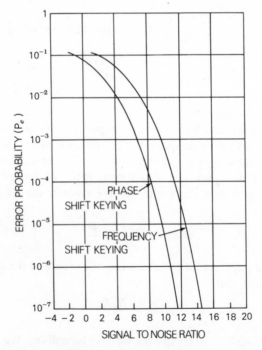

Figure 7.19 Bit error rate.

In phase shift keying (PSK), we indicated that the best choice of phase difference between the two signals is π. We can therefore use the same derivation and system we just finished analyzing. The only change is that to transmit a "1" we use $s(t) = A \cos (\omega_0 t + \pi) = -A \cos \omega_0 t$. The receiver can be simplified as the top leg of Fig. 7.13. This is redrawn as Fig. 7.20. The

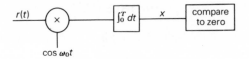

Figure 7.20   PSK receiver.

receiver therefore compares x to zero, where x is given by

$$x = \int_0^T r(t) \cos \omega_0 t \, dt \tag{7.19}$$

If x is greater than 0, we decode as a transmitted "0." If x is less than 0, we decode as a transmitted "1."

As before, we find that if "0" is transmitted, $r(t) = A \cos \omega_0 t + n(t)$, and

$$x = A \int_0^T \cos^2 \omega_0 t \, dt + \int_0^T n(t) \cos \omega_0 t \, dt \tag{7.20}$$

This is a Gaussian random variable. Its mean value is $AT/2$. The variance is found as in the previous case:

$$\sigma_x^2 = E \left\{ \int_0^T \int_0^T n(t)n(\tau) \cos \omega_0 t \cos \omega_0 \tau \, dt \, d\tau \right\}$$
$$= \frac{N_0}{2} \int_0^T \cos^2 \omega_0 t \, dt = \frac{N_0 T}{4} \tag{7.21}$$

Therefore, x is Gaussian with mean $AT/2$ and variance $N_0 T/4$. When we compare this to x_2 for the previous FSK analysis, we see that the mean values are identical and the variances differ by a factor of 2. The bit error rate is given by

$$\text{BER} = \int_0^\infty \frac{1}{\sqrt{N_0 T \pi / 2}} \exp \left\{ -\frac{(x - AT/2)^2}{N_0 T/2} \right\} dx \tag{7.22}$$

Making a change of variables, $y = (x - AT/2)/\sqrt{N_0 T/4}$, we have

$$\text{BER} = \frac{1}{\sqrt{2\pi}} \int_{-\sqrt{A^2 T/N_0}}^\infty e^{-y^2/2} \, dy \tag{7.23}$$

Equation (7.23) should be compared to Eq (7.18). We see that for FSK to achieve the same performance as PSK, the S/N must be twice as large. Equation (7.23) is plotted as part of Fig. 7.19. Note that the 3-dB separation of the two curves substantiates the two-to-one S/N observation just made.

It should further be noted that the above analysis assumed perfectly coherent PSK. That is, we assumed that the phase of the received signal was precisely matched by that of the local oscillator in the receiver.

PROBLEMS

7.1. Find the output of the 3-bit ADC of Fig. 7.1 for the following voltage inputs:
(a) 0.327 V
(b) 0.631 V
(c) 0.751 V
(d) 1.000 V

7.2. A ramp generator ADC is to be constructed for normal speech (assume a maximum frequency of 3 kHz). Choose an appropriate period and slope for the converter.

7.3. Design an OP AMP DAC for 8-bit digital signals and analog input amplitudes from 0 to 6 V.

7.4. You are given a time signal $f(t) = \sin t$. This signal is transmitted via 4-bit PCM. Find the PCM wave for the first 20 sec, and sketch the result.

7.5. A PCM wave is shown below where voltages of $+1$ and -1 are used to send a 1 and a 0, respectively. Two-bit quantization has been used. Sketch a possible analog information signal $f(t)$ that could have resulted in this PCM wave. Now assume that a pulse of $+1$ V represents a binary 0 and -1 V represents a 1. Sketch a possible $f(t)$.

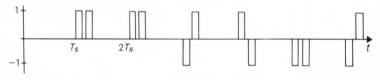

7.6. Two voice signals (3-kHz upper frequency) are transmitted using PCM with eight levels of quantization. How many pulses per second are required to be transmitted, and what is the minimum bandwidth of the channel?

7.7. The input to a delta modulator is $f(t) = 5t + 1$. The sampler operates at 10 samples/sec, and the step size is 1 V. Sketch the output of the delta modulator.

7.8. Delta modulation is used to transmit a voice signal with maximum frequency of 3 kHz. The sampling rate is set at 20 kHz. The maximum amplitude of the analog voice signal can be assumed to be 1 V. Discuss the choice of appropriate step size.

7.9. Find the mean-square quantization noise for the PCM modulation described in Problem 7.4.

7.10. Verify the entries in Fig. 7.19 for a signal to noise ratio of 6 dB. (Use the table of error functions given in Appendix III.)

7.11. An adaptive delta modulator is to be used to transmit $f(t) = 5t + 1.1$. The sampler operates at 10 samples/sec. The step size depends upon the number of 1's in the 4 most recent sampling points. The step size is either $\frac{1}{2}$ V, 1 V, or $1\frac{1}{2}$ V depending upon the number of 1's (i.e., if the number of 1's is 0 or 4, the maximum step size is used; if the number of 1's is 1 or 3, the middle step size is used; if the number of 1's is 2, the minimum step size is used). Sketch the staircase approximation to $f(t)$ for the first 4 sec and also sketch the transmitted digital waveform. Assume that all binary transmissions prior to $t = 0$ were "0."

REFERENCES

JAYANT, N. S., ed., *Waveform Quantization and Coding*, IEEE Press, New York, 1976. (*Note:* Contains many relevant papers.)

LINDSEY, W. C., and SIMON, M. K., *Telecommunication Systems Engineering*, Prentice-Hall, Englewood Cliffs, N.J., 1973.

SAKRISON, D. J., *Communication Theory*, Wiley, New York, 1968.

SPILKER, J. J., JR., *Digital Communications by Satellite*, Prentice-Hall, Englewood Cliffs, N.J., 1977.

TAUB, H., and SCHILLING, D. L., *Principles of Communication Systems*, McGraw-Hill, New York, 1971.

ZIEMER, R. E., and TRANTER, W. H., *Principles of Communications*, Houghton Mifflin, Boston, 1976.

INFORMATION THEORY AND CODING

8

We have spent a great deal of time in previous chapters investigating techniques for transmitting a particular *signal* from one point to another. In this chapter we shall go right back to the very beginning and return to basics. Instead of thinking of a message as described by a particular time function, we shall stop and think what it is we are trying to communicate from one point to another. Before deciding upon a technique to use in transmitting a message, we should first investigate the *information content* of that message. Let us motivate this discussion by getting ahead of the game with a few specific practical examples.

Suppose you enjoy Mexican cuisine. You enter your favorite restaurant to order the following: one cheese enchilada, one beef taco, one order of beans, one order of rice, two corn tortillas, butter, a sprig of parsley, chopped green onions, and a glass of water. Instead of this entire list being written out longhand, the person taking the order probably communicates something of the form "One #2 Combination." Stop and think about this. Rather than attempting to decide whether to transmit the specific words via English spelling or by some form of shorthand, a much more basic approach has been taken. Let's present another example.

A popular form of communication is the telegram. Have you ever called Western Union to send a wedding congratulations telegram? If so, you found that they try to sell you a standard wording such as "Heartiest congratulations upon the joining of two wonderful people. May your lives together be happy and productive. A gift is being mailed under separate cover." In the early days of telegraphy, the company noticed that many wedding telegrams sounded almost identical to that presented above. A primitive approach to transmitting this might be to send the message letter by letter. However, since the wording is so predictable, why not have a short code for this message. For example, let's call this particular telegram, "#79." The operator would then simply type the name of the sender and of the addressee and then #79. At the receiver, form #79 is pulled from the file and the address and signature are filled in. Think of the amount of transmission time saved.

As yet a third example, consider the (atypical, I hope) university lecture in which a professor stands before the class and proceeds to read directly from the textbook for 2 hours. A great deal of time and effort could be saved if the professor instead read the message "pages 103 to 120." In fact, if the class expected this and knew they were past page 100, the professor could shorten the message to "0320." The class could then read these 18 pages on their own and get the same information. The professor could dismiss the class after 20 seconds and return to more scholarly endeavors.

I think that by now the point should be clear. A basic examination of the information content of a message has the potential of saving considerable effort in transmitting that message from one point to another.

8.1 INFORMATION AND ENTROPY

The modest examples given at the start of this chapter indicate that, in an intuitive sense, some relatively long messages do not contain a great deal of information. Every detail of the restaurant order or the telegram is so commonplace that the person at the receiving end can almost guess what comes next.

The concept of information content of a particular message must now be formalized. After doing this, we shall find that the less information in a particular message, the quicker we can communicate the message.

The concept of information content is related to predictability. That is, the more probable a particular message, the less information is given by transmitting that message. If, for example, today were Tuesday and you phoned someone to tell them tomorrow would be Wednesday, you would certainly not be communicating any information. This is true since the probability of that particular message is "1."

The definition of information content should be such that it monotonically decreases with increasing probability and goes to zero for a probability of unity. Another property we would like this measure of information to have is that of additivity. If one were to communicate two (independent) messages in sequence, the total information content should be equal to the sum of the individual information contents of the two messages. Now if the two messages are independent, we know that the total probability of the composite message is the product of the two individual probabilities. Therefore, the definition of information must be such that when probabilities are multiplied together, the corresponding informations are added. The logarithm satisfies this requirement. We thus *define* the information content of a message x as I_x in Eq. (8.1).

$$I_x = \log \left(\frac{1}{P_x} \right) \qquad (8.1)$$

In Eq. (8.1), P_x is the probability of occurrence of the message x. This definition satisfies the additivity requirement, the monotonicity requirement, and for $P_x = 1$, $I_x = 0$. Note that this is true regardless of the base chosen for the logarithm.

We usually use base 2 for the logarithm. To understand the reason for this choice, let us return to the restaurant example. Suppose that there were only two selections on the menu. Further assume that past observation has shown these two to be equally probable (i.e., each is ordered half of the time). The probability of each message is therefore $\frac{1}{2}$, and if base 2 logarithms are used in Eq. (8.1), the information content of each message is

$$I = \log_2 \left(\frac{1}{1/2} \right) = \log_2 2 = 1$$

Thus, one unit of information is transmitted each time an order is placed.

Now let us think back to digital communication and decide upon an efficient way to transmit this order. Since there are only two possibilities, one binary digit (bit) could be used to send the order to the kitchen. A "0" could represent the first dinner, and a "1," the second dinner on the menu.

Suppose we now increase the number of items on the menu to 4, with each having a probability of $\frac{1}{4}$. The information content of each message is now $\log_2 4$ or 2 units. If binary digits are used to transmit the order, 2 bits would be required for each message. The various dinners would be coded as 00, 01, 10, and 11. We can therefore conclude that if the various messages are equally probable, the information content of each message is exactly equal to the minimum number of bits required to send the message. This is the reason for commonly using base 2 logarithms, and, in fact, the unit of information is called the *bit of information*. Thus one would say that in the last example each menu order contains 2 bits of information.

When all of the possible messages are equally likely, the information content of any single message is the same as that of any other message. In cases where the probabilities are not equal, the information content depends upon which particular message is being transmitted.

Entropy is defined as the average information per message. To calculate the entropy, we take the various information contents associated with the messages and weight each by the fraction of time we can expect that particular message to occur. This fraction is the probability of the message. Thus, given n messages, x_1 through x_n, the entropy is defined by

$$H = \sum_{i=1}^{n} P_{x_i} I_{x_i} = \sum_{i=1}^{n} P_{x_i} \log\left(\frac{1}{P_{x_i}}\right) \tag{8.2}$$

By convention, the letter "H" is used for entropy.

Illustrative Example 8.1

A communication system consists of six possible messages with probabilities $\frac{1}{4}$, $\frac{1}{4}$, $\frac{1}{8}$, $\frac{1}{8}$, $\frac{1}{8}$, and $\frac{1}{8}$, respectively. Find the entropy.

Solution

The information content of the six messages is 2 bits, 2 bits, 3 bits, 3 bits, 3 bits, and 3 bits, respectively. The entropy is therefore given by

$$\begin{aligned} H &= \tfrac{1}{4} \times 2 + \tfrac{1}{4} \times 2 + \tfrac{1}{8} \times 3 + \tfrac{1}{8} \times 3 + \tfrac{1}{8} \times 3 + \tfrac{1}{8} \times 3 \\ &= \tfrac{5}{2} \text{ bits/message} \end{aligned} \tag{8.3}$$

It proves instructive to examine Eq. (8.2) for the binary message case. That is, we consider a communication scheme made up of two possible messages, x_1 and x_2. For this case, $P_{x_2} = 1 - P_{x_1}$, and the entropy is given by

$$H_{\text{binary}} = P_{x_1} \log\left(\frac{1}{P_{x_1}}\right) + (1 - P_{x_1}) \log\left\{\frac{1}{1 - P_{x_1}}\right\} \tag{8.4}$$

Equation (8.4) is sketched as a function of P_{x_1} in Fig. 8.1.

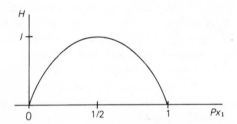

Figure 8.1   Entropy for the binary case.

Note that as either of the two messages becomes more likely, the entropy decreases. When either message has probability 1, the entropy goes to zero. This is reasonable since, at these points, the outcome of the transmission is certain without having to go through the trouble of sending any message. Thus, if $P_{x_1} = 1$, we know message x_1 will be sent all of the time (with probability 1). If $P_{x_1} = 0$, we know x_2 will be sent all of the time. In these two cases, no information is transmitted by sending the message.

If we were able to show a three-dimensional plot for the three-message case, we would find a similar result with a peak in entropy for message probabilities of $\frac{1}{3}$.

The entropy function is symmetrical with a maximum for the case of equally likely messages. This is an important property that is used in later coding studies.

8.2 CHANNEL CAPACITY

In this section, we investigate the rate at which information can be sent through a channel and the relationship of this rate to errors. Before we can do this, we must define the rate of information flow.

We assume that a source can send any one of a number of messages at a rate of r messages/sec. For example, if the source is a Teletype machine, the rate would be the number of symbols per second. Once we know the probability of each individual message, we can compute the entropy H in bits per message. If we now take the product of the entropy H with the message rate r, we get the information rate in bits per second, denoted R. Thus, $R = rH$ bits/sec. For example, if the messages given in Illustrative Example 8.1 were sent at 2 messages/sec, the information rate would be 5 bits/sec.

One can probably intuitively reason that, for a given communication system, as the information rate increases, the number of errors per second will also increase. As an example, suppose that a PCM signal were being sent. The faster the bits are transmitted, the greater the intersymbol interference and therefore the greater the probability of mistaking a "0" for a "1" (and vice versa).

C. E. Shannon has shown that a given communication channel has a maximum rate of information C known as the *channel capacity*. If the information rate R is less than C, one can approach arbitrarily small error probabilities by intelligent coding techniques. This is true even in the presence of noise, a fact that probably contradicts the reader's intuition.

The negation of the above theorem is also true. That is, if the information rate R is greater than the channel capacity C, errors cannot be avoided regardless of the coding technique used.

We shall not prove this fundamental theorem here. We shall, however, discuss its application to several commonly encountered cases.

We consider the band limited channel operating in the presence of additive white Gaussian noise. In this case, the channel capacity is given by

$$C = B \log_2 (1 + S/N) \tag{8.5}$$

In Eq. (8.5), C is the capacity in bits per second, B is the bandwidth of the channel in hertz, and S/N is the signal to noise ratio. Equation (8.5) makes intuitive sense. As the bandwidth of the channel increases, it should be possible to make faster changes in the information signal, thereby increasing the information rate. As the S/N increases, one would expect to be able to increase the information rate while still preventing errors due to the noise. Note that with no noise at all, the signal to noise ratio is infinity and an infinite information rate would be possible regardless of the bandwidth. Equation (8.5) might lead one to conclude that if the bandwidth approaches infinity, the capacity also approaches infinity. This last observation is not correct. Since the noise is assumed to be white, the wider the bandwidth, the more noise is admitted to the system (unless the noise is identically equal to zero). Thus as B increases in Eq. (8.5), S/N decreases.

It is instructive to expand upon the last observation. That is, suppose the signal and noise power were fixed and we were trying to design the best possible channel. There is expense associated with increasing the system bandwidth, so we should attempt to find the maximum channel capacity as a function of bandwidth. Suppose that the noise is white with power spectral density $N_0/2$. Further assume that the signal power is fixed at a value S. The channel capacity is given from Eq. (8.5) as

$$C = B \log_2 \left(1 + \frac{S}{N_0 B}\right) \tag{8.6}$$

Note that B is in hertz and that $N_0/2$ is the power spectral density in watts per radian per second. Therefore the factor of 2π cancels. We now find the value of channel capacity approached as B goes to infinity:

$$\begin{aligned}
C &= \lim_{B \to \infty} B \log_2 \left(1 + \frac{S}{N_0 B}\right) \\
&= \lim_{B \to \infty} \frac{S}{N_0} \log_2 \left(1 + \frac{S}{N_0 B}\right)^{N_0 B/S} \\
&= \frac{S}{N_0} \log_2 e = 1.44 \frac{S}{N_0}
\end{aligned} \tag{8.7}$$

Equation (8.7) shows the maximum possible channel capacity as a function of signal power and noise spectral density. In an actual system design, the channel capacity will be compared to this figure and a decision will be made whether further increase in bandwidth is worth the expense.

Suppose now that you were asked to design a binary communication system (e.g., send a PCM encoded signal) and you cranked the bandwidth of your communication channel, the maximum signal power, and the noise spectral density into Eq. (8.5). You come up with a maximum information rate C. You plug in the sampling rate and the number of bits of quantization, and you arrive at a certain information transmission rate, R bits/sec. Aha, you observe. R is less than C, so Shannon tells you that if you do enough work encoding this binary data train, you can achieve arbitrarily low probabilities of error. But you have already coded the original signal into a train of binary digits. What do you do next?

The next section of this text discusses answers to this question. As applied to the above example, most procedures would entail grouping combinations of bits and calling the result a new word. For example, we can group by twos to yield four possible messages, 00, 01, 10, and 11. Each of these four possible messages can be transmitted using some code word. By so doing, the probability of error can be reduced.

As an example, if you wished to transmit this textbook to a class of students, you could read each letter aloud as, for example,

r-e-a-d-e-a-c-h-l-e-t-t-e-r-a-l-o-u-d

Alternatively, you could group letters together and send the groups as a code word from an acceptable dictionary. For the above example, we send

read-each-letter-aloud

Suppose that in reading the individual letters, you slur the "d" sound and one student receives it as "v." That student receives the erroneous message

r-e-a-v-e-a-c-h-l-e-t-t-e-r-a-l-o-u-d

However, in the word coding, the student would receive

reav-each-letter-aloud

and since "reav" is not an acceptable code word, the student could correct the error to interpret the received message as

read-each-letter-aloud

and thus make no error. The essence of coding to achieve arbitrarily small errors amounts to grouping into longer and longer code words. The longer the code words, the more different the dictionary entries will be, and the less likely that individual errors cannot be corrected.

8.3 PRINCIPLES OF CODING

We begin with a basic statement of the problem. Given M possible messages, we wish to convert these into M possible code words. The code words are chosen to achieve objectives such as error correction and security. In a basic binary communication system, M is equal to 2, thus requiring two possible code words. One possible approach was introduced in the previous section. We can start with a binary communication system and group bits to form any desired number of possible messages. For example, by combining groups of 3 bits, we can form eight possible message words: 000, 001, 010, 011, 100, 101, 110, and 111. Each of these eight possible message words can now be coded into one of eight different code words. The code words need not necessarily be of the same length as the message words. In fact, to achieve error correction capabilities, redundancy must be introduced by making the code words longer than the message words.

In order to analyze the error correction capabilities of various codes, we will first define the concept of *distance* between two binary code words. The distance between two equal length binary code words is defined as the number of bit positions in which the two words differ. Thus, for example, the distance between 000 and 111 is 3; the distance between 010 and 011 is 1. The distance between any one code word and the word formed by changing 1 bit is "1."

Suppose now that we transmit one of the eight possible 3-bit words. Assume that the channel is noisy and that one bit position is incorrectly received. Since every possible 3-bit combination is used as a message, the received 3-bit combination will be identical to one of the code words. Thus, for example, if 101 is transmitted and an error occurs in the third bit, 100 is received. There is no way for the receiver to know that 100 was not the transmitted word. An error is therefore going to occur.

Suppose now that the dictionary of code words is such that the distance between any two words is at least 2. The following eight code words have this property:

$$0000, \quad 0011, \quad 0101, \quad 0110, \quad 1001, \quad 1010, \quad 1100, \quad 1111$$

The reader should verify that the minimum distance between code words is 2.

Now suppose that we transmit one of these eight words and that a single bit error occurs during transmission. Since the distance between the received word and the transmitted word is 1, the received word cannot be identical to any of the dictionary words. For example, suppose 0101 is transmitted and an error occurs in the third bit. 0111 would be received. This is not one of the eight acceptable words, and the system can tell that an error was made.

It is instructive to think of distance in terms of a multidimensional space. In the previous example of three-bit code words, we can plot this in three dimensions. Each code word is a point at the corner of a unit cube, as shown in Fig. 8.2. Starting at each corner, if a single bit error is made, we move along one of the edges to another corner a distance of 1 unit away.

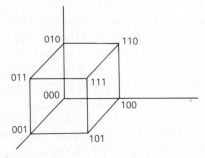

Figure 8.2 Three-bit words in three-dimensional space.

In the 4-bit code word case, we need an abstract figure with eight points representing the four-dimensional code words. This requires a figure in four dimensional space. An error of one bit position moves us away from the code word along a particular direction corresponding to the bit in error. Since the distance between code words is 2, we must change at least 2 bits to go from one code word to another. We can draw four-dimensional spheres of radius 1 around each code word point. Figure 8.3 shows three representative code

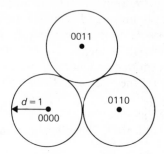

Figure 8.3   Three code words in four-dimensional space.

word points and their associated spheres. A 1-bit error will take us to some point on the sphere. Such a point on the sphere can be reached by changing one bit in each of four possible code words. Using the 4-bit example presented before as an illustration, assume 0111 is received. This could have resulted from 1111 being transmitted and an error occurring in the first bit position. It could also result from 0011, 0101, or 0110 with errors occurring in the second, third, or fourth bit positions, respectively. Thus a single error can be detected but not corrected.

Suppose now that we increased the minimum distance between code words to 3 units. We then see that if a single error is made, the received word will be distance 1 from the correct code word and at least 2 units of distance from every other code word. An intelligent receiver can therefore correct the single bit error.

As an example, suppose the system consisted of three code words, 01111, 10011, and 01000. The minimum distance between code words is 3 units. If, for example, 01111 were transmitted and the second bit were in error, 00111 would be received. The distance from this received word to the second code word is 2, and to the third it is 4. We could therefore easily decide that this should be decoded as the first code word, 01111.

The student should become convinced that, in general, if the minimum distance between code words is D_{min}, errors involving up to $D_{min} - 1$ bits can be detected. If D_{min} is an even number, errors up to $D_{min}/2 - 1$ can be corrected. If D_{min} is odd, errors up to $D_{min}/2 - \frac{1}{2}$ can be corrected. Carefully note the difference between error detection and error correction. Thus, for example, if D_{min} were 6, up to two bit errors can be corrected or up to five bit errors can be detected. One must decide in advance whether error correction or detection is the desired mode of operation. For example, if error correction were being performed in the $D_{min} = 6$ case, an error in five bit positions would place us only 1 unit away from another code word, and the receiver could erroneously decode the received message as that code word. If error detection were being performed, any received word which is not identical to a code word will be rejected.

The obvious next question is, "How do we increase distance between code words to achieve greater error correction?" The simplest way to increase distance is by repetition.

Suppose that we again have eight message words to transmit: 000, 001, 010, 011, 100, 101, 110, and 111. Let us repeat each of these 3-bit groupings to yield the code words 000000, 001001, 010010, 011011, 100100, 101101, 110110, and 111111. The minimum distance between code words is 2, and we can therefore detect (but not correct) single errors. This repetition process is not a very efficient method to achieve a distance of 2. We were able to present eight code words with minimum distance 2 using only 4-bit words (see the first example in this section). It would be a waste of transmission time to use 6-bit words if 4-bit words can accomplish the same error detection. In fact, the longer word would also have a greater probability of a single bit error. The following sections present several common techniques for choosing a code to achieve error detection and/or correction capabilities.

8.4 ALGEBRAIC CODES

The previous section presented several possible codes which achieve some level of error identification. The purpose of the present section is to present an organized technique for formulating code words and a complementary organized technique for recovering the original word and identifying errors at the receiver. The class of codes we will consider is known as *algebraic codes*.

The first coding example presented in Section 8.3 was a simple form of algebraic code known as a *single parity bit check code*. We repeat the message and code words for that example:

Message	Code Word
000	0000
001	0011
010	0101
011	0110
100	1001
101	1010
110	1100
111	1111

Note that the coding process can be described as an addition of a fourth bit
to the message word such that the total number of 1's is even. The receiver
simply checks to see that the total number of 1's is even and if so, ignores the
fourth bit in each code word. If the number of 1's is odd, the receiver rejects
the word since a bit error is present. This rejection is known as an *erasure*
and would normally require that the word be retransmitted. We thus can
detect but not correct one bit error. Note that if two bit errors occur, the
parity will remain even and the receiver would incorrectly decode the word.
The probability of two bit errors is often small enough to ignore.

Illustrative Example 8.2

Three-bit message words are encoded by adding a fourth parity check bit.
The coded words are sent via FSK, and the signal to noise ratio at the receiver
is 8 dB. Find the probability of an undetected error at the receiver.

Solution

An undetected error will occur in a message if a particular four bit code
word experiences either two or four bit errors. If either one or three bit errors
occur, the parity test will fail, and the receiver will detect an error. The bit
error rate is approximately 5×10^{-3} (*see* Fig. 7.19). The probability of any
particular 2 bits being in error is therefore 25×10^{-6} (we assume independence). Since there are six possible combinations of 2 bits in any 4-bit word
(i.e., first and second bit, first and third bit, first and fourth bit, second and
third bit, second and fourth bit, and third and fourth bit), the probability
of two bit errors is $6 \times (5 \times 10^{-3})^2 \times (1 - 5 \times 10^{-3})^2$. The probability of
four bit errors is 625×10^{-12}. The probability of either two or four bit
errors is therefore approximately 1.49×10^{-4}. In the above system, we can
expect about 3 incorrectly decoded messages in every 10,000 transmitted
messages.

In a similar manner, we can calculate the probability of a detected error
(i.e., 1- or 3-bit errors) at the receiver. The student should verify that this
probability is approximately 2.8×10^{-2}. We can therefore expect about 3
detected errors out of every 100 transmitted messages. Putting these two
results together, we see that in any group of 10,000 transmitted messages, we
can expect about 3 messages to be incorrectly decoded, about 280 messages
to result in parity errors (erasures), and the remaining 9,717 messages to be
correctly decoded.

We now generalize this type of coding. Suppose that the message words
consist of m bits (in the above example, $m = 3$). We therefore have a set of
up to 2^m distinct message words. We consider code words which add n parity
bits to the m message bits to end up with code words of length $m + n$ bits.
Thus, if a_i is an original message bit and c_i is a parity check bit, each code

word will be of the form

$$a_1 a_2 a_3 \cdots a_m c_1 c_2 c_3 \cdots c_n$$

Note that of the 2^{m+n} possible code words, only 2^m are used.

Each check bit is chosen to achieve even parity when combined with specific message bits. Thus, for example, c_1 can be chosen to yield an even number of 1's when combined with a_1, a_3, a_4, and a_6; c_2 can be chosen to yield an even number of 1's when combined with a_1, a_2, and a_5; and so on. Some elementary Boolean algebra and matrix skills permit us to express this relationship in a general format. The check bits are chosen in such a way as to satisfy

$$\bar{H}\bar{T} = 0 \tag{8.8}$$

In Eq. (8.8), $\bar{T}$ is an $m + n$ column vector as shown in

$$\bar{T} = \begin{bmatrix} a_1 \\ a_2 \\ a_3 \\ \cdot \\ \cdot \\ \cdot \\ a_m \\ c_1 \\ c_2 \\ \cdot \\ \cdot \\ \cdot \\ c_n \end{bmatrix} \tag{8.9}$$

and $\bar{H}$ is a $n \times (n + m)$ matrix as shown in

$$\bar{H} = \begin{bmatrix} h_{11} & h_{12} & \cdots & h_{1m} & 1 & 0 & 0 & 0 & \cdots & 0 \\ h_{21} & h_{22} & \cdots & h_{2m} & 0 & 1 & 0 & 0 & \cdots & 0 \\ \cdot & \cdot & \cdots & \cdot & \cdot & \cdot & \cdot & \cdot & & \cdot \\ h_{n1} & h_{n2} & \cdots & h_{nm} & 0 & 0 & 0 & 0 & \cdots & 1 \end{bmatrix} \tag{8.10}$$

Note that the $n \times n$ matrix formed by partitioning the right part of $\bar{H}$ is an identity matrix. $h_{11}, h_{12}, \ldots, h_{nm}$ are each either equal to 0 or 1. The sums in Eq. (8.8) are taken using modulo-2 addition. Thus, if the number of 1's is even, the sum is zero. To more fully appreciate Eq. (8.8), we expand the first two rows:

$$h_{11}a_1 + h_{12}a_2 + \cdots + h_{1m}a_m + c_1 = 0$$
$$h_{21}a_1 + h_{22}a_2 + \cdots + h_{2m}a_m + c_2 = 0 \tag{8.11}$$

Illustrative Example 8.3

Given the algebraic code with 4-bit message words and 3 parity check bits, $\bar{H}$ is defined as

$$\bar{H} = \begin{bmatrix} 1 & 1 & 0 & 1 & 1 & 0 & 0 \\ 1 & 0 & 1 & 1 & 0 & 1 & 0 \\ 0 & 1 & 1 & 1 & 0 & 0 & 1 \end{bmatrix}$$

Find the code words corresponding to the 16 possible message words.

Solution

The three equations corresponding to the matrix equation $\bar{H}\bar{T} = 0$ are

$$\begin{aligned} a_1 + a_2 \quad\quad + a_4 + c_1 &= 0 \\ a_1 \quad\quad + a_3 + a_4 + c_2 &= 0 \\ a_2 + a_3 + a_4 + c_3 &= 0 \end{aligned} \quad\quad (8.12)$$

Thus, c_1 is chosen to achieve even parity when combined with the first, second, and fourth bits; c_2 is chosen to achieve even parity when combined with the first, third, and fourth bits; c_3 achieved even parity when combined with the second, third, and fourth message bits. The following code words result:

Message Word	Code Word
0000	0000000
0001	0001111
0010	0010011
0011	0011100
0100	0100101
0101	0101010
0110	0110111
0111	0111001
1000	1000110
1001	1001001
1010	1010101
1011	1011010
1100	1100011
1101	1101100
1110	1110000
1111	1111111

We now transmit the code words. The receiver forms the product of the received word with the matrix $\bar{H}$, and if this product is not equal to zero, the receiver knows that at least one error was made in transmission. That is, we know that $\bar{H}\bar{T} = 0$ for all acceptable code words. Therefore if the product of $\bar{H}$ with the received vector is not zero, we know the received word is not one of the acceptable code words. Of course, we hope that this elaborate system will do more for us than simple error detection. We were able to accomplish this with only 1 parity bit.

The value of algebraic coding arises in multiple error detection and in error correction. Suppose that the transmitted vector is denoted as $\bar{T}$. We define an error vector $\bar{E}$ which contains a "1" in each bit position in which an error occurs. The received vector is therefore of the form

$$\bar{R} = \bar{T} + \bar{E} \tag{8.13}$$

where, again, the addition is modulo-2. Error correction is possible if we can isolate the vector $\bar{E}$ at the receiver. If we multiply the received vector by the matrix $\bar{H}$, we have

$$\bar{H}\bar{R} = \bar{H}(\bar{T} + \bar{E}) = \bar{H}\bar{T} + \bar{H}\bar{E} = \bar{H}\bar{E} \tag{8.14}$$

In order to arrive at the final term in Eq. (8.14), we used the fact that $\bar{H}\bar{T} = 0$ since $\bar{T}$ is an acceptable code word. We define the product $\bar{H}\bar{E}$ as $\bar{S}$, the *syndrome*. The dictionary defines syndrome as "a number of symptoms occurring together and characterizing a specific disease or condition." In fact, the error syndrome characterizes the specific bit error. Let us first assume that a single bit error occurs. $\bar{E}$ would then be a vector composed of a single "1," and the remaining positions would be zero. If we take the product of this with $\bar{H}$ to form the syndrome, the result is a vector which is actually one column of $\bar{H}$, that column being the one corresponding to the bit position in error.

Illustrative Example 8.4

Given the $\bar{H}$ defined in Illustrative Example 8.3, a message word 1010 is transmitted. A bit error occurs in the fourth bit position. Find the syndrome.

Solution

The transmitted code word is 1010101, and the received word is 1011101. We form the product $\bar{H}\bar{R}$ to get

$$\bar{H}\bar{R} = \begin{bmatrix} 1 & 1 & 0 & 1 & 1 & 0 & 0 \\ 1 & 0 & 1 & 1 & 0 & 1 & 0 \\ 0 & 1 & 1 & 1 & 0 & 0 & 1 \end{bmatrix} \begin{bmatrix} 1 \\ 0 \\ 1 \\ 1 \\ 1 \\ 0 \\ 1 \end{bmatrix} = \begin{bmatrix} 1 \\ 1 \\ 1 \end{bmatrix} \tag{8.15}$$

We note that the result is equal to the fourth column of $\bar{H}$, thus identifying an error as having occurred in the fourth bit position.

We see that in order to correct a single error using this technique, it is necessary that $\bar{H}$ contain no duplicate columns. It is also necessary that no column of $\bar{H}$ be composed of all 0's since if the syndrome were a vector composed of all 0's, we would not know whether we were seeing that column or the result of no errors.

If an error of more than one bit error occurs, the syndrome will be equal to the sum of the corresponding columns of $\bar{H}$. Thus if, for example, errors occurred in the third and fourth bits of the above example, the syndrome would be

$$\bar{S} = \begin{bmatrix} 1 \\ 0 \\ 0 \end{bmatrix}$$

This would be incorrectly interpreted as a single bit error in the fifth bit position. For the present, we shall restrict our discussion of algebraic decoders to the single error correction case.

Algebraic codes are not difficult to implement. Calculation of the various vector products is equivalent to summing combinations of bit positions. This is accomplished by entering the codes into shift registers, tapping off at the appropriate positions, and feeding these outputs into a summation device. Those readers familiar with digital processing know that modulo-2 addition of two inputs is performed in the EXCLUSIVE OR logic block.

Figure 8.4 shows the implementation of the vector product of (1011010) and a received word. This represents the second of three such multiplications required for the receiver in Illustrative Example 8.4.

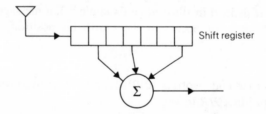

Figure 8.4 Implementation of algebraic code receiver.

As each bit enters a storage block in Fig. 8.4, the previous contents of that block are shifted one position to the right. Thus, for example, if the shift register contains the binary number 1011001 and a binary zero enters the left-hand block, the new contents of the register would be 0101100.

We have seen that a single error can be located, and therefore corrected, by using the syndrome. Correction of more than one error would involve finding code words with minimum distances greater than 2 units. If we are successful in devising such codes, the basic decoding technique involves finding the distance between a received word and each of the acceptable code words. The received word is then decoded into the code word which is the minimum distance from the received word, i.e., the closest word.

Further studies of coding center upon three considerations. The first is the design of codes with the maximum possible distance between words. The second is the design of codes which are simple to implement, both at the transmitter and at the receiver. Finally, one must consider whether the particular code permits the user to approach the channel capacity while achieving the required error performance. The reader is referred to the references at the end of this chapter for more detailed discussion.

8.5 PSEUDONOISE CODE

A *cyclic code* is a parity check code with the property that a cyclic shift of any code word results in another acceptable code word. A cyclic shift is a shift of each bit by one position, where the end bit is folded around to the other end. Thus, one possible cyclic shift of 1011011 would be 0110111.

A useful cyclic code is the so-called *PN* code. PN is an abbreviation of *pseudonoise*. We shall see that the binary sequence generated according to the rules for this code bears a relationship to white noise (i.e., the autocorrelation is similar to an impulse).

The pseudonoise code sequences are generated using shift registers. Figure 8.5 illustrates a PN generator for a 7-bit PN cyclic code. We initialize the

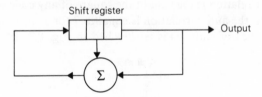

Figure 8.5   Seven-bit PN generator.

system with any one of the eight possible 3-bit sequences. Suppose, for example, we start with 010. The two inputs to the summing circuit are 0 and 1, so the sum, 1, is fed to the input of the shift register. Following this operation, the shift register contains 101, and a 0 has been fed out of the output (right end). The summer then sums 1 and 0 to yield a 1. After this is fed into the input, the register contains 110, and a 1 is fed out. Continuing this operation, we find that the output is of the form

$$0\ 1\ 0\ 1\ 1\ 1\ 0\ 0\ 1\ 0\ 1\ 1\ 1\ 0\ 0\ \ldots$$

Since the generator operates upon the three most recent bits, once the generating sequence (010) repeats in the above example, the entire remaining sequence will also repeat. The above code is therefore periodic with a period of 7 bits.

We now examine the first 9 bits of the output sequence shown above. If we examine these in groups of three adjacent bits, we see that every possible

3-bit combination occurs (with the exception of 000). That is, the first 3 bits are 010; the second, third, and fourth bits are 101; the third, fourth, and fifth bits are 011; and so on. Therefore each of the seven possible 7-bit code sequences is a cyclic shift of every other sequence. The PN code is cyclic.

The code word composed of all 0's is usually not used. Excluding this, the seven possible 7-bit PN code words are

$$0111001, \quad 1110010, \quad 1100101, \quad 1001011, \quad 0010111,$$
$$0101110, \quad 1011100$$

The PN code words have some very useful properties. We observe that each of the words contains exactly four 1's. We further observe that the distance between any pair of code words is 4 units (the reader should check this). We now find the autocorrelation[1] of a PN sequence. If the autocorrelation is taken over the 7-bit period of the sequence, we see that the value of the autocorrelation at any non-zero shift is simply related to the distance between two code words. That is, suppose we wish to evaluate the autocorrelation at a shift of 2 units. We multiply a 7-bit PN sequence by the 7-bit sequence resulting from a cyclic shift of 2 units. We then sum the product. This sum represents the number of bit positions in which the two words are the same. This number is equal to 7 minus the distance between the two words. At a shift of zero, the autocorrelation is the sum of the square of any code word, which is equal to 7. Thus, the autocorrelation is equal to 7 for a shift of zero and to 3 for all other values of shift. This is sketched in Fig. 8.6. PN sequences of

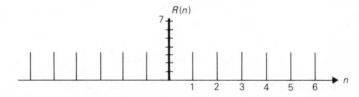

Figure 8.6 Autocorrelation of 7-bit PN sequence.

length greater than 7 bits are generated by using initiating sequences longer than 3 bits and by changing the taps feeding from the shift register into the summer. In general, for a generator length of n bits, the code word length will be $2^n - 1$. For $n = 3$, the length is $2^3 - 1$ or 7 bits, which checks with the above example. The distance between code words will always be 2^{n-1}. Without elaborating on details, we caution that certain rules must be followed in choos-

---

[1] *Autocorrelation*, as applied to binary codes, is closely related to the concept of distance. The autocorrelation of a code sequence is a measure of the number of bits in agreement between the sequence and its shifted versions. The product in the definition of autocorrelation is therefore redefined as:

$$0 \times 0 = 1 \times 1 = 1 \qquad ; \qquad 0 \times 1 = 1 \times 0 = 0$$

ing the taps used to generate the PN code. Not all combinations will result in PN sequences.

Observe that the autocorrelation of the PN sequence resembles a digital impulse. This is the reason the PN type of code is known as pseudonoise. In applications where deterministic "white" noise is required, the PN sequence is often used.

8.6 CONVOLUTIONAL CODES

Another important class of codes is the *convolutional code*. For this type of code, we no longer consider individual blocks of bits as code words. Instead, we operate upon a continuing stream of information bits. The source generates a continuing message sequence of 1's and 0's, and the transmitted sequence is generated from this source sequence. The technique of generating the transmitted sequence is to convolve the source sequence with a fixed digital function. Thus, a particular transmitted bit t_n is generated from a combination of source bits $s_n, s_{n-1}, s_{n-2}, \ldots, s_{n-k}$ according to the convolution equation

$$t_n = \sum_{m=0}^{k} h_m s_{n-m} \tag{8.16}$$

The h's in Eq. (8.16) are each either 1 or 0. The equation can be implemented with a shift register and a modulo-2 adder. Figure 8.7 shows a general implementation of Eq. (8.16). The switches in Fig. 8.7 are closed if the associated h in Eq. (8.16) is 1 and open if it is 0.

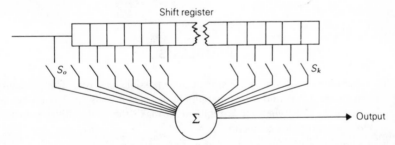

Figure 8.7   Implementation of convolutional code.

In the application of convolutional coding we often transmit more than 1 bit between each pair of input bits. Thus, referring to Fig. 8.7, we may shift in 1 input bit, set the switches to correspond to a particular set of h's, and generate the first output bit. Then, before feeding in another input bit, we could reset the switches to correspond to a second set of h's and transmit a second bit. If we send 2 output bits for each input bit, the code is referred to as a *rate-$\frac{1}{2}$ convolutional code*. In sending a rate-$\frac{1}{2}$ convolutional code, one often chooses the first bit of each transmitted pair to be identical to the information (input) sequence. This would result from setting $h_0 = 1$ and $h_n = 0$ for $n \neq 0$. In such cases, the code is known as a *systematic* code.

As an example of a rate-$\frac{1}{2}$ non-systematic convolutional code, consider a 3-bit shift register with the first set of h's given by 0101 and the second set by 0110. Suppose further that the input sequence is 1101001. The output would then be given by

$$1\,1\,1\,0\,1\,1\,0\,1\,0\,1\,1\,0\,1\,1\,0\,1\,1\,0\ldots$$

To arrive at this output sequence, we have assumed that the particular input sequence was surrounded by 0's. We find the first, third, fifth, etc., bits in the output sequence by adding the most recent input bit to the input bit two periods in the past (corresponding to an h of 0101). To find the second, fourth, sixth, etc., output bits, we add the most recent bit and the previous bit together (corresponding to an h of 0110).

Decoding of a convolutional code involves comparing received groupings of bits to the output of the encoder for various input sequences. We choose the input sequence which results in an output which is minimum distance from the received sequence. Since a continuous bit stream is received, the decoding decision must be made bit by bit. As each bit is received, a decision must be made between two alternatives, 1 or 0.

The student is referred to the references for techniques of decoding convolutional codes in order to achieve minimum error. Progress has recently been made in devising decoding techniques which are simple to implement while approaching the optimum performance.

PROBLEMS

8.1. A communication system consists of three possible messages. The probability of message A is p, and the probability of message B is also p. The probability of message C is whatever is required to make the probabilities sum to 1. Plot the entropy as a function of p.

8.2. A communication system consists of four possible messages. $Pr(A) = Pr(B)$ and $Pr(C) = Pr(D)$. Plot the entropy as a function of $Pr(A)$.

8.3. The probability of rain on any particular day in Las Vegas, Nevada is 0.01. Suppose that a weather forecaster in that city decides to save effort by predicting no rain every day of the year. What is the average information content of each forecast (in bits per day). Make the (simplistic) assumption that the information content of an incorrect prediction is zero. Repeat your calculation if the forecaster now decides to predict rain every day.

8.4. Find the channel capacity C if the S/N = 6 dB and a standard broadcast AM channel is used. Repeat for a broadcast TV channel and compare your answers.

8.5. Find the minimum distance for the following code consisting of four code words:

$$0111001, \quad 1100101, \quad 0010111, \quad 1011100$$

How many bit errors can be detected? How many bit errors can be corrected?

8.6. Find the minimum distance for the following code consisting of four code words:

$$0111001, \quad 1100101, \quad 0010111, \quad 0101001$$

How many bit errors can be detected? How many bit errors can be corrected?

8.7. Redo Illustrative Example 8.2 for a signal to noise ratio of 11 dB, and find the probability of an incorrectly decoded message.

8.8. Redo Illustrative Example 8.2 for a signal to noise ratio of 8 dB, and use PSK.

8.9. What is the form of $\bar{H}$ for the single parity bit code formed from 3 message bits, as described in Section 8.4?

8.10. What is the form of $\bar{H}$ for the code in which a 3- bit message word is repeated twice, as described in Section 8.3 (i.e., 101 is coded as 101101)?

8.11. Suppose two rows in the $\bar{H}$ matrix are switched. Discuss the changes that would occur. Repeat for the case of two columns being interchanged.

8.12. Suppose the probability of bit error is 10^{-3}. We transmit each bit by repeating it four times. That is, we transmit 0 as 0000 and 1 as 1111. Find the probability that a transmitted information bit is decoded as the incorrect bit at the receiver.

8.13. An algebraic code is formed using the following $\bar{H}$:

$$\bar{H} = \begin{bmatrix} 1 & 1 & 1 & 1 & 1 & 0 & 0 \\ 1 & 0 & 1 & 1 & 0 & 1 & 0 \\ 0 & 1 & 0 & 1 & 0 & 0 & 1 \end{bmatrix}$$

Can single errors be detected? Corrected?

8.14. An algebraic code is formed using the following $\bar{H}$:

$$\bar{H} = \begin{bmatrix} 1 & 1 & 1 & 1 & 0 & 0 & 0 \\ 0 & 0 & 1 & 0 & 1 & 0 & 0 \\ 0 & 1 & 0 & 0 & 0 & 1 & 0 \\ 1 & 0 & 0 & 0 & 0 & 0 & 1 \end{bmatrix}$$

Find the code words corresponding to each of the eight possible message words. Find the minimum distance. How many errors can be detected? Corrected?

8.15. A 4-bit PN code is formed using the following shift register:

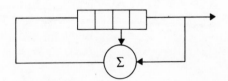

(a) What is the length (period) of the PN sequence?
(b) What is the distance between code words?
(c) Sketch the autocorrelation.
(d) If this is used as a code with the length found in part (a), how many errors can be detected? Corrected?

8.16. A rate-$\frac{1}{2}$ convolutional code is constructed with constraint length 4. The first set of h's is given by 10000 and the second set by 01011. Find the output sequence if the input sequence is given by 1011010110.

REFERENCES

ABRAMSON, N., *Information Theory and Coding*, McGraw-Hill, New York, 1963.

ASH, R., *Information Theory*, Wiley-Interscience, New York, 1965.

BERLEKAMP, E. R., *Algebraic Coding Theory*, McGraw-Hill, New York, 1968.

BERLEKAMP, E. R., ed., *The Development of Coding Theory*, IEEE Press, New York, 1974. (*Note:* Contains many relevant papers.)

PIERCE, J. R., *Symbols, Signals and Noise*, Harper & Row, New York, 1961.

REZA, F. M., *Introduction to Information Theory*, McGraw-Hill, New York, 1961.

SLEPIAN, D., ed., *The Development of Information Theory*, IEEE Press, New York, 1973. (*Note:* Contains many relevant papers.)

APPENDIX

I

In approximate order of appearance

SYMBOL   *NAME*

$\bar{a}_x, \bar{a}_y, \bar{a}_z$ Three rectangular unit vectors

$f^*(t)$ Complex conjugate of $f(t)$

$|x|$ Magnitude of the complex number x

$\underline{/x}$ Phase angle of the complex number x

$\triangleq$ Equal by definition

$\approx$ Approximately equal to

$\mathfrak{F}[\ ]$ Fourier Transform of the quantity in brackets

$\mathfrak{F}^{-1}[\ ]$ Inverse Fourier Transform of the quantity in brackets

$\leftrightarrow$ Quantities on left and right end of the arrow form a Fourier Transform pair

$f(t) * g(t)$ $f(t)$ convolved with $g(t)$

$U(t)$ The unit step function

$Re\{\ \}$ Real part of the term in braces

$Im\{\ \}$ Imaginary part of the term in braces

Hz	Hertz (cycles per second)
kHz	Kilohertz = 1,000 hertz
MHz	Megahertz = 1,000,000 hertz
$\delta(t)$	The unit impulse function
$sgn(t)$	Sign function. Equal to $+1$ for positive argument and to -1 for negative argument
ω_m	Upper cutoff frequency above which the Fourier Transform is zero
$f(t) \rightarrow g(t)$	$g(t)$ is the output when $f(t)$ is the input
DFT	Discrete Fourier Transform
FFT	Fast Fourier Transform
W	Complex exponential used with the DFT and FFT
$F(z)$	z-Transform
$h(t)$	Impulse response of a linear system
$H(\omega)$	System function. The Fourier Transform of $h(t)$
t_r	Rise time
E_f	Energy in a signal $f(t)$
P_f	Average power in a signal $f(t)$
$\psi_f(\omega)$	Energy spectral density of $f(t)$
$\phi_f(t)$	Energy autocorrelation of $f(t)$
$f_T(t)$	Truncated version of $f(t)$ limited to values of t between $-T/2$ and $+T/2$
$S(\omega)$	Power spectral density
$R(t)$	Power autocorrelation
$F(x_0)$	Probability distribution function
$p_x(x)$	Probability density function
m	Mean value of a random variable
σ^2	Variance of a random variable
$m_x(t)$	Mean value of the process $x(t)$
$R_{xx}(t_1, t_2)$	Autocorrelation of the process $x(t)$
$P(A/B)$	Conditional probability
$x(t), y(t)$	Quadrature components of narrowband noise
S/N	Signal to noise (power) ratio
ΔSNR	Signal to noise improvement figure
dB	Decibel $= 10 \log_{10} (\ )$
$f_c(t)$	Unmodulated carrier signal
AM	Amplitude modulation
DSBSC	Double sideband suppressed carrier
$f_m(t)$	Either an AM or PAM modulated waveform
ω_c	Carrier frequency
AMTC	AM transmitted carrier
η	Power efficiency
$z(t)$	Pre-envelope (analytic signal)

r.f.	Radio frequency
i.f.	Intermediate frequency
SSB	Single sideband
$f_{usb}(t)$	Upper sideband SSB signal
$f_{lsb}(t)$	Lower sideband SSB signal
VSB	Vestigial sideband
FM	Frequency modulation
$\omega_i(t)$	Instantaneous frequency
$\lambda_{fm}(t)$	FM signal
$\Lambda_{fm}(\omega)$	Transform of $\lambda_{fm}(t)$
k_f	Constant associated with FM
$J_n(\beta)$	Bessel function of the first kind
$\Delta\omega$	Maximum frequency deviation
BW	Bandwidth
PM	Phase modulation
$\theta(t)$	Instantaneous phase
k_p	Constant associated with PM
$\lambda_{pm}(t)$	Phase modulated signal
$\Delta\theta$	Maximum phase deviation
PAM	Pulse amplitude modulation
PPM	Pulse position modulation
PWM	Pulse width modulation
TDM	Time division multiplexing
A/D	Analog to digital converter
D/A	Digital to analog converter
lsb	Least significant bit
msb	Most significant bit
PCM	Pulse code modulation
CODEC	Coder-decoder
DM	Delta modulation
ADM	Adaptive delta modulation
FSK	Frequency shift keying
PSK	Phase shift keying
DPSK	Differential phase shift keying
RZ	Return to zero
NRZ	Non-return to zero
BER	Bit error rate
I_x	Information content of message x
H	Entropy
C	Channel capacity
D_{min}	Minimum distance between code words
$\bar{T}$	Transmitted column vector
$\bar{H}$	Matrix used for algebraic coding

$\bar{E}$	Error vector
$\bar{R}$	Received vector
$\bar{S}$	Error syndrome
PN	Pseudonoise

APPENDIX
II

COMMON FOURIER TRANSFORM PAIRS

General Relationships

In the following, assume $f(t) \leftrightarrow F(\omega)$ and $g(t) \leftrightarrow G(\omega)$:

$$f(t - t_0) \leftrightarrow e^{-j\omega t_0}(\omega)$$

$$e^{j\omega_0 t} f(t) \leftrightarrow F(\omega - \omega_0)$$

$$\frac{df}{dt} \leftrightarrow j\omega F(\omega)$$

$$\int_{-\infty}^{t} f(\tau)\, d\tau \leftrightarrow \frac{F(\omega)}{j\omega}$$

$$f(t) * g(t) \leftrightarrow F(\omega) G(\omega)$$

$$2\pi f(t) g(t) \leftrightarrow F(\omega) * G(\omega)$$

Specific Transform Pairs

$$e^{-at} U(t) \leftrightarrow \frac{1}{a + j\omega} \qquad (a > 0)$$

$$te^{-at} U(t) \leftrightarrow \frac{1}{(a + j\omega)^2} \qquad (a > 0)$$

$$e^{-at^2} \longleftrightarrow \sqrt{\frac{\pi}{a}} \exp\left(-\frac{\omega^2}{4a}\right)$$

$$|t| \longleftrightarrow -\frac{2}{\omega^2}$$

$$\frac{\sin at}{\pi t} \longleftrightarrow$$

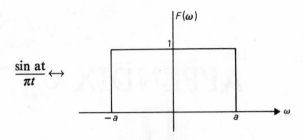

$$\longleftrightarrow \frac{\sin \omega T}{\omega}$$

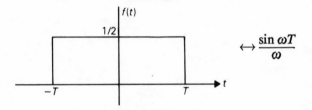

$$\longleftrightarrow \frac{\sin^2 \omega T/2}{T\omega^2/4}$$

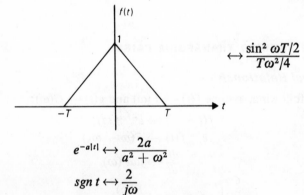

$$e^{-a|t|} \longleftrightarrow \frac{2a}{a^2 + \omega^2}$$

$$\text{sgn } t \longleftrightarrow \frac{2}{j\omega}$$

$$\delta(t) \longleftrightarrow 1$$

$$1 \longleftrightarrow 2\pi\delta(\omega)$$

$$e^{j\omega_0 t} \longleftrightarrow 2\pi\delta(\omega - \omega_0)$$

$$\cos \omega_0 t \longleftrightarrow \pi[\delta(\omega - \omega_0) + \delta(\omega + \omega_0)]$$

$$\sin \omega_0 t \longleftrightarrow j\pi[\delta(\omega + \omega_0) - \delta(\omega - \omega_0)]$$

$$f(t) \cos \omega_0 t \longleftrightarrow \tfrac{1}{2}[F(\omega - \omega_0) + F(\omega + \omega_0)]$$

$$U(t) \longleftrightarrow \pi\delta(\omega) + \frac{1}{j\omega}$$

APPENDIX
III

THE ERROR FUNCTION

$$\text{erf}(x) = \frac{1}{\sqrt{2\pi}} \int_{-\infty}^{x} e^{-y^2/2} \, dy$$

x	erf (x)	x	erf (x)
0 00	0.5000	0.05	0.5199
0.10	0.5398	0.15	0.5596
0.20	0.5793	0.25	0.5987
0.30	0.6179	0.35	0.6368
0.40	0.6554	0.45	0.6736
0.50	0.6915	0.55	0.7088
0.60	0.7268	0.65	0.7422
0.70	0.7580	0.75	0.7734
0.80	0.7881	0.85	0.8023
0.90	0.8159	0.95	0.8289
1.00	0.8413	1.05	0.8531
1.10	0.8643	1.15	0.8749
1.20	0.8849	1.25	0.8944
1.30	0.9032	1.35	0.9115
1.40	0.9192	1.45	0.9265
1.50	0.9332	1.55	0.9394

x	$erf(x)$	x	$erf(x)$
1.60	0.9452	1.65	0.9505
1.70	0.9554	1.75	0.9599
1.80	0.9641	1.85	0.9678
1.90	0.9713	1.95	0.9744
2.00	0.9772	2.05	0.9798
2.10	0.9821	2.15	0.9842
2.20	0.9861	2.25	0.9878
2.30	0.9893	2.35	0.9906
2.40	0.9918	2.45	0.9929
2.50	0.9938	2.55	0.9946
2.60	0.9953	2.65	0.9960
2.70	0.9965	2.75	0.9970
2.80	0.9974	2.85	0.9978
2.90	0.9981	2.95	0.9984
3.00	0.9987	3.05	0.9989

APPENDIX
IV

ANSWERS TO SELECTED PROBLEMS

Chapter 1

1.2. Squared error $= 16$

1.4. (a) $a_n = 0$ for n even; $a_n = 2/(\pi^2 n^2)$ for n odd

1.6. $f_p(t) = \dfrac{\tau}{T} + \displaystyle\sum_1^\infty \dfrac{1}{n\pi} \sin \dfrac{n\pi}{T} \cos \dfrac{2\pi n t}{T}$

1.12. $\dfrac{1}{2a} e^{-a|t|}$

1.25. $\hat{F}(\omega) = \dfrac{1}{\pi} F(\omega) * \dfrac{\sin \omega/2}{\omega} e^{-j\omega 3/2}$

1.28. $\dfrac{\pi}{2}[\delta(\omega - 2\omega_0) + \delta(\omega + 2\omega_0)] + \pi\delta(\omega)$

1.36. $F(0) = \tfrac{3}{4}; F(\Omega) = -j/4; F(2\Omega) = \tfrac{1}{4}; F(3\Omega) = j/4$

Chapter 2

2.5. $H(\omega) = j\omega/(1 + j\omega); h(t) = -e^{-t}U(t) + \delta(t)$

2.6. (a) $h(t) = \tfrac{1}{3}\delta(t) - \tfrac{2}{9}e^{-(2/3)t}U(t)$

2.13. (a) $\overline{v_{\text{out}}^2} = K/2$; (c) $\overline{v_{\text{out}}^2} = \tfrac{1}{4}$

329

2.14. $S(\omega) = 2\pi A^2 \delta(\omega)$

2.20. $H(\omega) = \dfrac{j\omega/R_1 C_1}{-\omega^2 + j\omega[(C_1 + C_2)/(R_3 C_1 C_2)] + [(R_1 + R_2)/(R_1 R_2 R_3 C_1 C_2)]}$

2.24. (a) $1, -1, 1, -1, 1, -1, 1, -1, \ldots$; (b) $0, 0, 0, 0, 0, 0, 0, 0, 0, 0, \ldots$

Chapter 3

3.2. $\frac{5}{6}$

3.4. $\frac{2}{9}$

3.5. $\frac{93}{256}$

3.6. $\frac{5}{32}$

3.10. $p_y(y) = \frac{1}{4}$ for $|y| < 2$

3.11. 0

3.15. (a) $A = 1$; (b) $1 - e^{-6}$; (c) $\frac{1}{2}$

3.16. $\text{erf}\left(\frac{1}{2}\right) = 0.6915$

3.20. $E\{y(t_1)y(t_2)\} = R_x(t_2 - t_1) + \sin 2t_1 \sin 2t_2$ assuming $E\{x\} = 0$

3.22. $P_{\text{out}} = \omega_m/\pi$

3.27. $R_v(\tau) = (A^2/2) \cos \omega_c \tau$

3.29. (a) 600π; (b) 0.011 dB; (c) 15.18 dB

Chapter 4

4.9. (a) $f(t)$; (b) $P_f/2$; (c) P_f

4.10. $\frac{1}{2}[F(\omega - \omega_c) + F(\omega + \omega_c)]$; $\frac{1}{2}[F(\omega - \omega_c)e^{j\pi/4} + F(\omega + \omega_c)e^{-j\pi/4}]$

4.18. 22.93 dB

4.19. 19.93 dB

4.20. (b) 22.93 dB; (c) $\Delta\text{SNR} = 2.22$ dB

4.21. DSBSC: 7.98 dB; DSBTC: 7.98 dB assuming large carrier; SSB: 4.98 dB

Chapter 5

5.5. $f(t) \simeq 1 - \dfrac{a^2 k_f^2}{4\omega_m^2} + \dfrac{a^2 k_f^2}{4\omega_m^2} \cos 2\omega_m t + \dfrac{3}{8} \dfrac{a^4 k_f^4}{24\omega_m^4} - \dfrac{a^4 k_f^4}{48\omega_m^4} \cos 2\omega_m t$

$\qquad + \dfrac{1}{8} \dfrac{a^4 k_f^4}{24\omega_m^4} \cos 4\omega_m t$

5.6. $\sin\left(\dfrac{ak_f}{\omega_m} \sin \omega_m t\right) \simeq \dfrac{ak_f}{\omega_m} \sin \omega_m t - \dfrac{a^3 k_f^3}{6\omega_m^3} \sin^3 \omega_m t + \dfrac{a^5 k_f^5}{120\omega_m^5} \sin^5 \omega_m t$

5.7. $\lambda_{fm}(t) = \cos\left\{\omega_c t \cos \dfrac{ak_f}{\omega_m} \sin \omega_m t\right\} - \sin \omega_c t \sin\left\{\dfrac{ak_f}{\omega_m} \sin \omega_m t\right\}$

5.10. (a) BW $= 20,048$ Hz; (b) BW $= 40,096$ Hz

5.11. BW $= 60,048$ Hz

5.13. BW $= 20$ kHz

5.14. BW $= 1,150$ kHz

5.15. BW $= 7000\pi$ rad/sec

5.17. (b) BW $= 1$ kHz

5.19. (a) 37.7 dB

Chapter 7

7.1. 010; 101; 110; 111

7.4. Choose $T_s = 3$ sec; 1000; 1001; 0101; 1011; 0011; 1101; 0001

7.6. 36 kbits/sec; minimum bandwidth $= 18$ kHz

7.7. Output sequence: 1 1 1 1 0 1 1 1 0 1 1 1 0 1 1 1 $\cdots$

7.9. 0.0013 V^2

7.11. Binary pulse train transmitted: 1 1 0 1 1 1 0 1 1 1 0 1 1 1 $\cdots$; quantized values after each sampling point: 1.5; 2.5; 2.0; 2.5; 3.5; 4.5; 3.5; 4.5; 5.5; 6.5; 5.5; 6.5; 7.5; 8.5 $\cdots$

Chapter 8

8.3. If rain predicted: 0.0664; if no rain predicted: 0.0144

8.4. $C = 11{,}610$ bits/sec

8.5. Detect 3; correct 1

8.12. $P \simeq 4 \times 10^{-9}$

8.15. (a) 15; (b) 8; (c) $R(0) = 15$, $R(n) = 7$; (d) correct 3, detect 7

INDEX